Fungicidal Activity

Wiley Series in Agrochemicals and Plant Protection

Series Editors:

Terry Roberts, *JSC International Ltd, Harrogate, England*
Junshi Miyamoto, *Sumitomo Chemical Company Ltd, Japan*

Series Advisory Editors:

Fitz Führ, *Institut für Radioagronomie, Jülich, Germany*
David H. Hutson, *Falmouth, Cornwall, UK*
Philip Kearney, *Water Resources Research Center, University of Maryland, USA*
Donald Mackay, *University of Toronto, Canada*

Previous Titles in the Wiley Series in Agrochemicals and Plant Protection:

The Methyl Bromide Issue (1996), ISBN 0 471 95521 3. *Edited by* C. H. Bell, N. Price *and* B. Chakrabarti
Pesticide Remediation in Soils and Water (1998), ISBN 0 471 96805 6. *Edited by* P. Kearney *and* T. R. Roberts
Chirality in Agrochemicals (1998), ISBN 0 471 98121 4. *Edited by* N. Kurihara *and* J. Miyamoto

Forthcoming Titles in the Wiley Series in Agrochemicals and Plant Protection:

Metabolism of Agrochemicals in Plants. *Edited by* T. R. Roberts
Optimizing Pesticide Use. *Edited by* M. F. Wilson

Fungicidal Activity

Chemical and Biological Approaches to Plant Protection

Edited by
DAVID HUTSON
Falmouth, Cornwall, UK

and

JUNSHI MIYAMOTO
Sumitomo Chemical Company, Osaka, Japan

JOHN WILEY & SONS
Chichester · New York · Weinheim · Brisbane · Singapore · Toronto

National 01243 779777
International (+44) 1243 779777
e-mail (for orders and customer service enquiries): cs-books@wiley.co.uk
Visit our Home Page on http://www.wiley.co.uk
or http://www.wiley.com

Other Wiley Editorial Offices

John Wiley & Sons, Inc., 605 Third Avenue,
New York, NY 10158-0012, USA

WILEY-VCH Verlag GmbH, Pappelallee 3,
D-69469 Weinheim, Germany

Jacaranda Wiley Ltd, 33 Park Road, Milton,
Queensland 4064, Australia

John Wiley & Sons (Asia) Pte Ltd, 2 Clementi Loop #02-01,
Jin Xing Distripark, Singapore 129809

John Wiley & Sons (Canada) Ltd, 22 Worcester Road,
Rexdale, Ontario M9W 1L1, Canada

Library of Congress Cataloging-in-Publication Data

Fungicidal activity : chemical and biological approaches to plant protection / edited by D. H. Hutson and Junshi Miyamoto.
p. cm. — (Wiley series in agrochemicals and plant protection)
Includes bibliographical references and index.
ISBN 0-471-96806-4 (hardback : alk. paper)
1. Fungicides. I. Hutson, D. H. (David Herd), 1935– .
II. Miyamoto, J. (Junshi) III. Series.
SB951.3.F8 1988 98-3121
632′.952—dc21 CIP

British Library Cataloguing in Publication Data

A catalogue record for this book is available from the British Library

ISBN 0 471 96806 4

Typeset in 10/12 pts Times by Thomson Press (India) Ltd, New Delhi.
Printed and bound in Great Britain by Biddles Ltd, Guildford and Kings Lynn.
This book is printed on acid-free paper responsibly manufactured from sustainable forestry, in which at least two trees are planted for each one used for paper production.

Contents

Contributors

J. M. Clough
ZENECA Agrochemicals, Jealott's Hill Research Station, Bracknell, Berkshire RG42 6EY, UK

H.-W. Dehne
Institut für Pflanzenkrankheiten, Universität Bonn, Nussallee 9, D-53115, Bonn, Germany

C. R. A. Godfrey
ZENECA Agrochemicals, Jealott's Hill Research Station, Bracknell, Berkshire RG42 6EY, UK

J. A. Gracia-Garza
Biocontrol of Plant Diseases Laboratory, USDA-ARS, BARC-West, Beltsville, MD 20705, USA

D. W. Hollomon
IACR-Long Ashton Research Station, Department of Agricultural Sciences, University of Bristol, Long Ashton, Bristol BS18 9AF, UK

S. J. Kendall
IACR-Long Ashton Research Station, Department of Agricultural Sciences, University of Bristol, Long Ashton, Bristol BS18 9AF, UK

R. P. Larkin
Biocontrol of Plant Diseases Laboratory, USDA-ARS, BARC-West, Beltsville, MD 20705, USA

E.-C. Oerke
Institut für Pflanzenkrankheiten, Universität Bonn, Nussallee 9, D-53115, Bonn, Germany

D. Roberts
Biocontrol of Plant Diseases Laboratory, USDA-ARS, BARC-West, Beltsville, MD 20705, USA

Y. Uesugi
1-28-9-903 Kaidori, Tama-shi, Tokyo 206, Japan

I. Yamaguchi
The Institute of Physical and Chemical Research (RIKEN), Wako, Saitama 3510106, Japan

K. Yoneyama
Faculty of Agriculture, Meiji University, 1-1-1, Higashimita, Tama, Kawasaki, Kanagawa 214, Japan

Series Preface

There have been tremendous advances in many areas of research directed towards improving the quantity and quality of food and fibre by chemical and other means. This has been at a time of increasing concern for the protection of the environment, and our understanding of the environmental impact of agrochemicals has also increased and become more sophisticated thanks to multi-disciplinary approaches.

Wiley has recognized the opportunity for the introduction of a series of books within the theme 'Agrochemicals and Plant Protection' with a wide scope that will include chemistry, biology and biotechnology in the broadest sense. This series will effectively be a replacement for the successful 'Progress in Pesticide Biochemistry and Toxicology' edited by Hutson and Roberts which has run to nine volumes. In addition, it will complement the international journals *Pesticide Science* and *Journal of the Science of Food and Agriculture* published by Wiley on behalf of the Society of Chemical Industry.

The scope of the new series encompasses all major areas of interest and will reflect the advances made during half a century after World War II and, more importantly, look to further successful developments in agricultural chemistry and biotechnology in the 21st century. The objective is to publish a series of books that will be complementary and will become a collectable series for those involved in research, development and registration of agrochemicals as well as more general aspects.

SCOPE OF 'AGROCHEMICALS AND PLANT PROTECTION'

As indicated, the scope is very broad and the following topics are illustrative:

- Discovery of new agrochemicals/QSAR approaches/other structure activity relationships
- Mode of action of all classes of agrochemicals and PGRs as well as natural products
- Synergy
- Safeners
- Chirality in agrochemical development
- Formulation technologies
- Biopesticides
- Resistance and resurgence/molecular mechanisms of resistance
- Toxicology/metabolism/human risk assessment/extrapolation from animals to man

- Environmental fate and effects/risk assessment/simulation modelling/ predicted environmental concentrations
- Residues analysis/formulations analysis/instrumentation
- Remediation/waste management
- Regulation of agrochemicals
- Modern biotechnology and agriculture/genetic modification/plant breeding/ regulatory implications of biotechnology
- IPM and ICM
- Plant protection in the 21st century.

There will also be scope for specific monographs on classes of agrochemical (e.g. triazole fungicides, sulphonylurea herbicides) and for broad tests on global issues of world food supply and environmental concerns.

THE SERIES EDITORS

Dr Terry Roberts is Director of Scientific and Regulatory Affairs at JSC International based in Harrogate, UK. He joined JSC in March 1996 and provides scientific and regulatory consulting services to the agrochemical, biocides and related industries with the emphasis on EU registrations.

Dr Roberts was formerly with Corning Hazleton as Director of Agrochemical and Environmental Services (1990–1996) and with Shell Research Ltd for the previous 20 years.

He has been active in international scientific organizations, notably OECD, IUPAC and ECPA, over the past 25 years and was recently appointed Secretary to the Division of Chemistry and the Environment within IUPAC. He has published extensively and is now Editor-in-Chief of the new Wiley series on 'Agrochemicals and Plant Protection'.

Dr Junshi Miyamoto is Corporate Advisor to the Sumitomo Chemical Company, where he has worked since 1957 after graduating from the Department of Chemistry, Faculty of Science, Kyoto University. After a lifetime of working in the chemical industry, Dr Miyamoto has acquired a wealth of knowledge in all aspects of mode of action, metabolism and toxicology of agrochemicals and industrial chemicals. He was previously Director General of Takazuka Research Center of the Company covering the areas of agrochemicals, biotechnology as well as environmental health sciences. He is currently President of the Division of Chemistry and the Environment, IUPAC, and in 1985 he received the Burdick Jackson International Award in Pesticide Chemistry from the American Chemical Society and in 1995, the Award of Distinguished Contribution to Science from the Japanese Government. He has published over 190 original papers and 50 books in the area, and is on the editorial board of several international journals, including *Pesticide Science*.

Preface

Fungicides have the longest history of the three main groups of crop protection agents (insecticides, herbicides and fungicides). Some 150 different compounds are now in use with a global end-user market value of some $6 billion, accounting for almost 20% of the agrochemicals market. Fungicides appear to fall into rather less well-defined groups than do, for example, the insecticides and they encompass a wide range of varied chemical structures. A major development in the 1960s was the successful commercialization of systemic fungicides (absorbed and transported within the plant) allowing more effective protection than did the earlier compounds. More efficient spray technology and formulation have similarly improved the efficiency of the non-systemic materials. With systemic fungicides, and with very specific modes of action and with intensive use situations (e.g. high value glass house crops), resistance has appeared. This in turn has provided a stimulus for the search for new modes of action. The development of compounds with low environmental impact and favourable mammalian toxicology has similarly been an impetus to further research. The search is also increasingly being extended to exploit biological control methods.

The size and scope of the fungicides market are presented by H.-W. Dehne and E.-C. Oerke in the introductory Chapter 1. This is followed by Y. Uesugi's description (Chapter 2) of the current status of the various groups of chemical fungicides, their modes of action and their uses. The debt owed to natural products research in the discovery of fungicides is then described by I. Yamaguchi in Chapter 3. The problem of resistance is discussed in Chapter 4 by D. W. Hollomon and S. J. Kendall. This chapter emphasizes the prediction and avoidance of the problem and also deals with its management and amelioration of its effects when encountered.

The detailed description by J. M. Clough and C. R. A. Godfrey in Chapter 5 of the discovery and development of the strobilurins provides an excellent example of the development of such a chemical class based on a novel mode of action. The remaining three chapters concentrate on biological aspects of control. Biological control methods are already successful in insect control; the biological control of fungal diseases is a relatively recent development and there are many difficulties yet to be overcome. The prospects are assessed by R. P. Larkin, D. P. Roberts and J. A. Gracia-Garza (Chapter 6). Chapter 7 on Systemic Acquired Resistance (I. Yamaguchi) commences with the far-from-obvious statement: 'Disease is a rare outcome in the spectrum of plant–microbe interaction'. The author then describes this remarkable phenomenon, akin in ways to the mammalian immune system, and its manipulation by chemical and other stimulae. Such manipulation affords a novel approach to disease control. The enhancement of natural protection by breeding and by gene transfer techniques is another promising field of research which will become increasingly important. K. Yoneyama, in Chapter 8, describes the various endogenous protection systems and how these may be modulated to improve the resistance of plants to fungal attack. The use of these techniques raises further important issues such as the toxicological aspects of natural fungicides and the public acceptance of genetically engineered agricultural commodities. These latter aspects perhaps deserve treatment in a separate volume.

D. H. HUTSON
Falmouth, Cornwall, UK

J. MIYAMATO
Osaka, Japan

January 1998

1 Impact of Diseases and Disease Control on Crop Production

H.-W. DEHNE and E.-C. OERKE
Bonn University, Germany

INTRODUCTION

It has become evident over the centuries that diseases caused by fungi can lead to high, sometimes complete losses of crops. This is most striking when the epidemic is spread over large areas and if all plants are more or less susceptible to the devastating pathogens. Since plants have been cultivated, fungal pathogens have found opportunities to spread epidemically and can cause a

Fungicidal Activity. Edited by D. H. Hutson and J. Miyamoto

biological threat to entire crops. Furthermore, fungal diseases especially can result not only in a quantitative yield loss but also in qualitative losses due to lower food quality and to decreased potential for storage and mycotoxin contamination. Most crops are attacked by several fungal pathogens—some are adapted to more moist conditions, such as downy mildews or fruit rots, others need only limited periods of wetness, such as rust fungi, others are even independent of high moisture, such as powdery mildews or smut fungi. Besides the leaf pathogens, high risks of latent damage also arise from soil-borne diseases.

There are several examples of disease epidemics that have caused remarkable food deficiencies or the depletion of a complete crop out of a particular area. The late blight of potato caused by *Phythophthora infestans* has led to periods of starvation by epidemic spread in Ireland during 1845 and 1846; a similar severe epidemic caused big problems with the food supply in Germany during 1916 and 1917. Since then frequent outbreaks of this disease have led to repeated problems wherever potatoes are grown. Additionally, the outbreak of coffee rust caused by *Hemileia vastatrix* led to a depletion of coffee from India and Ceylon, which then became substituted by tea. Another example of an epidemic spread shows the influence of cultivation on these particular problems: the leaf spot disease caused by *Cochiobolus heterostrophus* (*Helminthosporium maydis*) was endemic in maize growing areas, but it became devastating when cytoplasmic male sterility—used for the production of hybrid seed maize—was introduced into almost all maize varieties. When this high susceptibility to the leaf spot pathogen was introduced, the epidemic spread of the disease led to high rates of damage during the early 1970s in the USA and later in South America.

The detrimental outbreak of diseases and the permanent presence of plant pathogens leading to losses in quantity and quality of plant production, have stimulated research to inhibit the development of fungal plant pathogens. The first plant protection chemicals were fungicides introduced as Bordeaux mixture against downy mildew of grapes. A milestone in cereal production was the introduction of seed dressings based on mercury compounds early this century. The organotin chemicals, due to their broad spectrum of activity, led to a general introduction of fungicides in plant production. Since the 1960s the development of fungicides has been improved by the invention of systemic, more specific fungicides that have also offered the development of integrated disease control.

IMPACT OF DISEASES IN MAJOR CROPS

The impact of plant diseases and the most important plant pathogens can be differentiated by crops and the crop losses they cause (Figure 1, Table 1). Key pathogens causing yield-limiting diseases, major pathogens regularly causing

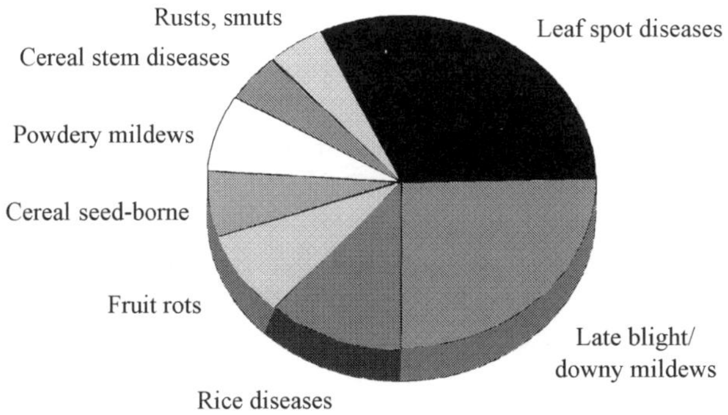

Figure 1 Relative importance of the major diseases based on market potentials (modified from Schwinn, 1992)

Table 1 Area harvested and production of the most important crops/crop groups in 1995

Crop	Area (1000 ha)	Production (1000 t)	Share (%) in area/production		
			Europe	North America/Oceania	East Asia
Wheat	220 605	541 120	12.0/22.9	20.8/18.8	13.2/19.0
Rice	149 151	550 193	0.3/0.4	0.9/1.6	23.4/38.0
Maize	136 245	514 506	7.9/10.7	20.1/37.9	17.3/22.3
Pulses	70 317	55 997	3.4/10.4	6.0/11.2	7.1/10.6
Barley	69 378	142 746	20.8/39.0	14.7/18.7	2.3/2.9
Soybeans	62 285	125 930	0.8/1.0	41.4/48.3	13.9/11.3
Oilseed rape	24 635	34 685	16.1/30.7	22.1/19.3	28.0/28.2
Cotton[a]	34 014	57 244	1.4/2.3	19.7/19.2	16.0/25.0
Peanuts[b]	22 476	27 990	0.1/0.1	2.9/5.8	17.2/37.0
Sunflower[c]	21 476	26 186	21.2/24.0	7.2/7.7	3.8/4.8
Potato	18 480	280 679	20.8/28.6	4.1/9.0	20.1/18.3
Coffee[d]	10 494	5 603	—/—	0.0/0.0	0.2/0.8
Sugarbeets	7 832	265 963	38.2/53.1	7.6/9.9	9.8/6.6
Vegetables	—	487 287	—/13.7	—/7.9	—/32.1
Fruits	—	396 873	—/15.7	—/8.9	—/11.9
Grapes	7 706	53 255	56.7/48.0	4.9/11.5	2.6/4.5

[a] Cottonseed [b] Groundnut in shell [c] Sunflower seed [d] Green coffee

(Source: FAO, 1996)

economic losses, and minor pathogens of lower importance have been differentiated. In some cases major pathogens from one region are not present in other areas. Regional differences for agricultural crops are given according to the intensity of production: Europe, dominated by the EU, with high productivity; North America and Oceania with lower productivity in cereals; East Asia with moderate to high productivity; Rest of world with low productivity. On a local basis, pathogens other than those listed may also be of economic importance. Yield levels and losses have been estimated from literature data as described earlier (Oerke *et al.*, 1994).

CEREALS

Cereals can be infected by a great number of pathogens, dominated by *Ascomycetes* and *Basidiomycetes*. Bacteria are of minor importance, however, they may reduce yields in some regions (e.g. *Xanthomonas campestris* pv *undulosa* in Brazil, etc.). Wheat and barley are highly susceptible to biotrophic pathogens such as rusts caused by *Puccinia* spp. and powdery mildew due to *Erysiphe graminis*. These diseases are especially important when they are

Table 2 Ranking of diseases in wheat, rice and maize in order of importance, by region

Europe	North America and Oceania	East Asia	Rest of world
Wheat			
Rusts	Rusts	Bunts	Bunts
Leaf and glume blotch	Common root rot	Rusts	Rusts
Powdery mildew	Bunts	Smuts	Smuts
Foot rot, eye spot	Take-all	Leaf and glume blotch	Foot rot, head blight
Smuts	Leaf and glume blotch	Powdery mildew	Leaf and glume blotch
Rice			
Rice blast	Sheath blight	Rice blast	Rice blast
Sheath blight	Rice blast	Sheath blight	Brown spot
Brown spot	Brown spot	Bacterial leaf blight	Sheath blight
Bacterial leaf blight	*Cercospora* leaf spot	Brown spot	*Cercospora* leaf spot
Leaf scald	Stem rot	Sheath rot	Bacterial leaf blight
Maize			
Foot rot, stalk rot	Leaf blight	Foot rot, ear rot	Foot rot, stalk rot
Head smut	Rusts	Rusts	Rusts
Common smut	Foot rot, stalk rot	Downy mildew	Downy mildew
Leaf blight	Head blight	*Aspergillus* spp.	Leaf blight
Head blight, ear rot	Head smut	Leaf blight	*Aspergillus* spp.

grown at high fertilization level (Table 2). In low input farming systems of wheat, smuts (*Ustilago* spp.), bunts (*Tilletia* spp.) and necrotrophic leaf pathogens such as *Mycosphaerella graminicola* (anam. *Septroria tritici*), *Phaeosphaeria nodorum* (anam. *Septoria nodorum*) are predominant. Seedling diseases due to *Fusarium* spp. and *Septoria* spp. can be controlled efficiently by seed dressing. The chemical control of soil-borne pathogens such as *Gaeumannomyces graminis* and *Cochliobolus sativus* causing take-all and common root rot/spot blotch, respectively, is often insufficient and production losses can only be reduced by long-term crop rotations. In barley, the pathogen spectrum is similar, however, leaf spot diseases such as net blotch (*Pyrenophora teres*) and scald (*Rhynchosporium secalis*) are among the most important yield-limiting factors in Europe, Oceania and North America. In South America *Puccinia striiformis* is very destructive.

Due to the high input-cultivation of wheat in Western Europe the loss potential of diseases is higher than in all other regions (Figure 2). Intensive disease control practices mainly in Western and Central Europe reduce losses by almost two-thirds to 7%. Soil-borne diseases still pose some control problems. In other regions disease control is rather low and, on a global level, the loss potential is reduced only from 16% to 12%. As world-wide yields currently average about 2.5 t/tha a higher use of fungicides would not be cost-effective in many areas.

RICE

More than 90% of actual production is harvested in Asia. Production systems vary from deepwater rice to upland rice, from intensive cultivation of irrigated lowland rice in East Asia to extensive production systems of upland rice. World-wide rice blast caused by *Magnaporthe grisea* (anam. *Pyricularia oryzae*) is considered to be the most damaging disease followed by sheath blight due to *Thanatephorus cucumeris* (anam. *Rhizoctonia solani*) (Table 2). Bacterial leaf blight caused by *Xanthomonas campestris* pv. *oryzae* has increased during the last decades especially in lowland rice. Besides brown spot (*Cochliobolus miyabeanus*), narrow brown leaf spot (*Cercospora oryzae*), stem rot (*Magnaporthe salvinii*), Bakanae (*Gibberella fujikuroi*), sheath rot (*Sarocladium oryzae*), kernel smut (*Tilletia barclayana*) and leaf scald (*Rhynchosporium oryzae*) and insect-transmitted viral diseaes are reported to cause economic losses.

In rice production the loss potential of pathogens exceeds 20% in Europe, North America/Oceania and East Asia where productivity is high. The infection pressure is lower in all other regions (Figure 2). Current disease control practices effectively reduce losses to an actual level of 8% in the USA, to about 12% in East Asia where China is the main rice producer. In Southeast Asia, South Asia, Latin America and Africa disease control is poor and current losses remain at a high level.

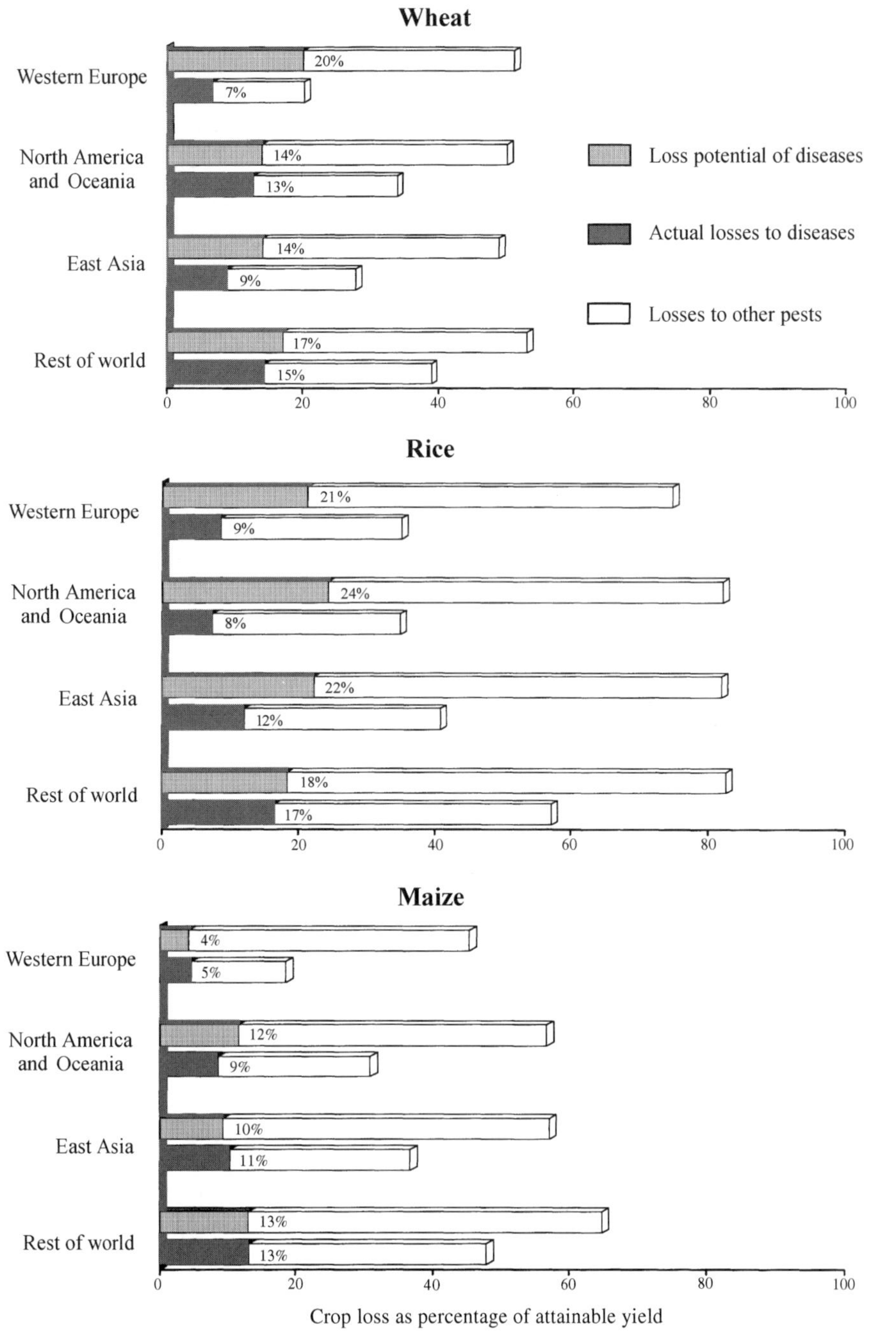

Figure 2 Loss potential and actual losses due to diseases in wheat, rice and maize, by region, in 1991–93

MAIZE

Diseases in maize production are of lower economic importance than weeds or arthropod pests. However, foot rot and stalk rot incited by *Fusarium moniliforme*, *Gibberella zeae* and *Diplodia* spp., rusts caused by *Puccinia* spp. and head smut due to *Sporosorium reilianum* (syn. *Sphacelotheca reiliana*) cause considerable losses when not controlled (Table 2). In the USA, the greatest producer of grain maize, leaf blights due to *Setosphaeria turcica* (Northern leaf blight) and *Cochliobolus* spp. are among the diseases regularly attacking maize. In the tropical growing regions of Asia, the downy mildews due to 10 species of *Sclerospora* spp. and *Sclerophthora* spp. are considered to be the most important diseases. Special emphasis has to be taken for ear rot caused by *Fusarium* spp. and *Aspergillus flavus* which are potent producers of mycotoxins. Control options for these pathogens are still unsatisfactory.

The loss potential varies considerably between regions because pathogens which are of major importance in the tropics and subtropics are only of minor importance in Europe and in North America where productivity is highest (Figure 2). Disease control is often limited to seed dressing. Only in the USA do control measures significantly reduce the loss potential. In all other regions actual losses are about the same or slightly higher than the potential because the control of weeds and insect pests is much more efficient than disease control.

POTATO

Production is most important in moderate climatic conditions, however, potatoes are also grown in the tropics and subtropics. More than 60 fungi and bacteria are pathogenic to potato. Under all growth conditions, late blight caused by *Phytophthora infestans* is considered to be the major yield-limiting disease. Without control, drastic yield reductions in the field and in storage have been reported. Similar to late blight, early blight incited by *Alternaria solani* occurs in all areas where potato is cultivated and causes crop losses especially in dry seasons. Other diseases of economic importance include black scurf (*Thanatephorus cucumeris*, anam. *Rhizoctonia solani*), potato wart (*Synchytrium endobioicum*), wilt, tuber rots (*Fusarium* spp., etc.) and black leg caused by the bacterium *Erwinia carotovora*. Under warmer conditions charcoal rot (*Macrophomina phaseolina*) and brown rot (*Pseudomonas solanacearum*) are known to cause crop losses. The bacterial disease has recently spread also to cultivation areas in cooler regions.

As major pathogens are spread world-wide, the loss potential does not differ significantly between regions. However, the intensity of disease control varies from very high in Western and Central Europe and North America to very low in some least developed countries. World-wide actual losses average 17%, ranging from 10% in West Europe to more than 25% in parts of South America and Africa.

FRUITS

Major fruit crops include citrus (grapefruit, lemon, limes, mandarins, pummelos, sour and sweet orange, tangelos), pome fruits (apple, pear), stone fruits (apricot, nectarine, peach, plums, sweet cherry and sour cherry), soft fruits (strawberry, raspberry, etc.), bananas, grapes, and nuts, especially peanut. The most important diseases of fruits are summarized in Table 3 and further details are given below.

Citrus

Important diseases are as follows: foot and root rot, gummosis (*Phytophthora* spp.), greasy spot (*Mycosphaerella citri*), melanose (*Diaporthe citri*), black spot (*Guignardia citricarpa*), scab (*Elsionë* spp. and *Sphaceloma* sp.), bacterial canker (*Xanthomonas campestris* pv. *citri*), mal secco (*Deuterophoma tracheiphila*) (confined to the Mediterranean countries and minor Asia), *Diplodia* stem-end rot (*Physalospora rhodina*, anam. *Diplodia natalensis*), green and blue mould (*Penicillium* spp.), grey mould (*Botroytinia fuckeliana*), anthracnose (*Glomerella cingulata*).

Pome Fruits

These are affected world-wide by scab (*Venturia* spp.), powdery mildew (*Podosphaera leucotricha*), crown, collar and root rot (*Phytophthora* spp.), canker (*Nectria galligena*, *Diaporthe* spp., *Valsa* spp.), fire blight (*Erwinia amylovora*), root rot (*Phymatotrichum omnivora*), rusts (*Gymnosporangium* spp.) and bitter rot (*Glomerella cingulata*). Blue mould (*Penicillium* spp.) and grey mould (*Botryotinia fuckeliana*) are the most important post-harvest diseases.

Table 3 Summary of the most important diseases in fruits: citrus, pome fruits, stone fruits, grapes and peanut

Citrus	Pome fruits	Stone fruits	Grapes	Peanut
Phytophthora diseases	Scab	Brown rot	Powdery mildew	Early and late leaf spot
Scab	Powdery mildew	Cherry leaf spot	Downy mildew	Rust
Black spot	Crown and root rot	Leaf curl	Grey mould	Stem rot
Melanose	Canker	Root and crown rot	Black rot	Web blotch
Mould diseases	Mould diseases	Powdery mildew	*Eutypa* dieback	Seed and seedling diseases

Stone Fruits

These are affected by the brown rot fungus *Monilinia* spp. causing blossom and twig blight as well as cankers and brown fruit rot, root and crown rot (*Phytophthora* spp.), powdery mildew (*Sphaerotheca* spp, *Podosphaera* spp.), cherry leaf spot (*Blumeriella jaapii*), leaf curl (*Taphrina deformans*), shot hole (*Wilsonomyces carpophilus* = anam. *Clasterosporium carpophilum*), bacterial canker (*Pseudomonas syringae*), anthracnose (*Glomerella cingulata*), rusts (*Tranzschelia discolor* and *T. pruni-spinosae*), *Cercospora* leaf spot (*Cercospora circumscissa*), *Leucostoma* canker (*Leucostoma* spp., syn. *Valsa* spp.), silver leaf disease (*Chondrostereum purpureum*, syn. *Stereum purpureum*) and *Eutypa* dieback of apricots (*Eutypa lata*).

Grapes

The following are the most important diseases of grapes: powdery mildew (*Uncinula necator*), downy mildew (*Plasmopara viticola*), grey mould fungus (*Botryotinia fuckeliana*) causing bunch rot and blight, black rot (*Guignardia bidwelli*), *Phomopsis* cane and leaf spot (*Cryptosporella viticola*, anam. *Phomopsis viticola*), *Eutypa* dieback (*Eutypa lata*), grape rust (*Physopella ampelopsidis*), *Dematophora* root rot (*Rosellinia necatrix*), anthracnose (*Elsinoë ampelina*) renewed outbreaks and Rotbrenner (*Pseudopezicula tracheiphila*).

Peanut

Diseases of peanut are early and late leaf spot (*Mycosphaerella spp.*, anam. *Cercospora arachidicola. Cercosporidium personatum*), rust (*Puccinia arachidis*), web blotch (*Phoma arachidicola*), stem rot (*Sclerotium rolfsii*), blight (*Sclerotinia* spp.), charcoal rot (*Macrophomina phaseoli*), wilt (*Verticillium dahliae*), seed and seedling diseases (*Pythium* sp., *Rhizoctonia solani*, *Aspergillus* spp, *Fusarium* spp.). The latter are especially important because *Aspergillus* spp. (yellow mould) and *Fusarium* spp. produce mycotoxins. Bacterial wilt (*Pseudomonas solanacearum*) is the only bacterial disease reported to cause economic losses.

Bananas and Soft Fruits

Black and brown sigatoka (*Mycosphaerella* spp.) and Panama disease (*Fusarium oxysporum* f.sp. *cubense*) are important diseases in banana production. Strawberries are the most important soft fruit and these are affected by grey mould caused by *Botryotinia fuckeliana*, *Phytophthora* spp., leaf scorch (*Diplocarpon earliana*), leaf spot (*Mycosphaerella fragariae*), anthracnose, black spot, fruit rot (*Glomerella cingulata*), *Rhizopus stolonifer*, and *Mucor* spp.

VEGETABLES

This very diversified group of vegetables includes cucurbits, tomato, cabbage, beans, peas, carrots, spinach, lettuce, etc. Their most important diseases are summarized in Table 4.

Cucurbits

These are affected by powdery mildew (*Sphaerotheca fuliginea*, *Erysiphe cichoracearum*), anthracnose (*Colletotrichum orbiculare*), downy mildew (*Pseudoperonospora cubensis*), damping-off and root rot (*Pythium* spp., *Phytophthora* spp., *Thanatephorus cucumeris*, etc.), gummy stem blight (*Didymella bryoniae*), angular leaf spot (*Pseudomonas syringae* pv. *lachrymans*), scab (*Cladosporium cucumerinum*), target leaf spot (*Corynespora cassiicola*) in greenhouse-grown cucumber and in the field and wilt (*Fusarium oxysporum*) which is destructive especially in melon and watermelon, less destructive in cucumber. *Phomopsis* black root rot (*Phomopsis sclerotioides*) is a serious disease of greenhouse cucumbers reported from Europe, Southeast Asia and Canada.

Tomato

Diseases of this crop include late blight, early blight, wilt diseases (*Fusarium oxysporum lycopersici*, *Verticillium* spp., *Clavibacter michiganensis*), stem blight (*Sclerotium rolfsii*), corky root rot (*Pyrenochaeta* spp.) and *Septoria* leaf spot (*S. lycopersici*). In most of the production areas powdery mildew caused by *Leveillula taurica* and *Erysiphe* sp., *Stemphylium* spp. and *Pseudomonas solanacearum* are only of minor importance.

Cabbage

Cabbage is affected by damping-off (*Pythium* sp., *Fusarium* spp., *Thanatephorus cucumeris*, *Olpidium brassicae*), downy mildew (*Peronospora* spp.), blackleg, canker caused by *Leptosphaeria maculans*, dark leaf spot (*Alternaria*

Table 4 Ranking of diseases in cucurbits, tomato, cabbage and other vegetable crops in order of importance

Cucurbits	Tomato	Cabbage	Others
Powdery mildew	Late blight	Damping-off	Damping-off
Black root rot	Early blight	Downy mildew	Downy mildew
Downy mildew	Wilt diseases	Blackleg, canker	Leaf spots
Damping-off, root rot	Grey mould	Dark leaf spot	Grey mould
Angular leaf spot	Corky root rot	Black rot	Powdery mildew

spp.), club root (*Plasmodiophora brassicae*) and black rot (*Xanthomonas campestris*).

Beans

The following are the most important diseases of beans: black root rot (*Thielaviopsis basicola*), Southern blight (*Rhizoctonia solani*, *Sclerotium rolfsii*), angular leaf spot (*Phaeoisariopsis griseola*), anthracnose (*Glomerella lindemuthiana*), grey mould (*Botrytis cinerea*), rust (*Uromyces appendiculatus*), white mould (*Sclerotinia sclerotiorum*), common bacterial blight (*Xanthomonas campestris* pv. *phaseoli*) and halo blight (*Pseudomonas syringae* pv. *phaseolicola*).

Other crops are attacked by downy mildews (lettuce, *Bremia lactucae*), seedling diseases, leaf spot diseases (celery, *Septoria* spp.), *Botrytis* spp., rusts *Uromyces* spp., powdery mildews (pea, *Erysiphe pisi*), etc.

INDUSTRIAL CROPS

Sugarbeet

Sugarbeet is grown under moderate and semi-arid climates for sugar production. Fungal and bacterial diseases are of minor importance when compared to weeds and insect pests which can also transmit important virus diseases. Pathogens attack sugarbeet plants during all stages of development, however, attack is most important in the seedling stage reducing plant density, and in the stage of carbohydrate translocation into the beet. *Cercospora* leaf spot is widespread and estimated to be the most important beet disease especially under warmer conditions considerably reducing sugar yields (Table 5, Figure 3). *Ramularia* leaf spot and powdery mildew (*Erysiphe betae*) have similar impacts under cooler and dry conditions, respectively. Seedling emergence and plant density is affected by a diseases complex caused by *Aphanomyces* sp., *Pythium* sp. and *Thanatephorus cucumeris*. The yield impact of beet rust (*Uromyces betae*) and downy mildew (*Peronospora farinosa*) is limited to some hot spots and specific weather conditions, respectively.

Oil Crops

Soybean is the most important oil-producing crop, although soybeans also account for nearly 60% of protein meals used for animal feed. More than 30 pathogens attacking soybean are of economic importance. The principal diseases injuring the vegetative parts of the plants are root rot (*Phytophthora megasperma* f.s.p. *glycinea*), seedling diseases caused by *Pythium* spp., *Thanatephorus cucumeris*, *Fusarium* spp., etc., *Cercospora* leaf spots, downy mildew (*Personospora manshurica*) and *Diaporthe* spp. (stem canker) which

Table 5 Summary of the most important diseases of industrial crops: sugarbeet, oilseed rape, sunflower, cotton and coffee

Sugarbeet	Oilseed rape	Sunflower	Cotton	Coffee
Cercospora leaf spot	Stem rot	Downy mildew	Seedling diseases	Rust
Ramularia leaf spot	Blackleg, canker	Grey mould	Foot and stalk rot	Coffee berry disease
Seedling diseases	Downy mildew	Foot and stem rot	Boll rots	Leaf spot
Powdery mildew	Dark/light leaf spot	*Verticillium* wilt	Macrosporiosis	Brown eye spot
Downy mildew	*Verticillium* wilt	Rust	Bacterial blight	Bacterial blight

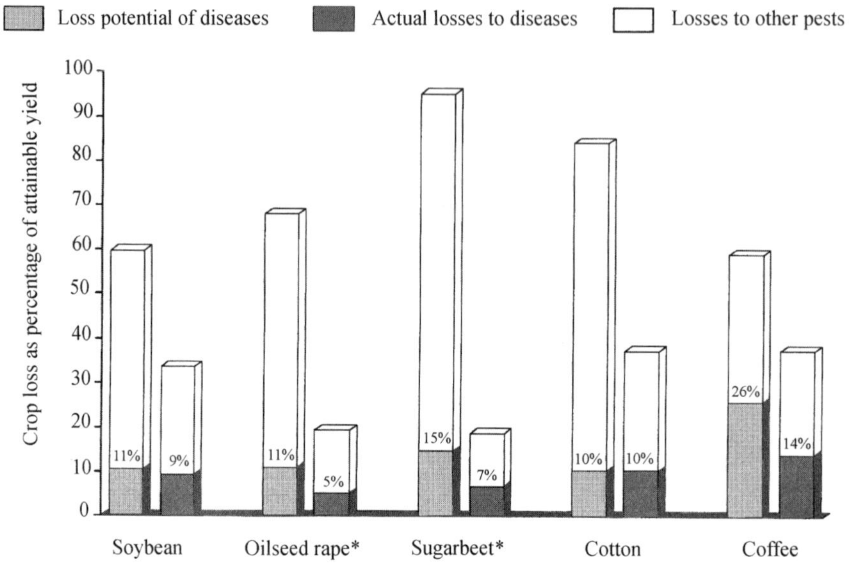

Figure 3 World-wide loss potential and actual losses due to diseases in soybean, oilseed rape, sugarbeet, cotton and coffee, in 1991–93 (*Figures for EU-15 only)

also infects the pods. Soybean rust (*Phakopsora pachyrhizi*), a very destructive pathogen, is limited to the eastern hemisphere, as is red leaf blotch (*Pyrenochaeta glycines*) to Africa. *Septoria* leaf spot (*Septoria glycines*) is prevalent in the Americas. Important bacterial diseases include *Pseudomonas syringae* pv. *glycinea* and *Xanthomonas* sp.

Oilseed rape and sunflower are grown in temperate climates for the production of plant oils. Oilseed rape is a ubiquitous crop in Europe and North America, East and South Asia. As with sugarbeet, the control of weeds and insect pests is more important than disease control. Especially in high-input production, oilseed rape can be attacked by stem rot (*Sclerotinia sclerotiorum*), blackleg and canker (*Leptosphaeria maculans*), downy mildew (*Peronospora parasitica*) and dark leaf spot (*Alternaria* spp.). Club root (*Plasmodiophora brassicae*), a typical disease of short crop rotations, and light leaf spot (*Pyrenopeziza brassicae*, anam. *Cylindrosporium concentricum*) are of minor importance.

Sunflower is produced under warmer climates, especially in southern Europe, Russia and India. Sunflower is attacked by the downy mildew fungus *Plasmopara halstedii* systemically infecting the plants, grey mould (*Botryotinia fuckeliana*, anam. *Botrytis cinerea*), the stem rot fungus *Sclerotinia sclerotiorum*, foot and stem rot (*Fusarium* spp., *Macrophomina phaseolina*) and the wilt fungi *Verticillium* spp. Sunflower rust incited by *Puccinia helianthi* can also be of local importance.

Cotton

Cotton is grown in the warmer parts of temperate zones and in the tropics and subtropics for fibre production. It is a major cash crop also for developing countries and, in general, crop protection is intensive. Arthropod pests and weeds are the most important yield-limiting factors. Diseases of economic impact are the seedling disease complex caused by *Thanatephorus cucmeris*, *Pythium ultimum* and *Glomerella gossypii*, wilt diseases due to *Fusarium* spp. and *Verticillium* spp., boll rots (*Phytophthora* spp., *Glomerella gossypii*), foot and stalk rot by *Thielaviopsis basicola* and *Phymatotrichum* sp. Macrosporiosis (*Alternaria macrospora*) causes major problems in the CIS and other Asian regions as well as in Africa, whereas bacterial blight (*Xanthomonas campestris* pv. *malvacearum*) is widespread in Africa.

Other industrial crops include totacco and hop which are regularly attacked by destructive pathogens. In tobacco, blue mould (*Peronospora tabacina*,) black shank caused by *Phytophthora nicotianae*, black root rot (*Thielaviopsis basicola*), powdery mildew (*Erysiphe cichoracearum*) and leaf spot diseases due to *Alternaria alternata* and *Cercospora nicotianae* (frogeye disease) are of economic importance. Hop production is concentrated in Central Europe where cultivation is endangered especially by downy mildew (*Pseudoperonospora humuli*) and powdery mildew (*Sphaerotheca humuli*).

COFFEE

Coffee is produced especially in the poorer countries of Latin America and Africa where it is the major export crop and a key factor in the development of these countries. Diseases are of major importance in coffee production and can be highly destructive when not controlled. Coffee rust (*Hemileia vastatrix*) and coffee berry disease (*Colletotrichum kahawae*) are prevalent especially in East

Africa, whereas the strains occuring in South America are of lower virulence. Bacterial blight of coffee (*Pseudomonas syringae*) is limited to some high quality producing areas of East Africa. South American leaf spot (*Mycena citricolor*) is confined to Latin America. Brown eye spot (*Cercospora coffeicola*) and tracheomycosis (*Gibberella xylariodes*) locally cause losses.

The intensity of disease control varies considerably between regions. Many growers prefer broad-spectrum copper products which also have side effects on bacterial blight. World-wide crop losses to diseases are estimated to be an average of 14% (Oerke *et al.*, 1994, Figure 3). Without control measures losses would exceed 25%.

ORNAMENTALS

Ornamentals comprise very different plant species. Important diseases include damping-off (*Pythium* spp., *Phytophthtora* spp., ect.), wilt diseases due to *Fusarium* spp. and *Verticillium* spp., *Botryotinia fuckeliana*, leaf spot pathogens such as *Diplocarpon rosae*, rusts (*Phragmidium mucronatum*, *Puccinia* spp.) and powdery mildews.

TURFGRASS

Important diseases are dollar spot (*Sclerotinia homoeocarpa*), grey leaf spot (*Magnaporthe grisea*), stripe smut (*Ustilago striiformis*), *Erysiphe graminis*, red thread (*Laetisaria fuciformis*), *Puccinia* spp., *Uromyces* spp., snow mould (*Coprinus psychromorbidus*), *Typhula* blight (*Typhula incarnata*), pink snow mould (*Microdochium nivale*), anthracnose (*Glomerella graminicola*) and leaf and spots caused by *Pyrenophora* spp., *Cochliobolus* spp., *Setosphaeria* spp., *Leptosphaeria* spp. (spring dead spot). Foot and root diseases such as *Fusarium* spp., *Pythium* spp., *Thanatephorus cucumeris* and Bermudagrass decline (*Gaeumannomyces graminis*) play an important role during seedling stages as well as in established crops.

CONTROL MEASURES

The control of diseases should be managed by integrated disease management including crop rotation, use of resistant cultivars, cultivation with proper soil preparation, irrigation and judicious fertilization. All of these preventive measures can be optimized in order to reduce inoculum density and the build up of epidemics. However, often seed dressing is necessary to safeguard the juvenile stages of plant development. Active disease control in a growing crop almost completely depends on the use of synthetic fungicides. Biocontrol agents limiting crop losses are at present only of minor importance—some due to low efficacy, some due to high prices.

Table 6 Development of the fungicide market and the total pesticide market in the period 1960–95

	Sales ($ million)				
	1960	1970	1980	1990	1995
Total	850	2700	11600	26150	30300
Fungicides	340	594	2181	5550	5848
% of total	40.0	22.2	18.8	21.2	19.3

Foliar applications as sprays, in some cases also as dusts, predominate in the application of fungicides. Seed treatments are estimated to account for 10% of the total fungicide market, whereas soil treatments and the application of fumigants are of minor importance and rather limited due to restricted registration.

Data available on the actual fungicide market cannot represent the total use of fungicides because of the unknown production figures for East and South Asia, especially the People's Republic of China and India. These countries are estimated to produce currently a large amount of organic and inorganic compounds for domestic demand. However, the data demonstrate the continuous increase of fungicide sales peaking in the early 1990s (Table 6). Subsequently fungicide use was slightly reduced, namely in European cereals, and recovered only recently. Sales are projected to increase to more than $6000 million by the year 2000.

The very diverse group of fruits and vegetables accounts for about 40% of the total fungicide market (Figure 4). Fungicides in cereals, namely wheat, account

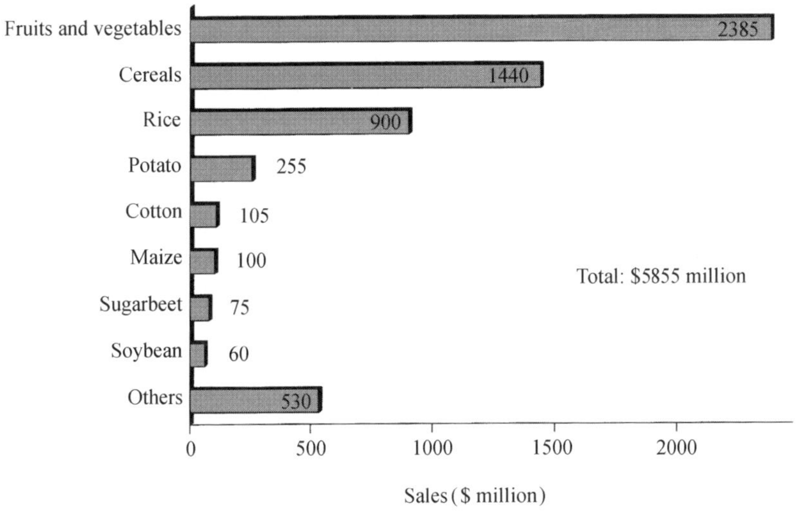

Figure 4 Fungicide use by crop in 1995 (data from County NatWest WoodMac, 1996)

Table 7 The fungicide market and the total pesticide market by crop in 1995

Crop	Total pesticides	Fungicides	% of total
Fruits and vegetables	6950	2385	34.3
Cereals	4560	1440	31.6
Rice	3650	925	25.4
Maize	3295	95	2.9
Cotton	2740	105	3.9
Soybeans	2600	50	2.0
Sugarbeet	875	75	8.6
Potato	825	255	31.1
Oilseed rape	505	40	8.1
Others	4345	300	6.9

for almost 25% and in rice for 15%. All other crops have only a small segment in the fungicide market totaling about 19% of annual sales. The importance of disease control is highly variable between crops. The share of fungicide costs in total expenditures on pesticides may be used as an indicator (Table 7). Disease control is of major importance in fruits and vegetables, wheat and barley, rice and potato. In most of the other crops fungicides account for less than 10% of total pesticide use.

Fruits and vegetables (grapes, citrus, pome fruits, nuts, stone fruits, tomato, cucurbits and cabbage) are the major crops receiving spray applications and/or seed dressings. With almost \$795 million spent on fungicides, grapes are by far the most important crop followed by citrus (\$285 million), apples (\$250 million) and nuts (\$200 million). Other fruits account for about \$310 million, while world-wide fungicide expenditure on vegetables is estimated to total \$540 million (County Nat West WoodMac, 1996). East Asia and West Europe are the major markets for fungicides in fruits and vegetables. In contrast to the market for cereal fungicides, North America and Latin America have also a higher share in this market (Figure 5).

In cereals, especially in wheat and to a significantly lower extent in barley, about \$1445 million are spent on fungicides. Western and Central Europe are by far the most important users of cereal fungicides safeguarding high productivity per area of land. Fungicide consumption is highest in France (\$565 million) followed by Germany and the UK. In all other regions the use is largely restricted to seed dressing, spray applications being rather rare (Figure 6).

The fungicide market in rice is dominated by Japan which accounts for two-thirds of world sales of \$900 million on rice fungicides, especially for the control of *Pyricularia oryzae* and *Rhizoctonia solani*. South Korea and China have a market share of about 10% and 5%, respectively. Despite having the world's greatest rice area, India accounts for less than 5% of fungicide sales. Fungicides in potato production are used regularly to control late blight and early blight. Because of the high damaging potential of these diseases,

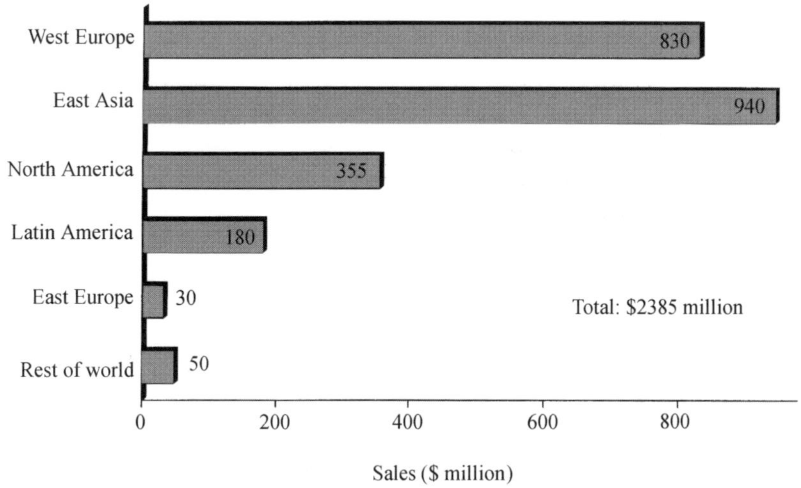

Figure 5 Fungicide use in fruits and vegetables, by region, in 1995 (data from County NatWest WoodMac, 1996)

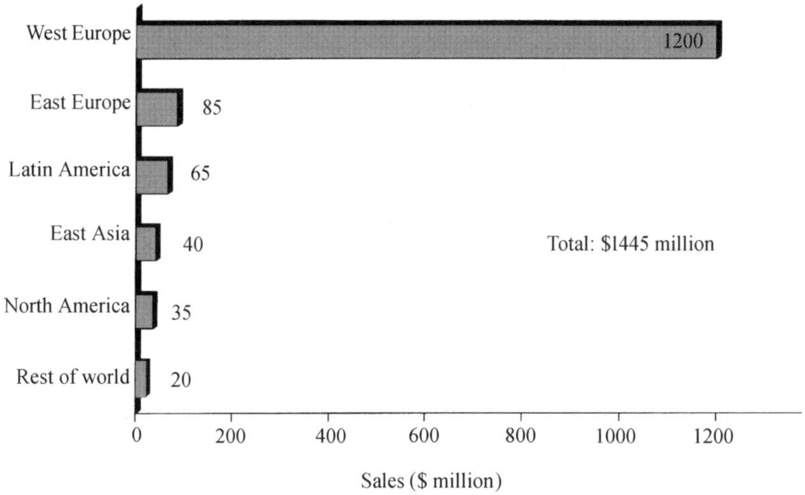

Figure 6 Fungicide use in cereals, by region, in 1995 (data from CountyNatWest WoodMac, 1996)

fungicide use is widespread, Nevertheless, East Asia and Western Europe account for about 60% of products used to control potato diseases.

Summarizing the fungicide sales by region, Western Europe has the highest intensity of fungicide use (Figure 7). A total of 41% of all fungicides are applied to only about 8% of agricultural land. The intensity is lower in East

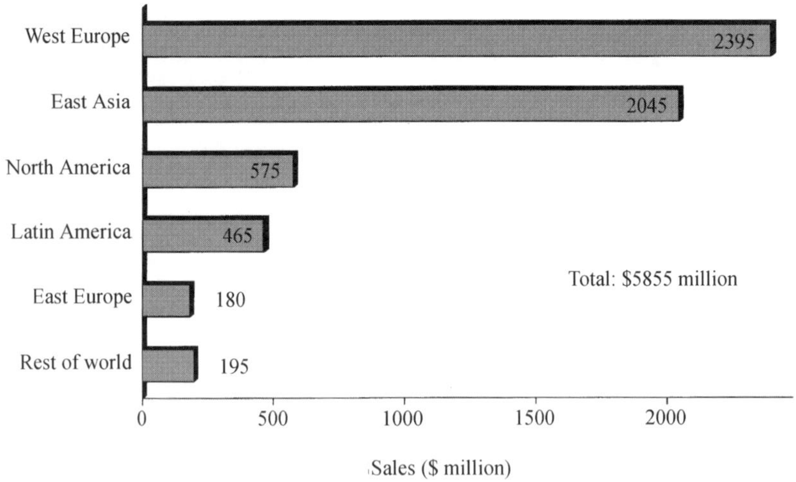

Figure 7 Total fungicide use by region in 1995, (data from CountyNatWest WoodMac, 1996)

Asia, where rice and fruits and vegetables account for the greatest market shares. In the Americas fruits, nuts and other perishable crops such as vegetables are the most important crops.

RECENT DEVELOPMENTS

Major developments in the use of fungicides include (a) the change to systemic fungicides with higher persistence, and (b) the reduction in the amount applied (Table 8). Almost all new introductions into the fungicide market have been

Table 8 Recent trends in the total sales of systemic and non-systemic fungicides in the period 1990–95

	Sales ($ million)			Sales ($ million)	
Group	1990	1995	Group	1990	1995
Systemic compounds			**Non-systemic compounds**		
Benzimidazoles	673	430	Dithiocarbamates	801	680
Triazoles	1172	1385	Inorganics	702	520
Substituted anilides	638	680	Other non-systemics	1033	780
Organophosphorus	232	205			
Morpholines	366	295			
Other systemics	818	880			
Total	3899	3875	**Total**	2536	1980

made by systemic fungicides. Their share in total sales simultaneously increased from 60% in 1990 to 66% in 1995. In this period of shrinking (EU-) markets, only triazoles representing the most important systemic fungicides and some other newer outlets also increased in absolute sales. Recent introductions in the fungicide market include strobilurins, anilinopyrimidines and quinolin-derivatives.

New techniques for the diagnosis and identification of plant diseases such as ELISA and PCR, and the development of systems for disease prognosis have contributed to a better, target-oriented use of fungicides. The introduction of curative compounds as well as the application of combination products have made possible the use of economic thresholds and a more cost-effective disease control. In the past decades, the amount of active ingredients applied per unit of area has been reduced, in some cases considerably. In many of the new introductions to the fungicide market high selectivity for specific pathogens, high efficiency due to fungitoxicity and low adverse effects on the environment are combined. Economic considerations have also resulted in a reduction in the dosage applied as well as in split applications and the use of combination products, respectively.

Apart from these positive trends, the rise in registration costs—especially due to stringent regulatory requirements and ecotoxicological studies—has led to a concentration of the research efforts on fungicides to be used in crops widely grown throughout the world. Less emphasis has been put on crops limited to developing countries and on minor crops susceptible to specific diseases. In some industrialized countries, too, shortcomings in disease control are evident for some crops. This trend is expected to continue.

Fungicides have been used for more than 100 years. Depending on crop and target pathogen(s), the application frequency varies between only one treatment per year and several treatments in spray schedules. The efficacy of broad spectrum, multisite fungicides has suffered hardly any change whereas resistance or reduced sensitivity have been reported for many systemic compounds, in some cases only a few years after introduction. For example, dicarboximide resistance in *Botryotinia fuckeliana* appeared in some areas after only about three years of intensive use of these fungicides which had replaced benzimidazoles because of resistance problems (Lorenz *et al.*, 1994). With phenylamide fungicides, used especially for the control of oomycetes, resistance was observed in 1980 reaching levels of over 60% of all *Phytophthora infestans* isolates in the UK and Netherlands (Clayton and Shattock, 1994; Staub, 1994). Resistance against DMI fungicides (azoles) was first noticed in cereal powdery mildew in 1981 and it is now known for at least 13 species in cereal and fruit crops (De Waard, 1994). For more detailed and up-to-date information see Chapter 4 of this volume.

In 1981, the Fungicide Resistance Action Committee (FRAC) was founded and fungicide resistance management strategies have been developed and successfully integrated. As a result, phenylamides are still a valuable tool to control oomycetes, dicarboximides are still widely used and DMI fungicides are growing in importance despite the occurrence of less sensitive strains of some target pathogens (Urech, 1994). Due to the introduction of new fungicide classes with other modes of action and a more sophisticated management of fungicide use, and the awareness that fungicide resistance is likely to appear for all single-site fungicides, crop losses due to fungicide-resistant pathogens have been largely prevented.

Research on genetically modified crops for the improvement of pathogen resistance is underway. However, at present only crops with modified food quality have been released. Only 3% of the field trials with genetically modified plants involved resistance to bacteria or fungi in the period 1986–93 (Ahl Goy and Duesing, 1995). Current efforts on higher levels for disease resistance in wheat, potatoes, fruits, etc. may result in introduction by the year 2000 or soon thereafter. In contrast to the state-of-the-art for insect-resistant crops, (gene) sources for resistance against pathogens are either not known or the transfer into plants has failed until now. Furthermore, the impact and success of conventional breeding for disease resistance are much higher than for other pest groups. Future release of genetically modified crops with disease resistance is likely to affect the fungicide market to a lesser extent than insect-resistant crops (maize, cotton, etc.). The latter may reduce the insecticide market especially in North America where public acceptance of genetic engineering is high at present.

REFERENCES

Ahl Goy, P. and Duesing, J. H. (1995). 'From pots to plots: genetically modified plants on trial', *Biotechnology*, **13**, 454–458.

Clayton, R. C. and Shattock, R. C. (1994). 'Strategies for phenylamide deployment: effects on phenylamide reistance in populations of *Phytophthora infestans*', in *Fungicide Resistance. BCPC Monograph No. 60* (eds S. Heaney, D. Slawson, D. W. Hollomon, M. Smith, P. E. Russel and D. W. Parry), pp. 179–181, British Crop Protection Council, Farnham, UK.

CountyNatWest WoodMac (1996). *Agrochemical Service*, County National West Wood Mackenzie, London.

De Waard, M. A. (1994). 'Resistance to fungicides which inhibit sterol 14α-demethylation, an historical perspective', in *Fungicide Resistance. BCPC Monograph No. 60* (eds S. Heaney, D. Slawson, D. W. Hollomon, M. Smith, P. E. Russel and D. W. Parry), pp. 3–10, British Crop Protection Council, Farnham, UK.

FAO (1996). *FAO Production Yearbook 1995*, Vol. 49, FAO Rome, Italy.

Lorenz, G., Becker, R., and Schelberger, K. (1994). 'Strategies to control dicarboximide-resistant *Botrytis* strains in grapes', in *Fungicide Resistance. BCPC Monograph No. 60* (eds S. Heaney, D. Slawson, D. W. Hollomon, M. Smith, P. E. Russel, and D. W. Parry), pp. 225–232, British Crop Protection Council, Farnham, UK.

Oerke, E.-C., Dehne, H.-W., Schoenbeck, F., and Weber, A. (1994). *Crop Production and Crop Protection—Estimated Losses in Major Food and Cash Crops*, Elsevier Science, Amsterdam.

Schwinn, F. J. (1992). 'Significance of fungal pathogens in crop production', *Pesticide Outlook*, **3**, 18–25.

Staub, T. (1994). 'Early experiences with phenylamide resistance and lessons for continued successful use', in *Fungicide Resistance. BCPC Monograph No. 60* (eds S. Heaney, D. Slawson, D. W. Hollomon, M. Smith, P. E. Russel, and D. W. Parry), pp. 131–138, British Crop Protection Council, Farnham, UK.

Urech, P. A. (1994). 'Fungicide resistance management: Needs and success factors', in *Fungicide Resistance. BCPC Monograph No. 60* (eds S. Heaney, D. Slawson, D. W. Hollomon, M. Smith, P. E. Russel, and D. W. Parry), pp. 349–356, British Crop Protection Council, Farnham, UK.

2 Fungicide Classes: Chemistry, Uses and Mode of Action

Y. UESUGI
Tokyo, Japan

INTRODUCTION

Although inorganic materials such as sulfur, copper sulfate, mercuric chloride and arsenic compounds were reported to be effective in controlling plant diseases even before 1800, use of these compounds as agricultural fungicides was not popular until their utility was widely recognized in the nineteenth century by spraying boiled lime-sulfur, which was originally used as an insecticide, and Bordeaux mixture, made from lime and copper sulfate.

The first synthetic organic chemicals used as agricultural fungicides may be the organomercurials introduced in the early twentieth century for seed treatment. In the 1930s and thereafter a variety of organic fungicides was

Fungicidal Activity. Edited by D. H. Hutson and J. Miyamoto

introduced. The fungicides developed in the early stages were selected by screening simple organic compounds by rather simple test methods such as the spore germination test which evaluates the inhibition of germination of spores of test fungi in a drop of solution of test chemical on a slide glass under the microscope. Among the organic fungicides thus developed, one of the most important groups is the dithiocarbamates. They are dimethyldithiocabamic acid derivatives such as ferbam, ziram and thiram, and ethylenebis(dithiocarbamates) (EBDC) such as zineb and maneb (Figure 1). Other important groups are quinones, such as chloranil and dichlone, *N*-haloalkylthiodicarboximides, such as captan, and dinitrophenol derivatives, such as dinocap (Figure 1).

These organic fungicides are usually far more effective than inorganic fungicides. Organic molecules are generally more compatible with fungal cells which are surrounded by walls and membranes in which a lipid layer is important in exchanging substances through the layer and preserving indispensable constituents within the cells. Most organic fungicides developed in the early stages have chemical reactivity which is another reason for their higher

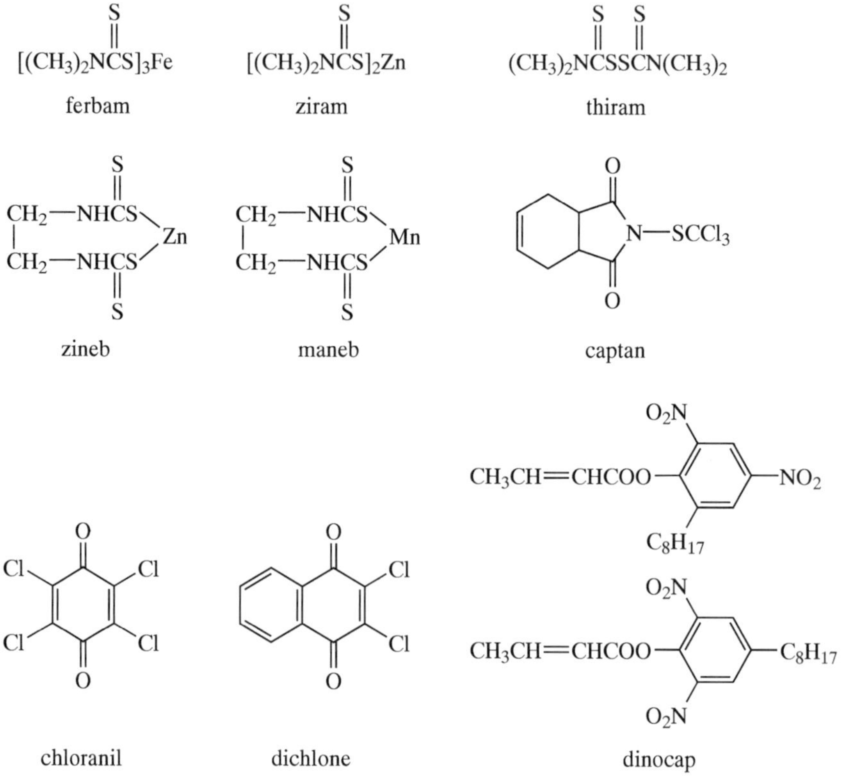

Figure 1 Chemical structures of multisite inhibiting and/or non-selective fungicides

activity. They react with functional groups of molecules of enzyme proteins and other important constituents of fungal cells and disturb the fungal physiology. Each of such fungicides acts on various sites in fungal physiology. This type of multiple-site inhibiting fungicides, which are also called multisite inhibitors, are liable to act on organisms other than their targets.

Even though some old types of fungicides act on specific sites within the fungal cells, their action is often non-selective when their site of action is common among a variety of organisms. An example of site-specific but non-selective action is the case with chlorinated phenol and dinitrophenol derivatives which uncouple oxidative phosphorylation in the process of aerobic respiration, and thus they are more or less toxic to a wide range of aerobic organisms. Non-selective action of the conventional fungicides sometimes caused undesirable effects on crop plants, mammals and other beneficial organisms. Since avoidance of the undesirable effects of pesticides was emphasized in and after the 1960s when environmental problems became a matter of concern, fungicide screening programmes were conducted to discover fungicides with higher effectiveness on plant diseases and lower undesirable effects on the environment. This effort resulted in the development of novel types of fungicides which are specifically active to plant diseases but not toxic to crop plants, mammals and other beneficial organisms. To discover these novel fungicides, various tests have been necessary for toxicological and environmental safety as well as for higher activity on host plants.

Specific activities in these novel fungicides are generally caused by their limited actions in the fungal physiology. Each of them acts on a single specific site in the fungal cells which is peculiar to the target pathogenic fungi. Since these sites are within the fungal cells living in or on the host plants, systemic properties are often necessary. Thus, most novel fungicides are systemic fungicides, while most conventional non-selective fungicides usually protect crop plants from pathogens by adhering to surfaces of the plants and they are often called protective fungicides. Sites of action of the novel fungicides are generally specific to the pathogenic fungi. Various types of specific-site inhibiting fungicides have been developed. Among them, inhibitors of biosynthesis of indispensable constituents in fungal cells form one of the most important groups of fungicides. The fungal cell constituents or their process of biosynthesis may be often specific to the fungi. Modes of action of some fungicide classes have not been elucidated but they appear to act on specific sites in fungal cells as their actions are specific.

This character in the novel fungicides may be closely correlated with the emergence of fungicide-resistant strains of the pathogenic fungi, which is sometimes a serious problem for the novel site-specific fungicides. The resistant mutants may arise by a mutation on just a single gene related to the site of action in the case of site-specific fungicides, while the mutations for marked resistance to multiple-site inhibiting fungicides need changes on many genes related to the multiple sites of the fungicides and therefore the problem of fungal resistance to multiple-site inhibiting fungicides is quite rare in practice.

Another problem with the novel selective fungicides is their limited range of efficacy among plant diseases. Although a lot of novel fungicides have been developed, control of some plant diseases is still unavailable because of lack of good novel fungicides and other control methods. Even though high efficacy is not necessarily expected with conventional multiple-site inhibiting fungicides, they are still important to cope with the resistance problem and to control diseases which are not controlled by other methods and they are still being used where their environmental effects are not hazardous.

BENZIMIDAZOLES

One of the most important specific-site inhibiting and systemic fungicide groups introduced in the 1960s was benzimidazole fungicides. The first compound developed was thiabendazole, 2-(1,3-thiazol-4-yl)benzimidazole, but it was introduced originally as an anthelmintic and was later used also as an agricultural fungicide. Though a variety of compounds were tested and introduced, a representative of this class of fungicides is benomyl. Another group of fungicides, seemingly different from benzimidazoles in their chemistry, was introduced during this period. They are thiophanate, diethyl 4,4'-(*o*-phenylene)bis(3-thioallophanate), and its dimethyl homologue, thiophanate-methyl. Biological activities of thiophanate fungicides were, however, quite similar to the benzimidazole fungicides especially to benomyl. A fungitoxic derivative easily derived from technical samples of benomyl was reported to be methyl benzimidazolylcarbamate (Clemons and Sisler, 1969). Later, conversions of thiophanates were studied and it was found that both benomyl and thiophanate-methyl are converted under natural conditions to a common active product, methyl benzimidazol-2-ylcarbamate, carbendazim (Figure 2) (Selling *et al.*, 1970; Vonk and Kaars Sijpestejn, 1971). This may be the reason why thiophanates exhibit disease control effects similar to benzimidazoles. Thus, thiophanate fungicides are now generally classified as benzimidazole fungicides. Carbendazim, the common active principle of benomyl and thiophanate-methyl, is also used as a fungicide. However, its physical properties demonstrated by (lower) solubility in ordinary organic solvents, for instance, are somewhat different from those of benomyl and thiophanate-methyl, and such differences in physical properties may be the reason for somewhat different effectiveness in plant disease control compared with its precursors.

Benzimidazole fungicides are active against various fungi especially *Ascomycetes* and their relative *Fungi Imperfecti* such as *Botrytis*, *Cercospora*, *Fusarium*, *Sclerotinia*, *Venturia* and *Verticillium*. These fungi include a wide range of pathogens of crop plants especially of vegetables, fruit crops and

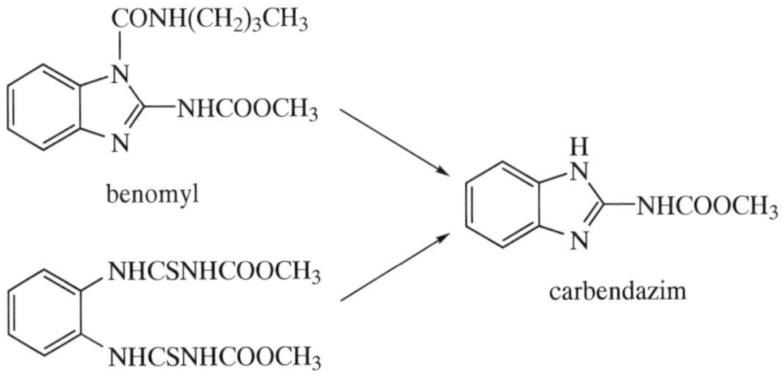

Figure 2 Conversions of benomyl and of thiophanate-methyl to a common product, carbendazim

ornamentals. It is interesting to note that benzimidazoles are effective in the control of diseases of mushrooms caused by *Trichoderma* and *Verticillium*. The host crops are fungi belonging to *Basidiomycetes*, to which benzimidazoles are less toxic, and thus the fungicides are used without phytotoxicity problems.

Concerning the mode of action of carbendazim, its inhibition of the process of mitosis in synchronized cultures of *Saccharomyces cerevisiae* and of *Ustilago maydis* was observed prior to its effects on DNA biosynthesis (Hammerschlag and Sisler, 1973). Davidse (1973) observed effects of carbendazim on mitosis of *Aspergillus nidulans* and suggested that its mode of action was via interference with spindle formation. Davidse and Flach (1977) further studied binding of radio-labeled carbendazim to fungal protein, presumably identical to tubulin. Tubulin is the building block of microtubules which are the components of spindles appearing in the process of mitosis. They compared the results with those obtained with other mitosis inhibitors and with other test organisms. They also tested mutants of the test fungus having varied sensitivity to carbendazim, which van Tuyl (1977) obtained and genetically analyzed. Differential binding of carbendazim was confirmed among the mutants. From these experimental results the mode of action of benzimidazole fungicides was concluded to be the inhibition of fungal mitosis by binding to the fungal tubulin. Benzimidazole fungicides are thus concluded to be multiplication inhibitors acting on fungal mitosis. At the site of action, however, they act by inhibiting biosynthesis of microtubules through the polymerization of tubulin. They may be regarded, therefore, also as a type of biosynthesis inhibitor.

N-ARYLCARBAMATES

N-Arylcarbamates are a group of fungicides having interesting selectivity in their action. The first development of this class was triggered by an observation

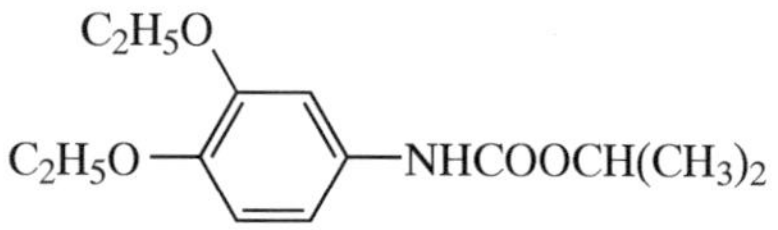

Figure 3 Chemical structure of diethofencarb

by Leroux and Gredt (1979) that some *N*-phenylcarbamate herbicides such as barban exhibit selective fungitoxicity especially against some strains of *Botrytis cinerea* resistant to benzimidazole fungicides. Since *N*-phenylcarbamate herbicides were known to be inhibitors of mitosis, this observation suggested a difference in selectivity in antimitotic activity between benzimidazoles and *N*-phenylcarbamates. The time was just when benzimidazole resistance in plant pathogenic fungi was a serious problem, but *N*-phenylcarbamate herbicides were unable to be used to control fungal strains resistant to benzimidazoles due to their phytotoxicity. Screening of derivatives of *N*-phenylcarbamate was conducted for a novel type of fungicide effective to the resistant fungal strains but not toxic to crop plants. Several compounds were found to be fungitoxic to benzimidazole-resistant strains without phytotoxicity. Among them diethofencarb (Figure 3) was developed as an agricultural fungicide (Takahashi *et al.*, 1988). Although this compound is effective in the control of various plant diseases caused by benzimidazole-resistant strains of pathogenic fungi, it is inactive to some types of resistant strains. In the case of *Botrytis*, most benzimidazole-resistant strains in the field are sensitive to diethofencarb, so it is used to control the resistant *Botrytis*. Since it is, however, less active to the original wild type of *Botrytis* sensitive to benzimidazoles, it is generally used as mixed preparations with a dicarboximide fungicide, procymidone, or with benzimidazoles such as thiophanate-methyl and carbendazim which are active to the wild type of the fungus.

Recently fungitoxicity seemingly due to antimitotic action of a derivative of propargyl *N*-2-pyridylcarbamate was reported (Mitani *et al.*, 1995). Selectivity of its fungitoxicity is different from that of diethofencarb and also from that of the benzimidazoles. It is effective not only to benzimidazole-resistant strains of *Botrytis* but also to the original wild-type strains sensitive to benzimidazoles. Various selectivity in antimitotic activity among *N*-arylcarbamates is an interesting fact and novel fungicides having other selectivities might be expected in this class.

A morphological observation was reported that an *N*-phenylcarbamate fungicide, MDPC, inhibits mitosis of a benzimidazole-resistant isolate of *Botrytis cinerea* in a similar manner to the action of carbendazim on a wild-type isolate (Suzuki *et al.*, 1984). Studies have been conducted on the increased sensitivity to *N*-phenylcarbamates accompanied by resistance to benzimidazoles. As stated above, the increased sensitivity to *N*-phenylcarbamates is not always observed with benzimidazole resistance. There seem to be several genes related to benzimidazole resistance and some or most of them may be also related with increased sensitivity to *N*-phenylcarbamates. This relationship has been recently clearly elucidated by analyses of genes corresponding to fungal β-tubulin which is the site of action of benzimidazoles and maybe of *N*-phenylcarbamates. Positions 198 and 200 in the amino acid sequence of β-tubulin, as analyzed by nucleic acid sequencing of the corresponding gene, were found to be important in benzimidazole resistance and *N*-phenylcarbamate sensitivity in *Venturia inaequalis* and other plant pathogenic fungi (Koenraadt *et al.*, 1992; Koenraadt and Jones, 1992). Changes at position 198 from glutamic acid, which is in the original wild-type fungus, to glycine and to alanine give rise to strains having increased sensitivity to diethofencarb and moderate and very high levels of resistance to benomyl, respectively. However, a change at position 198 from original glutamic acid to lysine and a change at position 200 from the original phenylalanine to tyrosine give rise to strains having unchanged low sensitivity to diethofencarb and high and moderate levels of benomyl resistance, respectively, and these two changes are unfavorable for control of the fungi by benzimidazoles and *N*-phenylcarbamates.

STEROL BIOSYNTHESIS INHIBITORS

This class of fungicides acts on fungal cells by inhibiting biosynthesis of fungal sterols. Since ergosterol was believed to be the main sterol in most fungi, this group of fungicides was originally called ergosterol biosynthesis inhibiting (EBI) fungicides. However, in some of the main targets of this fungicide group, such as powdery mildews of barley, apple and cucurbit and stem rust of wheat, the main fungal sterols were proved to be not ergosterol but related sterols produced through similar biosynthesis pathways (Pontzen *et al.*, 1990; Loeffler *et al.*, 1992). Powdery mildew and rust are caused by obligate parasites and isolation and identification of their fungal sterols had been difficult. Now agricultural fungicides of this class are called sterol biosynthesis inhibiting (SBI) fungicides rather than EBI fungicides.

SBI fungicides are effective in the control of plant diseases caused by *Ascomycetes*, *Basidiomycetes* and *Fungi Imperfecti* but they are not effective against *Oomycetes*. The spectrum of their fungitoxicity varies within the class and seems dependent on their chemical structures. This selectivity in their action signals the importance of sterols in the fungal cells as well as fitness of the chemical structures of the fungicides to bind the target sites in the respective fungi. SBI fungicides are practically useful to control diseases such as powdery mildews and rusts which were rather difficult to control with other fungicides or by other methods. Among them, powdery mildew of cereals is an economically important target of this class. Resistance problem with other classes of potent fungicides, such as benzimidazole and hydroxypyrimidine fungicides, is partly the reason for this. Although SBI fungicides themselves have resistance problems, avoidance of the problem seems often possible by switching to another SBI with a low level of cross-resistance or by changing the application method. The merit of this class is its utility in the control of a wide range of diseases having resistance problem with other classes of fungicides.

There are several groups in terms of their mode of action, though most agricultural SBI fungicides belong to two main groups, one is the demethylation inhibitor (DMI) group and the other includes derivatives of morpholine and its analogues. DMI fungicides inhibit a demethylation step in fungal sterol biosynthesis. Morpholines and analogues are assumed to inhibit an isomerization step and a reduction step in fungal sterol biosynthesis. These modes of action will be described in more detail below.

DEMETHYLATION INHIBITORS

This class includes triazole and imidazole fungicides which are important in practice and sometimes called azole fungicides. The DMI class also includes fungicides having other aryl heterocyclic rings such as pyrimidine and pyridine. Another interesting compound in the class is triforine which has no aryl ring but has a saturated heterocyclic ring, piperazine, instead. This fungicide class is therefore rather varied in terms of chemistry. Chemical structures of these DMI fungicides are shown in Figures 4, 5, and 6. In the case of azole fungicides (Figure 4 and 5), stereo-isomeric configuration especially around the second (β) atom in the side chain attached to nitrogen atom of the azole ring often affects the activity. The configurations around the first (α) and the third (γ) atoms sometimes also affect the activity seemingly when they affect the configuration at the second atom. This fact suggests that their fungitoxic action is due to binding of the azole ring to a stereo-specific target site in the fungal cells.

The mode of action of the DMI fungicides was first elucidated with a pyrimidine fungicide triarimol and with a piperazine fungicide triforine (Figure 6). The effect of triarimol on fungal sterol biosynthesis from radio-labeled acetate in *Ustilago maydis* was tested and the inhibition of C-14 demethylation of 24-methylenedihydrolanosterol in the process was observed (Ragsdale and Sisler, 1973; Ragsdale, 1975). Similarities were observed between triforine and triarimol in fungitoxic action and also in cross-resistance in *Cladosporium cucumerinum* and its mutants resistant to triarimol (Sherald *et al.*, 1973). Sterol biosynthesis inhibition by triforine at the C-14 demethylation step in *Aspergillus fumigatus* was also observed (Sherald and Sisler, 1975). During the period of these studies, the mode of action of a pyridine fungicide buthiobate (Figure 6) was also investigated. Inhibition of fungal sterol biosynthesis from radio-labeled acetate in *Monilinia fructigena* (Kato *et al.*, 1974), probably at the demethylation step (Kato *et al.*, 1975) was reported. Inhibition of C-14 demethylation in sterol biosynthesis by buthiobate was confirmed in an experiment with cell-free extracts of *Saccharomyces cerevisiae* by using radio-labeled intermediates of sterol biosynthesis (Kato and Kawase,

1976). The modes of action of azole fungicides triadimefon (Buchenauer, 1977) and triadimenol (Buchenauer, 1978) (Figure 4) in *Ustilago avenae* were studied somewhat later and their inhibition of C-14 demethylation in the process of fungal sterol biosynthesis was suggested. Since triadimefon was reported to be reduced to the more active triadimenol (Gasztonyi and Josepovits, 1979), the mode of action of the former might proceed through the formation of the latter.

A common mode of action among pyrimidine, piperazine, pyridine and triazole fungicides is astonishing. Among them the piperazine fungicide triforine has a somewhat exceptional chemical structure as stated above. Concerning the action of this fungicide, Langcake *et al.* (1983) cited an interesting

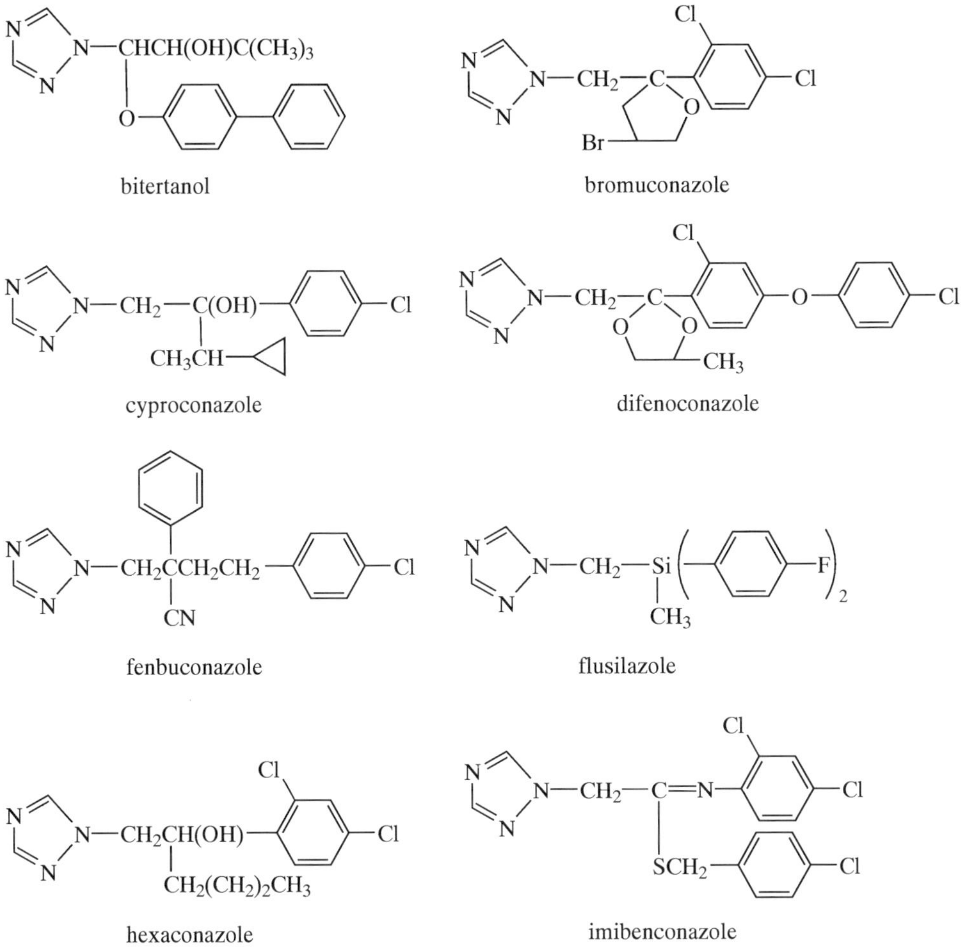

Figure 4 Chemical structures of triazole fungicides

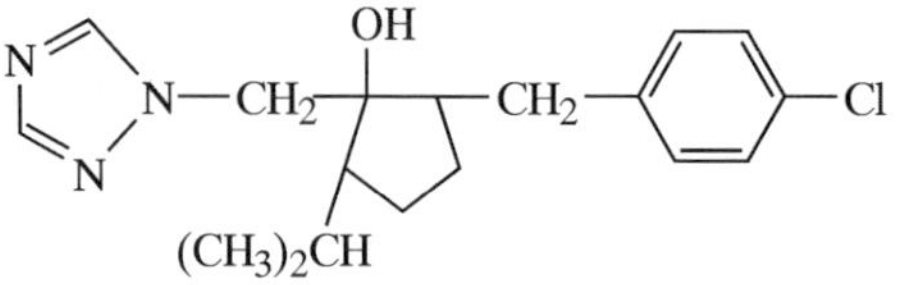

ipconazole

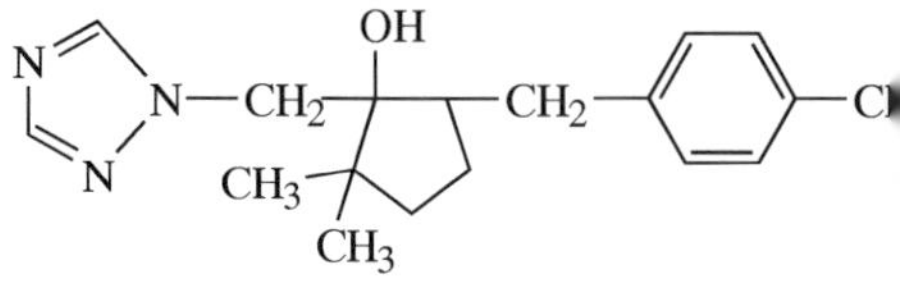

metconazole

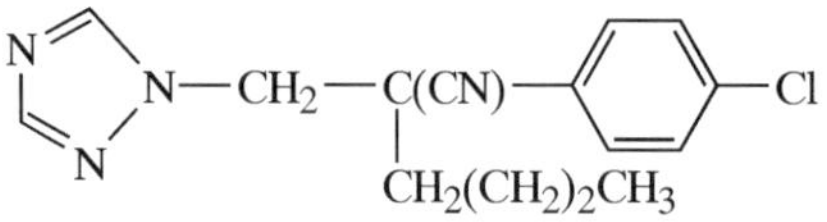

myclobutanil

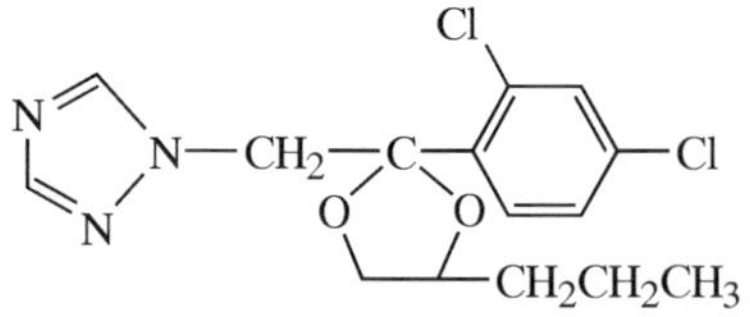

propiconazole

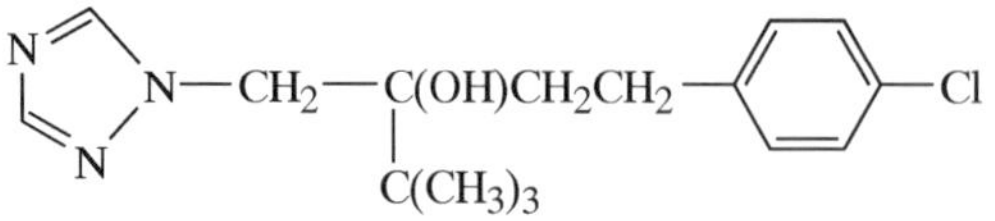

tebuconazole

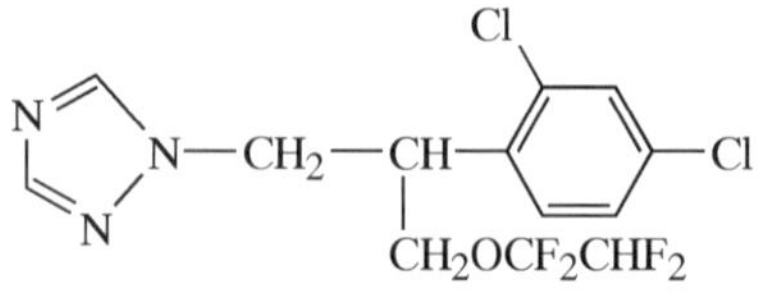

tetraconazole

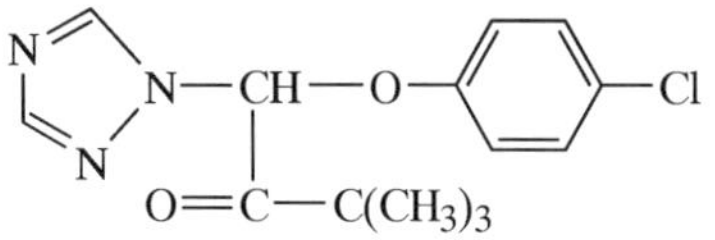

triadimefon

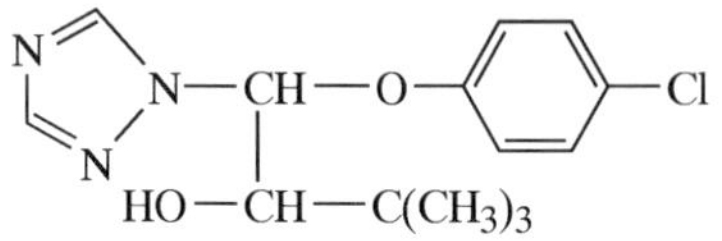

triadimenol

Figure 4 *(continued)*

unpublished observation that triforine is inactive on sterol demethylation when tested with a cell-free enzyme system of yeast which is sensitive to other DMI fungicides. Activation or other additional mechanism(s) might be involved in the mode of action of triforine.

The C-14 demethylation step in sterol biosynthesis is generally mediated by a mixed-function oxygenase (mfo) cytochrome P-450, which is a kind of hemoprotein. In fungal cells, there may be a variety of cytochromes P-450 involved not only in sterol biosynthesis but also in metabolism of other lipids, xenobiotics, etc. DMI fungicides presumably bind to cytochrome P-450 involved in sterol demethylation. They are also reported to act as synergists and antagonists to other pesticides, indicating that they may act on enzymes involved in various functions other than sterol demethylation. Although it is just a speculation by the author, DMI fungicides may sometimes fail to bind to some of the cytochrome P-450 in sterol demethylation and sometimes they may bind

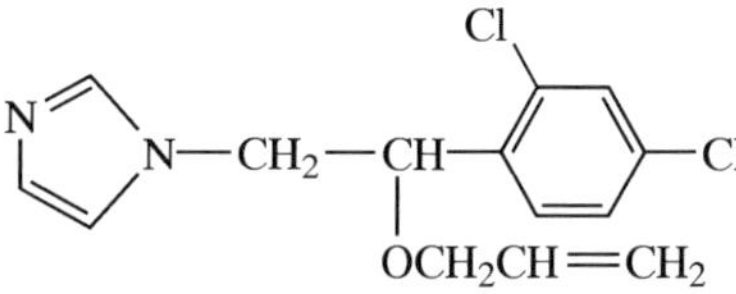

imazalil

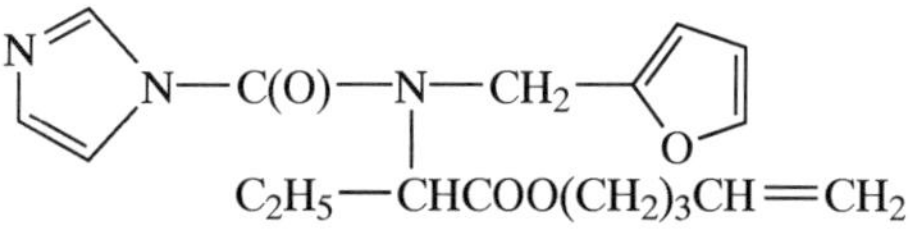

pefurazoate

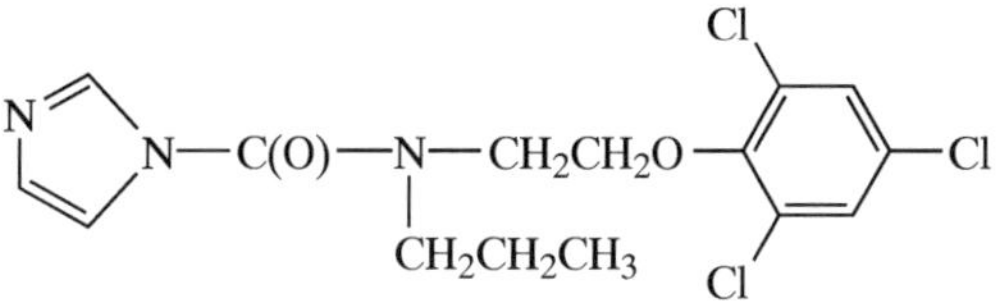

prochloraz

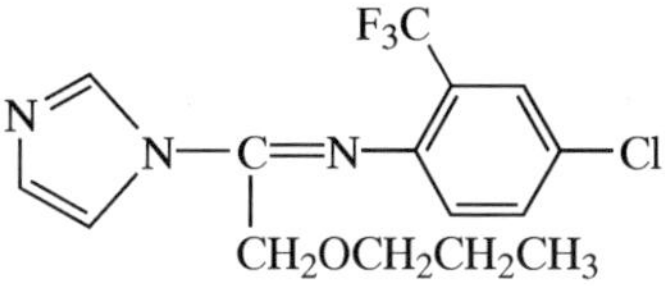

triflumizole

Figure 5 Chemical structures of imidazole fungicides

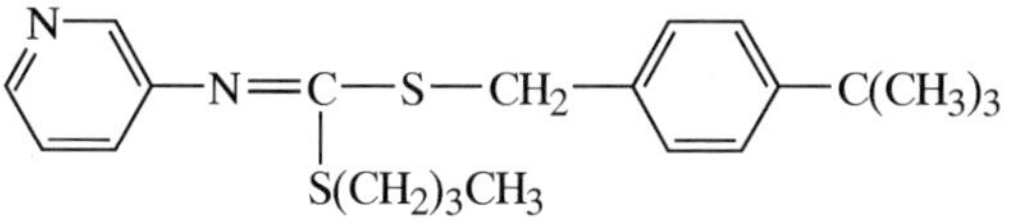

buthiobate

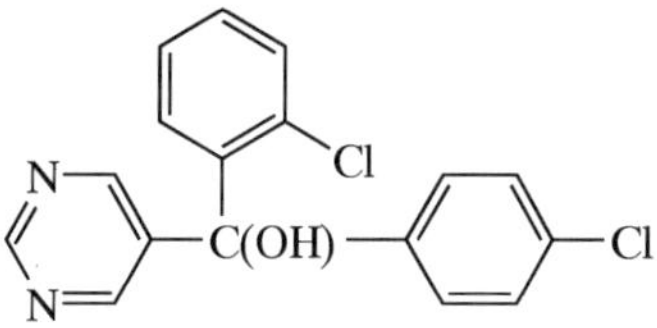

fenarimol

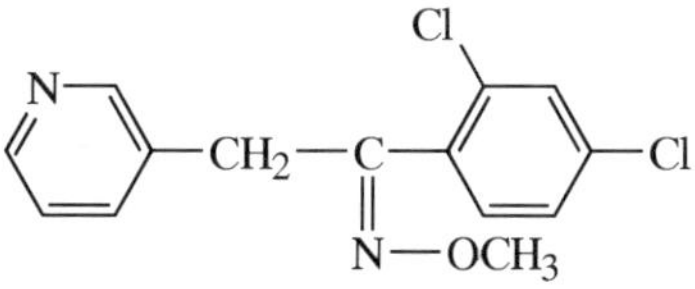

pyrifenox

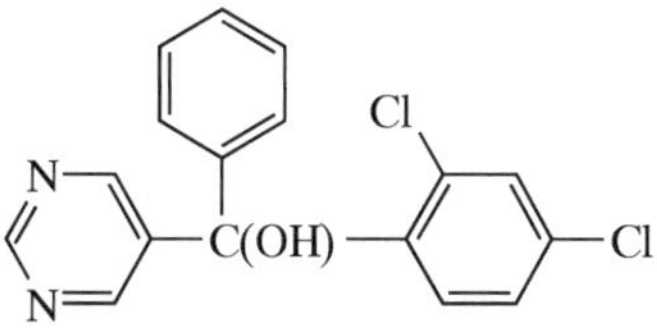

triarimol

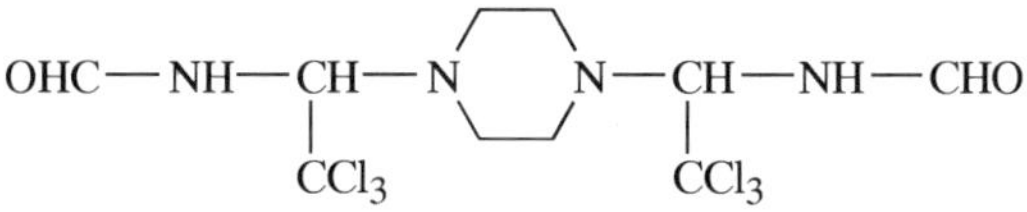

triforine

Figure 6 Chemical structures of DMI fungicides with six-membered nitrogen heterocycles

to P-450 and hemoprotein involved in other functions. This might be the reason for multiple aspects in the antifungal action of DMI fungicides which are exemplified by various antifungal spectra, various resistance factors in cross-resistance relationships and various interactions (i.e. synergism and antagonism) with other fungicides.

THE MORPHOLINES

The 4-substituted 2,6-dimethylmorpholines are the oldest group of SBI fungicides. Substituents were bulky hydrocarbon groups such as cyclododecyl (dodemorph) and long chain alkyl groups with around 13 carbons (tridemorph). Although economical advantage of the azoles and other DMI fungicides overtook the morpholines, the merit of morpholines was re-evaluated after resistance problems with DMIs became serious in practice, and novel morpholine fungicides and the analogous piperidine fungicides have been developed. Examples of these are illustrated in Figure 7.

The mode of action of the morpholine fungicide, tridemorph, in *Botrytis cinerea* was studied and the inhibition by the fungicide of isomerization by a shift of unsaturation $\triangle 8 \rightarrow \triangle 7$ in the fungal sterol biosynthesis pathway was suggested (Kato *et al.*, 1980). This isomerization seems a common site of action of morpholine fungicides and their analogues in general. Another site in the process of sterol biosynthesis, $\triangle$14-reduction, was also proposed for the action of the same fungicide in *Ustilago maydis* (Kerkenaar *et al.*, 1981). Inhibition of this site was also confirmed in *U. maydis* and *Saccharomyces cerevisiae* (Baloch *et al.*, 1984) and in a cell-free enzyme system of *S. cerevisiae* (Baloch and Mercer, 1987) in addition to the inhibition of isomerization $\triangle 8 \rightarrow \triangle 7$. The inhibition of reduction was rather weak in the case of tridemorph, but it was more remarkable with fenpropimorph and fenpropidin and suggested an additional action of the latter two fungicides. The dual site of action may be advantageous in combatting resistance to those fungicides.

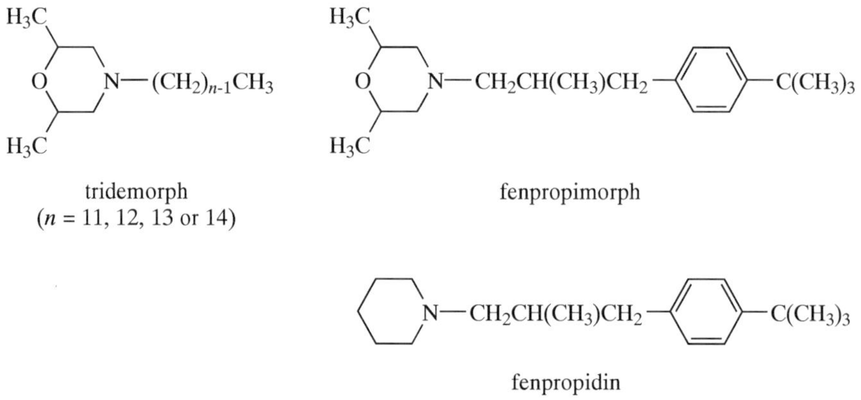

Figure 7 Chemical structures of morpholine and analogous fungicides

ACYLALANINES AND RELATED FUNGICIDES

This class of fungicides is often called the phenylamide fungicides. There are, however, several other classes having chemical structures regarded as phenylamides, e.g. aryl carboxanilides, dichlofluanid, tolylfluanid and flusulfamide. These fungicides differ from the acylalanines and related fungicides in mode of action, antifungal spectrum and other properties as plant protecting agents. To minimize the confusion, the name 'phenylamide' is not used here. Examples of this class are shown in Figure 8. The *N*-acyl-*N*-(substituted phenyl)alanine ester structure is common to benalaxyl, furalaxyl and metalaxyl, while the methyl propionate moiety of the structure is substituted with an oxazolidinone ring in oxadixyl.

This class holds an important position in controlling diseases caused by *Oomycetes* to which most other fungicide groups are less effective. Soil diseases caused by *Phytophthora* including late blight of potato and tomato, and downy mildew of vines and other crops are important targets of this class.

Their mode of action was investigated first with metalaxyl. Its interference with fungal biosynthesis of ribonucleic acid (RNA) was observed by several groups of workers. However, inhibition of the biosynthesis was not complete even at concentrations high enough to inhibit the fungal growth as observed by the effect on the incorporation of uridine, an RNA precursor. Later work by Davidse and his coworkers (1983) revealed selective inhibition of biosynthesis of ribosomal RNA (rRNA) by metalaxyl in comparison with the effect on biosynthesis of other classes of RNA such as transfer RNA and messenger RNA.

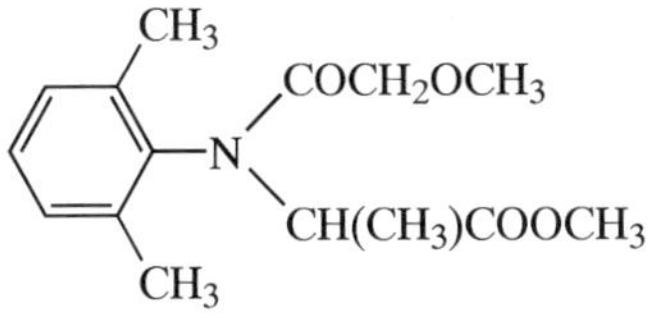

metalaxyl

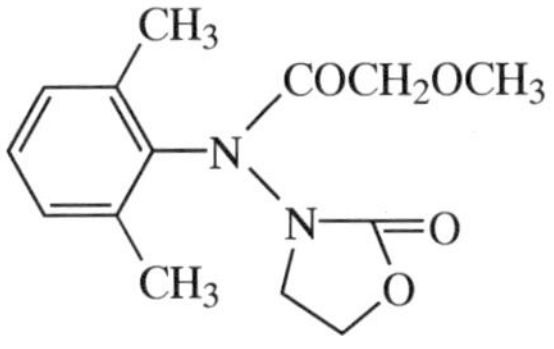

oxadixyl

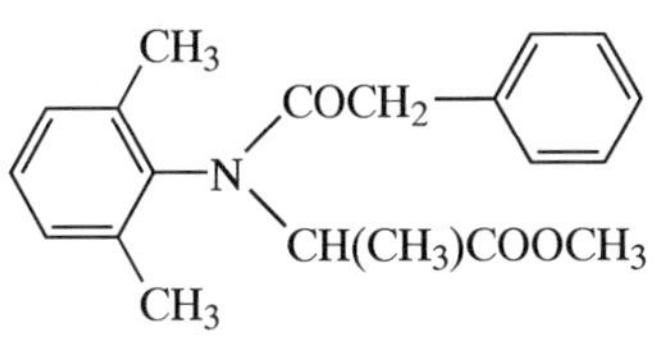

benalaxyl

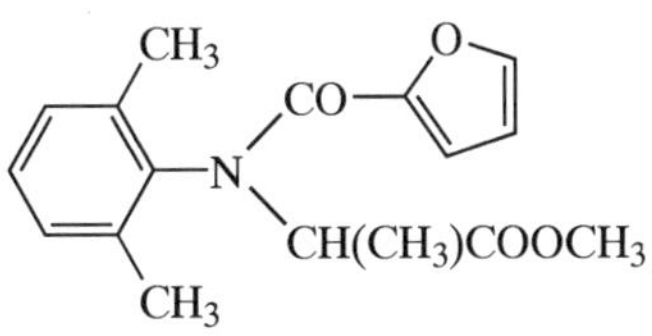

furalaxyl

Figure 8 Acylalanine and related fungicides

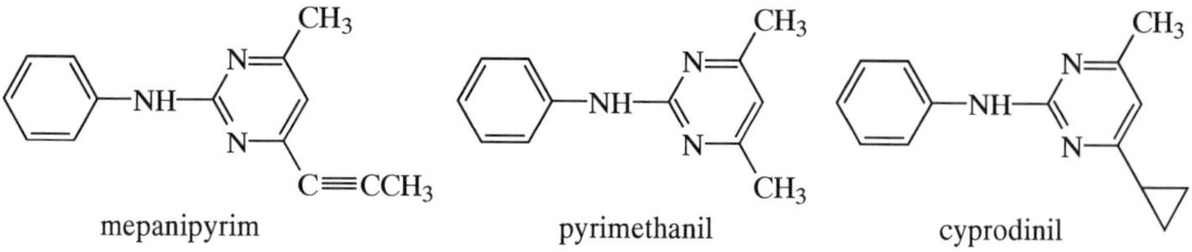

Figure 9 Anilinopyrimidine fungicides

ANILINOPYRIMIDINES

Although the chemical structures of this class of fungicides are relatively simple, they were developed rather recently. Examples are mepanipyrim, pyrimethanil and cyprodinil as shown in Figure 9. They are effective in the control of plant diseases caused by *Ascomycetes* and *Fungi Imperfecti*. Mepanipyrim and pyrimethanil exhibit good control of gray mold on vegetables, vines and pome fruits and cyprodinil was reported to be promising in controlling a wide range of diseases including powdery mildew and eye spot of cereals.

On the mode of action of these fungicides, mepanipyrim (Miura *et al*., 1994) and pyrimethanil (Milling and Richardson, 1995) were reported to inhibit the secretion of hydrolyzing enzymes such as cutinase, pectinase and cellulase from fungal cells. These hydrolyzing enzymes are necessary for the fungi to penetrate plant cell surfaces by digesting the plant surface. Thus the fungicides may inhibit the infection process without possessing a direct fungitoxic action. However, the mode of action of cyprodinil was studied in relation to that of mepanipyrim and pyrimethanil and another mechanism was proposed. Since fungitoxicities of cyprodinil as well as of mepanipyrim and pyrimethanil were reversed by the addition of methionine, inhibition of fungal biosynthesis of methionine by the fungicides was suggested as the mechanism of their action (Masner *et al*., 1994). Further studies may be necessary on the relation between the proposed two different mechanisms of action.

ORGANOPHOSPHORUS AND RELATED FUNGICIDES

Since phosphorus is an atom capable of forming a wide range of compounds having versatile biological functions and activities, many organophosphorus compounds have been proposed as fungicides and plant disease control agents. Among them, the phosphorothiolate fungicides (Figure 10) have been important for the control of rice blast caused by *Pyricularia oryzae*. Other organophosphorus fungicides now being used are pyrazophos, a pyrazolopyrimidinyl phosphorothionate, effective in controlling powdery mildews, tolclofos-methyl, a substituted-phenyl phosphorothionate, active on *Rhizoctonia* and other pathogens of soil-borne diseases, and fosetyl-Al, aluminum ethyl phosphonate,

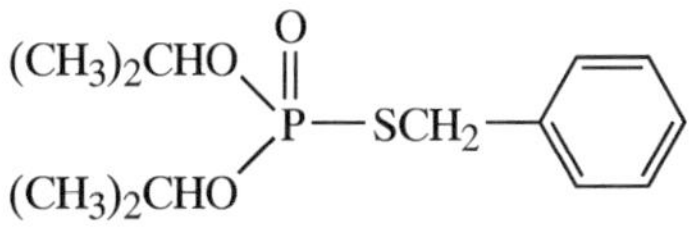

iprobenfos

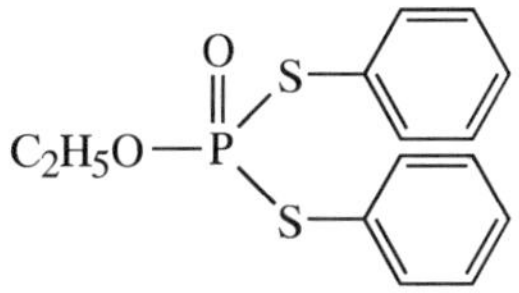

edifenphos

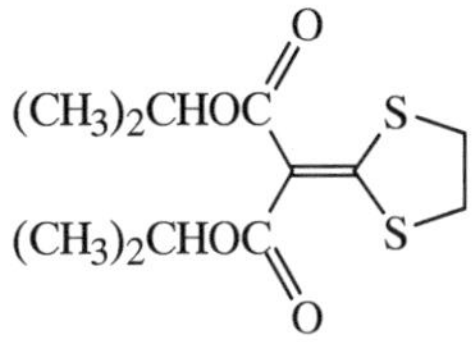

isoprothiolane

Figure 10 Phosphorothiolate fungicides and isoprothiolane

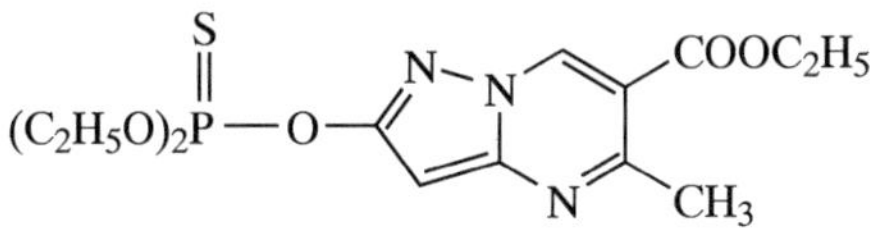

pyrazophos

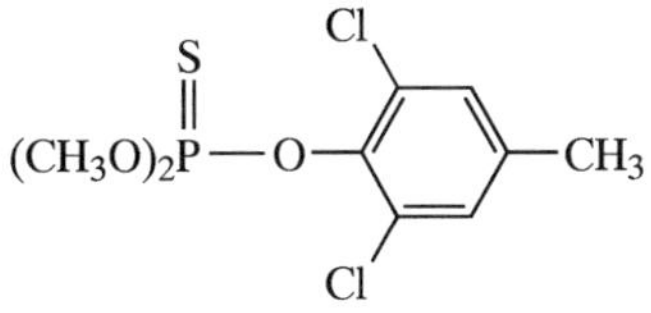

tolclofos-methyl

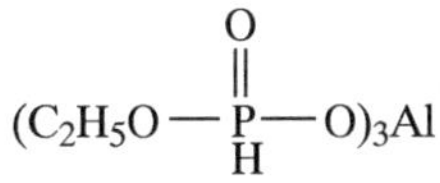

fosetyl-Al

Figure 11 Other organophosphorus fungicides

effective in controlling diseases caused by *Oomycetes* (Figure 11). These are effective to rather specific ranges of diseases, and there seems to be no common action among them.

Modes of action of phosphorothiolates have been intensively studied and the inhibition of chitin biosynthesis was first proposed. However, several groups of workers pointed out that the inhibition of chitin synthesis was unlikely or may be a secondary effect of their fungitoxic action. Finally Kodama and his coworkers discovered the inhibitions by iprobenfos (1979) and by edifenphos (1980) of fungal transmethylation from methyl-^{14}C-labeled methionine into phosphatidylcholine in *P. oryzae*. Thus the mode of action of the fungicides was concluded to be the phospholipid biosynthesis inhibition.

For the phosphatidylcholine biosynthesis, two major pathways are known. One is Greenberg's pathway in which transmethylation is the last step, i.e. the last intermediate, phosphatidylethanolamine, is methylated to produce phosphatidylcholine. The other pathway is Kennedy's pathway in which transmethylation is the first step or rather just before the pathway, i.e. ethanolamine is methylated to produce choline which is used as one of the building blocks of phosphatidylcholine. Since an inhibitor of Kennedy's pathway, hemicholinium-3, was not toxic to the fungus, the pathway was postulated as unnecessary to the fungus, and the inhibition site by phosphorothiolates was once suggested to be on the other route, Greenberg's pathway. Another approach to the mode of action of phosphorothiolates was investigated (Yoshida *et al.*, 1984). *P. oryzae* was incubated with [^{13}C-methyl] methionine and the change of the status of ^{13}C in the intact fungal cells was observed with ^{13}C NMR. Fungal transmethylation was shown in NMR spectra by the decreasing peak height at the chemical shift corresponding to the C–S bond and by an increasing peak height corresponding to the C–N bond. Inhibition of fungal transmethylation by phosphorothiolates was again confirmed. The ^{13}C–N compound produced at the initial stage of the incubation was, however, identified as choline. Thus it was concluded that the phosphorothiolates are inhibitors of fungal biosynthesis of choline and Kennedy's pathway may be important in the fungal biosynthesis of phosphatidylcholine.

Another fungicide, isoprothiolane, is also active on *P. oryzae* and its fungitoxic action was similar to the phosphorothiolates in many respects and cross-resistance was often observed between the phosphorothiolates and isoprothiolane. Although isoprothiolane has no phosphorus atom in its molecule, it has two sulfur atoms in thioether linkages in the molecule (Figure 10). The sulfur atoms are linked to a carbon atom having a double bond which is conjugated with carbonyl groups. This special position of sulfur atoms may be a cause of similarities with phosphorothiolates which have sulfur atoms in thioester linkages. In fact, it has also proved to be an inhibitor of the fungal choline biosynthesis (Yoshida *et al.*, 1984). Phosphorothiolate fungicides and isoprothiolane might be, therefore, better called organosulfur fungicides.

On the fungitoxic action of phosphorothiolate fungicides, a close connection between the fungitoxicity and fungal metabolism of the fungicides was suggested by comparative studies using sensitive and resistant strains (Uesugi and Takenaka, 1992). Normal wild types of *P. oryzae* metabolized phosphorothiolates by cleaving the P–S and S–C bonds. The metabolism seemed to be mediated by cytochrome P-450 which was induced by the fungicide itself. This metabolism was barely detected in laboratory mutants of the fungus selected for resistance to the fungicides. In other strains which were isolated from the field intensively sprayed with the fungicides and moderately resistant to them, the fungicides were degraded mainly by cleavage of the S–C bond and the P–S cleavage was hardly detected. These results suggest that activation of the fungicides through P–S cleavage occurs but the active metabolite has not been identified. The proposed activation of the phosphorothiolates was further supported by the joint action of *P. oryzae* with DMI fungicides which are inhibitors of constitutive cytochrome P-450 (Sugiura *et al.*, 1993). The DMIs inhibited the P–S cleavage and antagonized the fungitoxicity of the thiolates. The cytochrome P-450 induced by thiolates in this experiment seemed to bind with the DMIs and to decrease its availability for binding with the constitutive cytochrome P-450 involved in sterol biosynthesis. Thus fungitoxicity of the DMIs was also antagonized by the thiolates.

On the mode of action of pyrazophos, activation by oxidation to its oxone followed by hydrolysis of the oxone to produce a fungitoxic pyrazolopyrazole derivative, PP, was proposed by De Waard (1974). However, other possibilities were also suggested by other researchers. Although several investigations have been reported on the action of pyrazophos, fungitoxicities of pyrazophos and of PP were variable depending on the test fungi and on fungal growth stages and infection stages and no definite specific action has been identified especially on its main targets, powdery mildews. Multiple mechanisms might be involved in its fungitoxic action.

The mechanism of action of tolclofos-methyl was studied with *Ustilago maydis* by Nakamura and Kato (1984) who proposed the inhibition of fungal cytokinesis as a major mode of action. Disorder of fungal cell structure and morphology caused by this fungicide seems somewhat similar to those by aromatic hydrocarbons, dicarboximides and phenylpyrroles. Actually cross-resistance was often (but not always) observed among them in fungal strains and both tolclofos-methyl and aromatic hydrocarbons inhibit the swimming of zoospores of *Oomycetes* fungi showing the inhibition of flagellar movement which is thought to involve cytokinesis inhibition.

The mode of action of fosetyl-Al is still unsolved. It has little fungitoxic activity *in vitro* and several hypotheses have been proposed for its disease control effect. They are the induction of natural disease resistance in the plants by the fungicide, a direct action of the fungicide on fungal cells which is much higher *in vivo* than *in vitro*, and the reduction of the virulence of the pathogen by the fungicide.

MELANIN BIOSYNTHESIS INHIBITORS

The first melanin biosynthesis inhibiting (MBI) fungicide was 1,2,3,4,5-pentachlorobenzyl alcohol (PCBA) which was introduced in the 1960s though its mode of action was unknown at that time. It had little fungitoxic activity *in vitro* but it was effective as a preventative treatment for rice blast caused by *Pyricularia oryzae*. Derivatives of pentachlorobenzaldehyde, namely, the oxime and cyanohydrin were similarly effective. They are, however, not used now because their metabolites, i.e. penta- and tetrachlorobenzoic acids, caused indirect phytotoxicity on other sensitive crops such as vegetables. Fungicides of

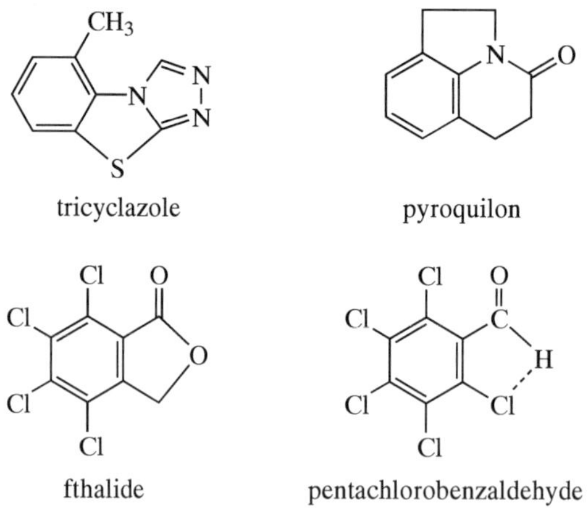

Figure 12 Chemical structures of MBI fungicides and pentachlorobenzaldehyde

various structures were later found to belong to this class. Those now used for rice blast control are fthalide (phthalide), tricyclazole and pyroquilon (Figure 12). The compounds effective as melanin biosynthesis inhibitors (MBIs) generally have fused bicyclic structures (Woloshuk and Sisler, 1982). In the case of PCBA and pentachlorobenzaldehyde mentioned above, their possible common metabolite, pentachlorobenzaldehyde, may form a fused bicyclic structure by an intramolecular hydrogen bond as shown by the dotted line in Figure 12. It is interesting to note that some intermediates of fungal melanin biosynthesis are naphthalene derivatives which have fused bicyclic structures and compounds having similar structures may act as competitors. The mode of action of the MBIs was first elucidated with tricyclazole by Tokousbalides and Sisler (1979) to be inhibition of fungal melanin biosynthesis at the reduction step from 1,3,8-trihydroxynaphthalene (1,3,8-THN) to vermelone in the biosynthetic pathway. At the same time, they suggested an important role for the melanin layer of the appressorial wall of *P. oryzae* in the epidermal penetration of the host plants by the fungus. The melanin layer may give physical rigidity to the appressorial wall to support and maintain the pressure within the fungal cells in the process of epidermal penetration of the plants.

Another MBI, carpropamid (Figure 13), was recently developed. Its chemical structure has, however, no fused rings. The mode of action of this fungicide was also inhibition of fungal melanin biosynthesis but the site of action was reported to be different from that of the above mentioned MBIs. Two dehydration processes were proposed as the sites for carpropamid, one is from scytalone to 1,3,8-THN and the other is from vermelone to 1,8-dihydronaphthalene in the fungal melanin biosynthetic pathway (Kurahashi *et al.*, 1996).

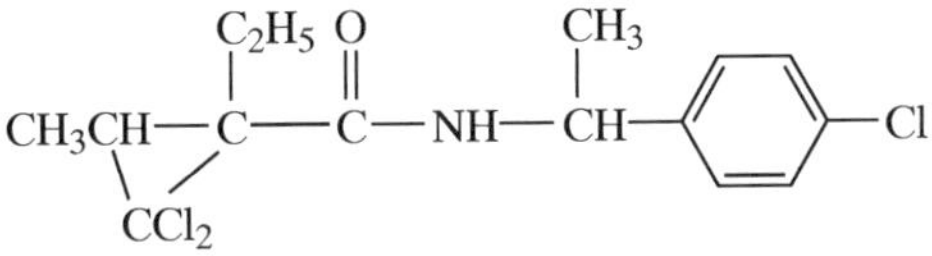

Figure 13 Chemical structure of carpropamid

Since melanin is unnecessary to *P. oryzae* except at the stage of the penetration of the surface of host plants, MBIs are usually not toxic to ordinary fungal growth. Although selective biosynthesis inhibitors generally have resistance problems in the field, there has been almost no problem with MBIs probably because the process of selection of resistant strains in the field is quite different from that of other fungicides and the chance of emergence of resistant strains is considered to be far less frequent with MBIs as compared with other biosynthesis inhibiting fungicides.

PROBENAZOLE

Probenazole (Figure 14) is also a plant protecting agent having little fungitoxicity *in vitro*, but it is effective in the prevention of rice blast, an important fungal disease, when the plants are treated prior to arrival of the pathogen as in the case of MBI fungicides. Its mode of action is, however, seemingly different from that of the MBI fungicides. Although the details are still unknown, it is supposed to induce resistance of the plants to the disease (Sekizawa and Watanabe, 1981). The rice plants treated with probenazole exhibit signs of increased resistance to the disease such as higher level of peroxidase activity. The acquired systemic resistance in the plants is also effective in the control of another disease caused by bacteria, bacterial leaf blight. Screening of test compounds for novel plant protecting agents having activity similar to that of probenazole has been intensively conducted and several novel candidates have been reported. Also some existing fungicides have been reported to induce acquired systemic resistance to diseases, but probenazole may be the only compound in current use belonging exclusively to this class of plant protecting agents.

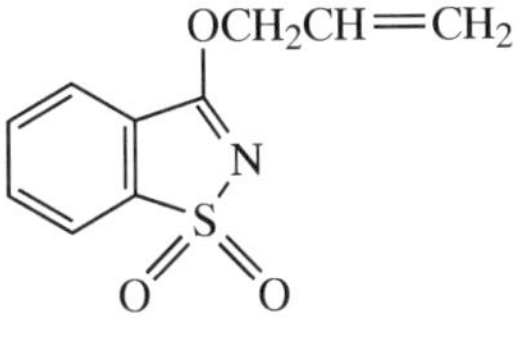

Figure 14 Chemical structure of probenazole

DICARBOXIMIDES

Fungicides having a common chemical structure, *N*-(3,5-dichlorophenyl)-dicarboximide, are called dicarboximide fungicides and they have a similar antifungal spectrum and activity in common. Fungicides of this class used now are chlozolinate, iprodione, procymidone and vinclozolin shown in Figure 15. There are other fungicides having the dicarboximide structure but no 3,5-dichlorophenyl group, such as captan and fluoroimide, but they differ somewhat from the *N*-(3,5-dichlorophenyl)dicarboximides in their antifungal spectra and actions and are usually not included in this class.

Dicarboximide fungicides are effective in controlling *Botrytis*, *Sclerotinia* and other plant pathogenic fungi on vines, vegetables, fruit crops, ornamental plants and other crops. Although details of their mode of action are still unknown, they cause disorder of fungal cell structure especially when the cells are growing and multiplying. Morphological changes to hyphae and germ tubes and leakage of fungal cell contents are often observed following treatment with the fungicides. Cross-resistance has been observed with aromatic hydrocarbons, with tolclofos-methyl, and with phenylpyrroles in fungal strains resistant to this class. The modes of action of these fungicide groups are also unknown but similarities in morphological changes of the test fungi treated with them suggest a similar action.

Recently an interesting result was reported on the molecular genetic analysis of a gene conferring a high level of resistance to vinclozolin and tolclofos-methyl in laboratory mutants of *Ustilago maydis*. The gene encoded a sequence quite similar to those corresponding to serine (threonine) protein kinases (Orth *et al.*, 1995). This indicates that protein kinase may be closely related with the

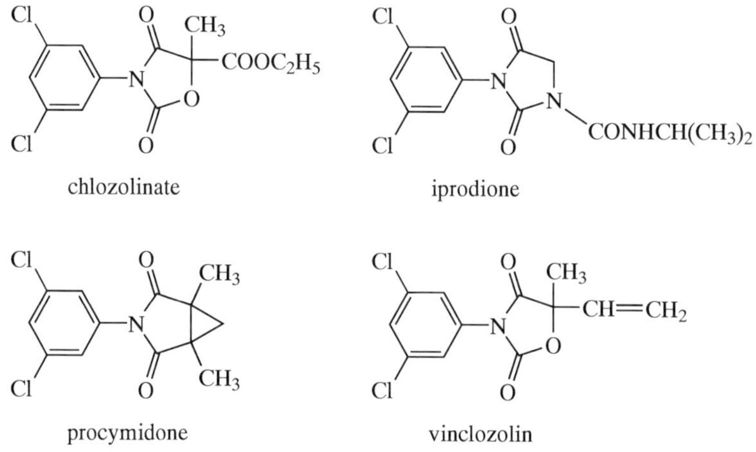

Figure 15 Chemical structures of dicarboximide fungicides

high level of resistance to vinclozolin and tolclofos-methyl and, therefore, somehow with the mode of action of these fungicides. Protein kinases are enzymes phosphorylating proteins by transferring a phosphate group from ATP to proteins. Phosphorylation and de-phosphorylation of proteins affects inactivation and activation of the functions of proteins (including various enzymes) and thus various physiological functions of organisms are controlled. Signal transduction, exchange of ions and substances through membranes, cell cycle, biogenesis of cell structures and other functions of organisms are controlled in this way. Thus the mode of action of carboximides might be related with some of these functions in fungal cells.

PHENYLPYRROLES

This class of fungicides, fenpiclonil and fludioxonil, have structures derived from that of an antifungal antibiotic pyrrolnitrin as shown in Figure 16. They have a rather wide range of antifungal spectrum including *Ascomycetes* and *Basdiomycetes* and are used as seed fungicides to control *Fusarium*, *Tilletia* and other seed-borne pathogens. Fludioxonil is also used as a foliar fungicide to control diseases caused by *Botrytis*, *Monilinia*, *Sclerotinia* and other fungi. The mode of action of pyrrolnitrin has been reported as inhibition of energy production by uncoupling oxidative phosphorylation in fungal respiration. The uncoupling was also observed with fenpiclonil in the mitochondrial respiration of *Botrytis cinerea* but the concentration needed for this action was much higher than that for its fungitoxicity and it seemed unlikely as its mode of action (Leroux *et al.*, 1992).

Uncoupling of oxidative phosphorylation is often reported to be caused by disorder of the mitochondrial membrane which is analogous to the cell membrane in both physiological and physical senses. The effects of these fungicides on fungal growth, multiplication, formation of cell structure and its maintenance seems to be related with the functions of the cell (cytoplasmic) membrane. Phosphorylation and de-phosphorylation of proteins by protein kinases and phosphatases stated earlier are often reported to be carried out on

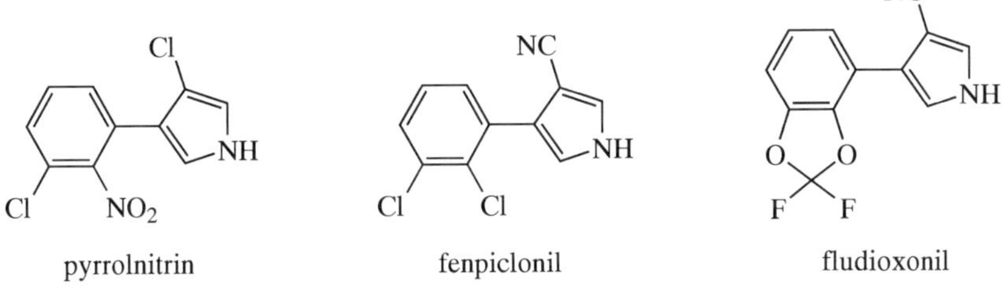

Figure 16 Chemical structures of pyrrolnitrin and phenylpyrrole fungicides

and in the membrane. The mode of action of these fungicides might be, therefore, related with the disorder of some function(s) of the membrane.

ARYL CARBOXANILIDES

The original fungicides of this class were carboxin and its sulfone, oxycarboxin, (Figure 17) introduced in the late 1960s. These fungicides were specifically effective on *Basidiomycetes* and were used to control rusts, smuts and other diseases caused by *Basidiomycetes* and related fungi on various crops. Their use was rather limited, however, for economic reasons. Around 1980 and thereafter, benzanilide and other aryl carboxanilide derivatives were developed to control sheath blight of rice caused by *Rhizoctonia solani* which is an important disease of rice and against which arsenic fungicides had been widely used. These were mepronil, flutolanil, thifluzamide and furametpyr as shown in Figure 17. They have fungitoxic action similar to that of carboxin. The primary action of carboxin, oxycarboxin and other arylcarboxanilides has been investigated by several research groups and it is concluded that they inhibit fungal respiration by inhibiting succinate dehydrogenase, i.e. succinate-ubiquinone reductase. The

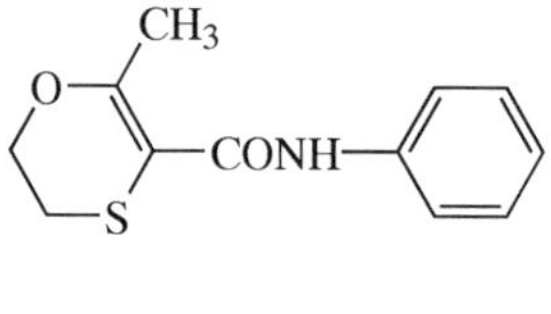

carboxin

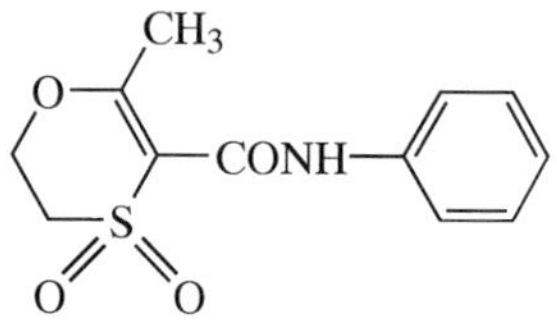

oxycarboxin

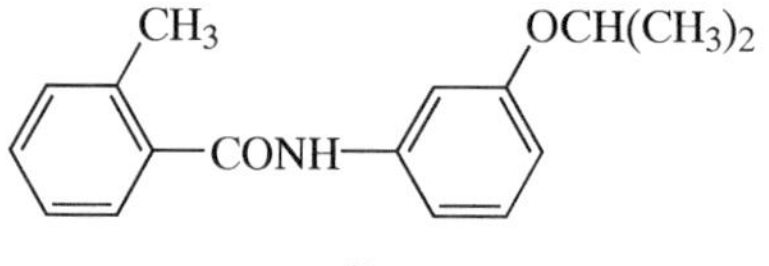

mepronil

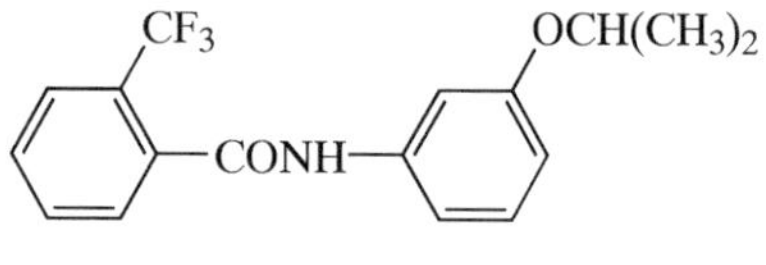

flutolanil

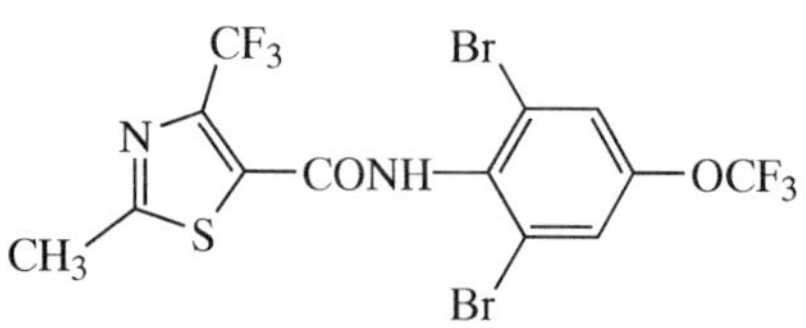

thifluzamide

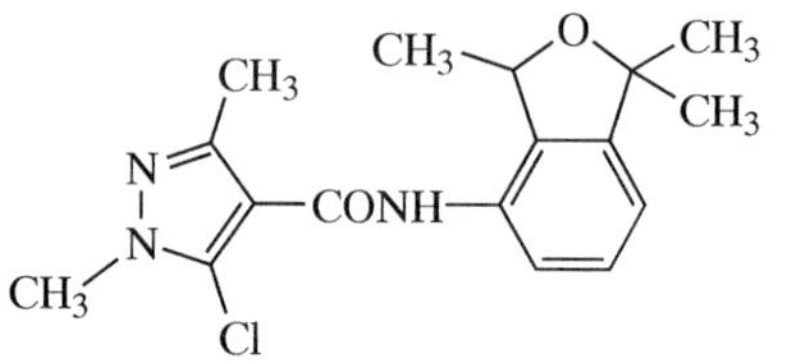

furametpyr

Figure 17 Chemical structures of aryl carboxanilides

binding of carboxin derivatives to this site of action seemed specific and the effects of the derivatives on succinate dehydrogenase complex (complex II) extracted from fungal cells were dependent on the molecular form of the inhibitors and strains and mutants of the test fungi. An interesting example was demonstrated by some laboratory mutants of *Ustilago maydis* moderately resistant to carboxin which were obtained *in vitro* by a single gene mutation. To this category of mutants, a derivative of carboxin, 4′-phenylcarboxin, was highly active in the fungal growth inhibition test and in the enzyme inhibition test, though the derivative was not necessarily an effective inhibitor to wild types of the fungus (White *et al.*, 1978). Thus, this single gene mutation seemed to bring about the change in binding of this series of inhibitors to the target enzyme of the fungus.

It is interesting to note that target organisms of the aryl carboxanilides developed so far are almost limited to the *Basidiomycetes*. Although there have been reports that some aryl carboxanilides inhibit cell-free succinate dehydrogenase extracted from fungi other than *Basidiomycetes* (such as *Ascomycetes*), there have been no practical fungicides of this class sufficiently effective towards them. This fact might suggest that this site of action is less vital to those fungi except *Basidiomycetes*, possibly because an alternative pathway, e.g. the glyoxylate cycle, bypasses or compensates for the succinate dehydrogenation step in the citric acid cycle of the fungal respiration system.

METHOXYACRYLATE AND RELATED FUNGICIDES

This class of fungicides originated from the discovery of the antifungal antibiotics, strobilurins and oudemansins, produced by fungi belonging to the *Basidiomycetes*. They have a basic chemical structure of *E*-β-methoxyacrylate in common. Since synthetic analogues of these antibiotics often exhibited considerable activity towards diseases caused by a wide range of fungi, trials to discover novel fungicides have been conducted. Among them, a methoxyacrylate, azoxystrobin (ICIA 5504), a methoxyiminoacetate, kresoxim-methyl (BAS 490F), and a methoxyiminoacetamide, SSF-126, are being developed and will probably be put into practice soon (Figure 18). Although these fungicides are effective towards a rather wide range of diseases, each has its specialty in control. Thus, kresoxim-methyl is highly effective against powdery mildews, azoxystrobin is a good fungicide against downy mildews and SSF-126 is especially effective towards rice blast.

The mode of action of these fungicides has been identified as inhibition of the respiratory chain at the cytochrome b/c_1 complex (complex III) site (Ammermann *et al.*, 1992; Baldwin *et al.*, 1995; Mizutani *et al.*, 1995). An interesting observation on an effect of this class of fungicides was that the inhibition of the cytochrome respiratory pathway gives rise to induction of an alternative respiratory pathway insensitive to cyanide in *Pyricularia oryzae*

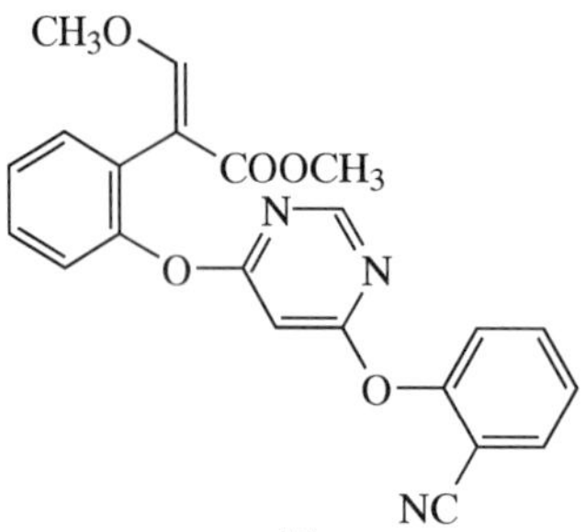

azoxystrobin

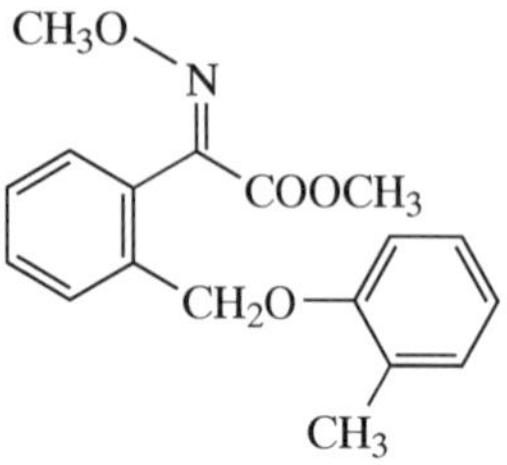

kresoxim-methyl

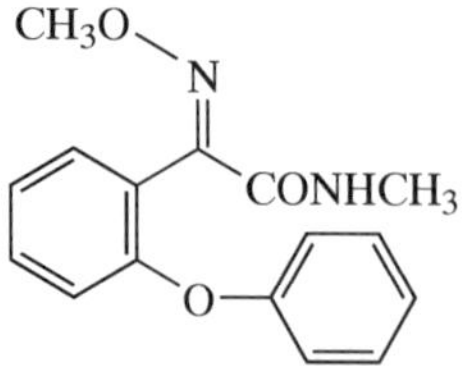

SSF - 126

Figure 18 Chemical structures of methoxyacrylate and related fungicides

(Mizutani *et al.*, 1995). Similar effects were also observed in *Botrytis cinerea* and other fungi (Hayashi *et al.*, 1996). The fungicides seem, therefore, not absolutely fungicidal, but their activity is probably due to cooperative inhibition of the alternative respiratory pathway and/or of induction of the pathway by natural antioxidants such as flavonoids and other phenolic substances present in the host plants. In fact, this class of fungicides generally exhibit rather low fungitoxicity *in vitro* in contrast to their good control effects on the fungal diseases in the field. The methoxyacrylates are described in more detail in Chapter 5.

OTHER FUNGICIDES

FLUAZINAM

This fungicide has a broad antifungal spectrum and is used for a wide range of fungal diseases of various crop plants including late blight on potatoes, downy mildew and gray mold on vines, apple scab, etc. It is an anilinopyridine having two nitro- two chloro- and two trifluoromethyl-substituents on the rings linked through the amino group in its structure (Figure 19). This rather special chemical structure may cause protonation of the amino group and give rise to a biological activity as a potent uncoupler of oxidative phosphorylation which

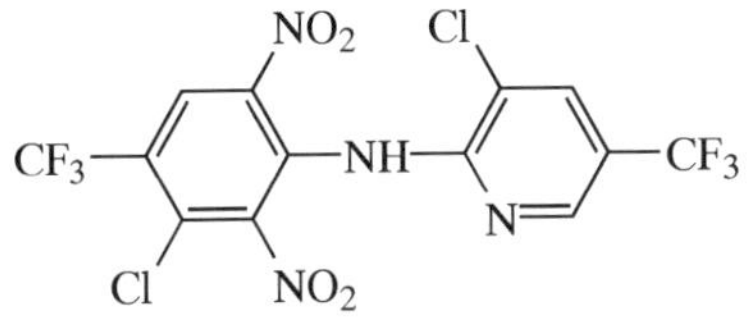

Figure 19 Chemical structure of fluazinam

was observed in mammalian mitochondria (Guo *et al.*, 1991). Uncoupling activity of the fungicide was rapidly decreased, however, in ordinary preparations of mammalian mitochondria probably due to detoxification by conjugation with glutathione.

Since uncoupling activity was not necessarily paralleled by antifungal activity in this series, structure-activity relationships were analyzed. In *Sphaerotheca fuliginea*, uncoupling of oxidative phosphorylation was suggested as the mode of fungitoxic action of fluazinam (Akagi *et al.*, 1996), while in *Botrytis cinerea*, another factor such as reactivity with sulfhydryl or other groups was suggested as its mode of action (Akagi *et al.*, 1995). Further study will be necessary for the elucidation of its mode of antifungal action.

FERIMZONE

This fungicide was developed to control rice blast caused by *Pyricularia oryzae*, but it is also effective in the control of *Helminthosporium oryzae*, *Cercospora oryzae* and other fungi which cause miscellaneous diseases on rice plants. Since its action is rather curative, it is usually used in combination with other blast fungicides with preventative action such as melanin biosynthesis inhibitors. Ferimzone is a pyrimidinylhydrazone of 2′-methylacetophenone (Figure 20). Its mode of action was studied and no or little effect was observed on fungal respiration or on biosynthesis of fungal cell constituents but fungilytic action and inhibition of the incorporation of leucine and acetate into fungal cells were observed. Thus the mode of action of ferimzone was suggested to be disruption of membrane function (Okuno *et al.*, 1989). This kind of action is often found with compounds relevant to uncouplers of oxidative phosphorylation, as in the case of phenylpyrroles stated earlier. In the case of ferimzone,

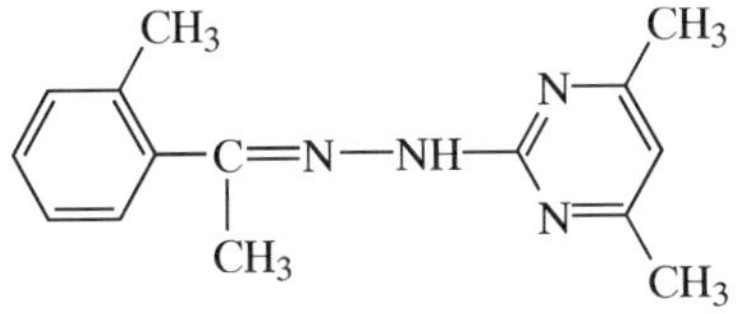

Figure 20 Chemical structure of ferimzone

potent uncouplers CCCP and FCCP have a basic chemical structure of phenylhydrazone in contrast to the pyrimidinylhydrazone structure of ferimzone. Disorder of functions of the mitochondrial membrane and of the cytoplasmic membrane caused by these inhibitors might have similar effects.

AROMATIC HYDROCARBONS

The substituted benzenes shown in Figure 21 are rather old fungicides. In contrast to other old types of fungicides such as dithiocarbamate, quinone and copper fungicides, they are rather inert in their ability to react with fungal cell constituents having sulfhydryl or other reactive groups. They are, therefore, not necessarily multiple-site inhibiting fungicides and have their specific niches in controlling plant diseases. Thus, quintozene is effective in controlling soil diseases caused by *Plasmodiophora*, such as club root of lettuce and cabbage. It is also effective in controlling *Sclerotinia* and *Botrytis* but not effective against *Fusarium* and *Pythium*. On the other hand, chloroneb is effective in controlling diseases caused by *Pythium*. It is also effective against *Rhizoctonia* and *Typhula* and it is used as a soil and seed fungicide. Dicloran is effective in controlling diseases of vegetables, stone fruits, berry fruits and other crops caused by *Botrytis*, *Sclerotinia*, *Monilinia* and other fungi. This spectrum is somewhat similar to dicarboximide fungicides developed later and now dicloran has been partially replaced by them. It is interesting to compare the chemical structures of dicloran and dicarboximides. Instead of the nitro group of dicloran, dicarboximides have dicarboximido groups which are somewhat similar to the nitro group in attracting electrons within the respective molecule, and the two chlorine atoms are substituted at similar positions in those molecules.

Cross-resistance was reported between dicloran and the dicarboximide fungicides in some field isolates of *Sclerotinia*, *Botrytis* and other fungi. Cross-resistance was also sometimes observed in laboratory mutants of various fungi among aromatic hydrocarbon fungicides as well as among dicarboximides, tolclofos-methyl, phenylpyrroles and aromatic hydrocarbon fungicides.

The mode of action of the aromatic hydrocarbon fungicides has long been investigated and their effects on multiplication, respiration, biosyntheses of protein, lipids and nucleic acids, formation and the maintenance of cellular

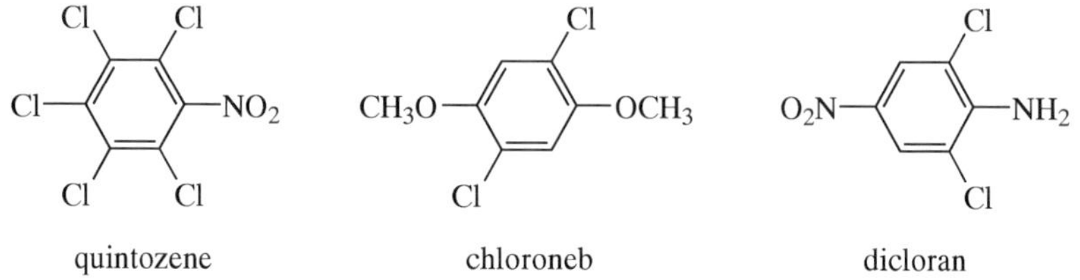

Figure 21 Chemical structures of aromatic hydrocarbon fungicides

and mitochondrial structures and other physiological profiles of fungal growth have been reported but no definite conclusion on the primary mode of action has been reached. It might be related with the modes of action of dicarboximides, tolclofos-methyl and phenylpyrroles as suggested by their cross-resistance relationships, but the actions of those fungicides are also still unknown.

FUNGICIDES CONTROLLING DISEASES CAUSED BY *OOMYCETES*

Plant diseases caused by *Oomycetes* have long been a serious problem in agriculture and the chemical control was initiated with Bordeaux mixture followed by ethylenebis(dithiocarbamate). Now fosetyl-Al, methoxyacrylates, fluazinam, acylalanine and related fungicides have been developed to control those diseases as stated earlier. However, due to the complex nature of the diseases, other fungicides shown in Figure 22 are also used.

Chlorothalonil is a fungicide developed early in the 1960s. It appears to belong to the aromatic hydrocarbon fungicide class but it has some reactivity with sufhydryl or other reactive groups and this was suggested as a probable mechanism of action (Vincent and Sisler, 1968), though a definite conclusion has not been drawn. It is effective in controlling not only diseases caused by

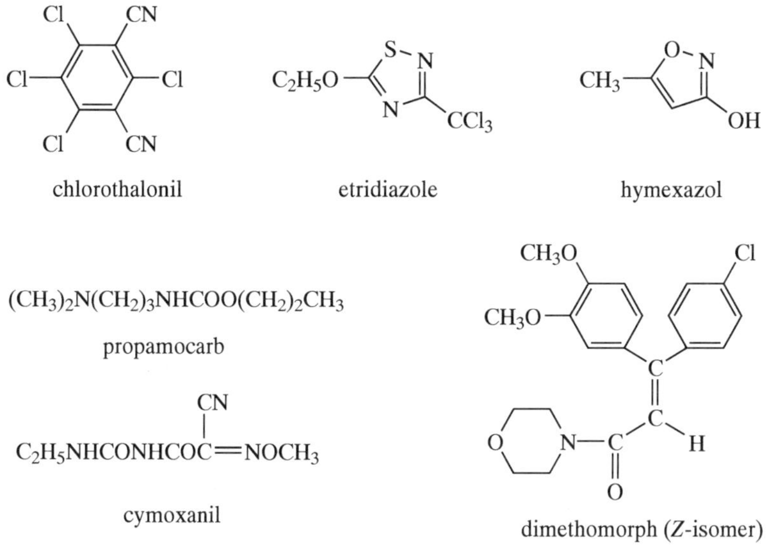

Figure 22 Chemical structures of fungicides controlling *Oomycetes*. Acylalanine and related fungicides (Figure 8), fosetyl-Al (Figure 11), methoxyacrylates (Figure 18) and fluazinam (Figure 19) are also effective in the control of *Oomycetes*

Oomycetes fungi such as downy mildews of cucumber, melons and cabbages and late blight of tomato and peppers, but also gray mold, powdery mildews and other diseases caused by fungi other than *Oomycetes*.

Etridiazole is a soil fungicide effective in controlling diseases caused by *Pythium* and *Phythophthora*. It is not necessarily a member of aromatic hydrocarbon fungicides chemically, but it is sometimes included as a member of the group because of its similarities in fungitoxic action (Lyr, 1995). It was once suggested to be an inhibitor of respiration blocking electron transfer between cytochromes b and c in *Pythium spp*. (Halos and Huisman, 1976).

Hymexazol is a soil fungicide used for the control of seedling blight of rice. This has been a problem that arose from the dense cultivation in seedling boxes since the introduction of transplanting machines into rice cultivation. The disease is caused by *Pythium*, *Fusarium*, *Rhizopus* and/or *Trichoderma*. The fungicide is effective on the former two pathogens, one is an *Oomycetes* fungus and the other is not. It is also effective against *Aphanomyces* but hardly or not effective against other important *Oomycetes* fungi, *Peronospora*, *Plasmopara* and *Phythophthora*. Its mode of action is thought to be via inhibition of fungal biosynthesis of DNA in *Fusarium oxysporum*, and disruption of permeability of cell membranes of *Pellicularia sasakii* (*Rhizoctonia solani*), but it has not been confirmed in spite of intensive studies with *Pythium aphanidermatum* (Nakanishi and Sisler, 1983).

Propamocarb is effective only on diseases caused by *Oomycetes* fungi, such as *Pythium*, *Phytophthora*, *Aphanomyces* and *Peronospora*. It is used by soil drench and by foliar spray. Its mode of action was suggested to be related with cell membrane function (Papavizas *et al*., 1978). Production of an unusual fatty acid was later reported in mycelial cells of *Pythium* and *Phytophthora* treated with the fungicide but it was not observed in *Ascomycetes* and *Basidiomycetes* (Burden *et al*., 1988).

Cymoxanil is especially effective against downy mildews of vine, hop and other crops caused by *Plasmopara*, *Peronospora* and *Pseudoperonospora*. It is also effective against late blight of potato and tomato caused by *Phytophthora infestans*. Its mode of action was studied by Ziogas and Davidse (1987) but it remains unknown.

Dimethomorph has a morpholine ring in its molecule but its fungitoxic action is different from that of the morpholine fungicides described earlier. Since it has an unsaturated double bond in its molecule, technical products contain *Z*- and *E*-isomers. Although the *Z*-isomer is the active principle, isomerization takes place rather easily especially in light, and therefore a mixture of both isomers is used in practice. It is effective against most diseases caused by *Oomycetes* except those by *Pythium*, and is used by foliar spray. Although the biochemical mode of action of dimethomorph has not been elucidated, the fungicide is thought to affect the fungal cell wall or the formation of the cell wall, on the basis of morphological observation.

FUNGICIDES CONTROLLING SHEATH BLIGHT OF RICE

Sheath blight is one of the important diseases of rice. The causal fungus of this disease is *Rhizoctonia solani* which is usually classified as *Deuteromycetes* or *Fungi Imperfecti*, a group unidentified in taxonomic classification, but the nature of the fungus is similar to *Basidiomycetes* in many respects. Fungicides specifically active to *Basidiomycetes* are often effective in controlling sheath blight of rice. Aryl carboxanilides already mentioned are effective in the control of this disease as well as other various diseases caused by *Basidiomycetes*. Two interesting antibiotics have also been developed to control this disease. One is polyoxorim (polyoxin D) which inhibits biosynthesis of chitin, an important constituent of the fungal cell wall. The other is validamycin A which inhibits utilization of trehalose, a sugar important for fungi as an energy source, by inhibiting trehalase, the enzyme hydrolyzing the fungal sugar.

Besides these, the following fungicides have recently been developed for control of this disease. Diclomezine, a pyridazinone fungicide (Figure 23), is a fungicide introduced late in the 1980s. The antifungal spectrum of this fungicide seems to be limited to *Rhizoctonia* and its relatives such as *Sclerotium*. Its mode of action has not been elucidated but it seems different from that of other fungicide classes as judged by morphological observation of the hyphal growth after treatment with the fungicide. It caused loss of septum in hyphal cells followed by leakage of cell contents.

Pencycuron a phenylurea fungicide (Figure 23) was developed and introduced in Japan in the 1980s for control of rice sheath blight. It is also effective against black scurf of potato and bottom rot of lettuce. Its antifungal spectrum is limited to *Rhizoctonia*, or rather a part of *Rhizoctonia*, and selectivity is often found within the class of *Rhizoctonia*. In spite of physiological and biochemical investigations into a wide range of its effects on fungal growth, no final conclusion was reached on its mode of action (Ueyama *et al.*, 1990), though its effects on the morphology of fungal growth resembled those caused by carbendazim.

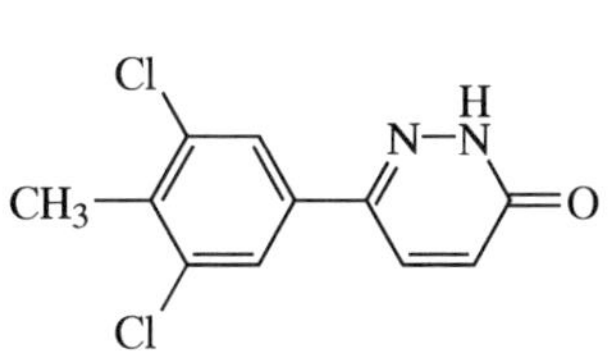

diclomezine

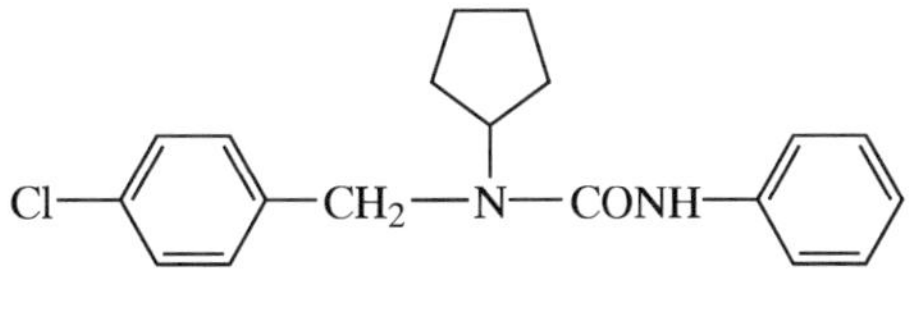

pencycuron

Figure 23 Chemical structures of fungicides controlling rice sheath blight. Aryl carboxanilides (Figure 17) are also effective in the control of the disease

MISCELLANEOUS FUNGICIDES

Hydroxypyrimidine fungicides, dimethirimol and ethirimol (Figure 24), were developed late in the 1960s as systemic fungicides for powdery mildews on various crops. Hydroxypyrimidines are thought to inhibit fungal biosynthesis of nucleic acids by inhibiting adenosine deaminase in the purine reutilization pathway (Hollomon and Chamberlain, 1981).

Dichlofluanid and tolylfluanid (Figure 24) are contact fungicides with protective action. They are rather old fungicides developed in the 1960s. They have a wide spectrum and are used for control of *Botrytis*, *Alternaria*, *Venturia*, *Phytophthora*, powdery mildews, downy mildews and other pathogens of vegetables, fruit crops, ornamentals and other crops by foliar application and seed dressing. Dichlofluanid is also used as a fumigant in smoke preparations in greenhouses. Their mode of action has not been clear, but they seem to be types of multiple-site inhibiting fungicides. Since specific-site inhibitor types of novel systemic fungicides have problems with fungal resistance, multiple-site inhibitor type contact fungicides are still important as alternatives or for use in combination with novel systemic fungicides.

A sulfonanilide fungicide, flusulfamide (Figure 24), is a soil fungicide developed recently. It is effective in the control of club root of cabbage, cauliflower, turnip and other *Brassica* plants caused by *Plasmodiophora*

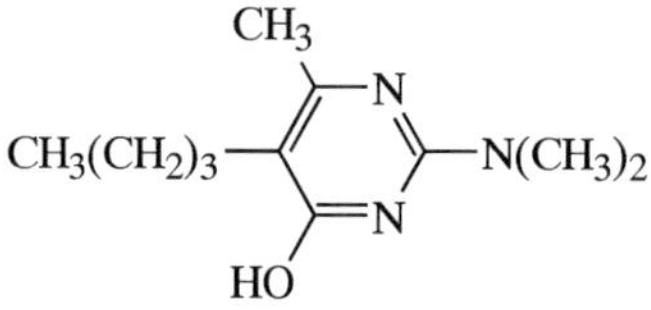

dimethirimol

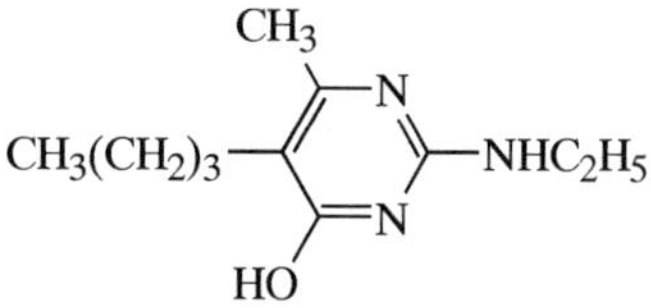

ethirimol

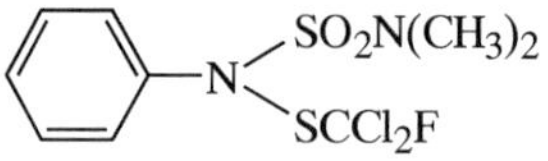

dichlofluanid

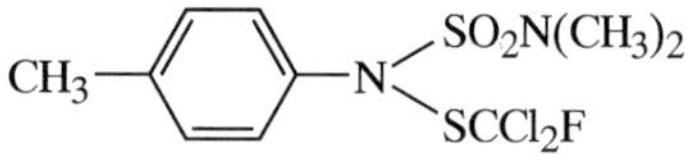

tolylfluanid

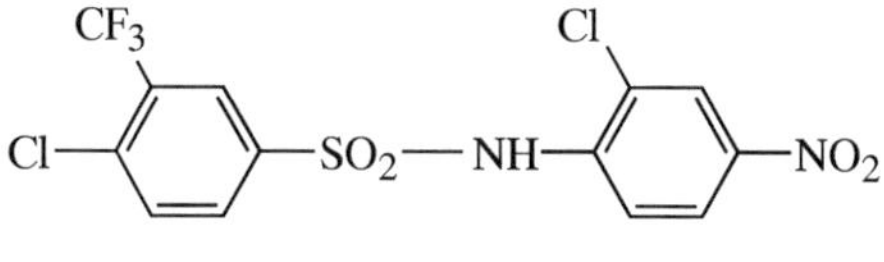

flusulfamide

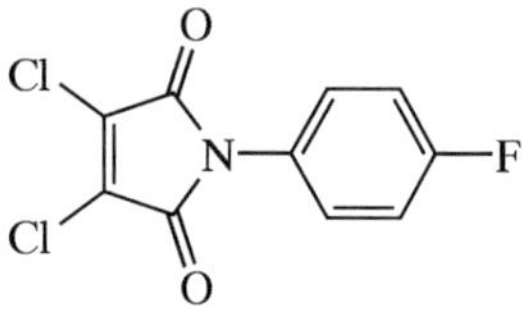

fluoroimide

Figure 24 Chemical structures of miscellaneous fungicides *brassicae*. It also controls other pathogenic soil fungi and fungal vectors of pathogenic viruses. Although its detailed mode of action has not been elucidated, it inhibits germination of resting spores of the soil fungi and thus halts the infection.

Fluoroimide (Figure 24) developed late in the 1970s is a kind of dicarboximide fungicide in chemical terms but it seems different from the so-called dicarboximide fungicides in its antifungal actions. It controls a wide range of diseases of apple and citrus caused by *Alternaria*, *Monilinia*, *Diaporthe* and other fungi. Although details of its mechanism of action have not been elucidated, antifungal action is clearly on spore germination rather than on mycelial growth and it is reversed by the addition of sulfhydryl compounds such as glutathione and dithiothreitol. This suggests that reactivity to sulfhydryl groups is important in its action. This might be expected from the chemistry of the molecule.

REFERENCES

Akagi, T., Mitani, S., Komyoji, T., and Nagatani, K. (1995). 'Quantitative structure-activity relationships of fluazinam and related fungicidal *N*-phenylpyridinamines: Preventive activity against *Botrytis cinerea*', *J. Pesticide Sci.*, **20**, 279–290.

Akagi, T., Mitani, S., Komyoji, T., and Nagatani, K. (1996). 'Quantitative structure-activity relationships of fluazinam and related fungicidal *N*-phenylpyridinamines: Preventive activity against *Sphaerotheca fuliginea*, *Pyricularia oryzae*, and *Rhizoctonia solani*', *J. Pesticide Sci.*, **21**, 23–29.

Ammermann, E., Lorenz, G., Schelberger, K., Wenderoth, B., Sauter, H., and Rentzea, C., (1992). 'BAS 490F—A broad-spectrum fungicide with a new mode of action, *Proceedings of Brighton Crop Protection Conf.*, pp. 403–410.

Baldwin, B. C., Clough, J. M., Godfrey, C. R. A., Godwin, J. R., and Wiggins, T. E. (1995). 'The discovery and mode of action of ICIA 5504', *Proceedings of 11th International Symposium on Systemic Fungicides and Antifungal Compounds*, pp. 69–77.

Baloch, R. I. and Mercer, E. I. (1987) 'Inhibition of sterol $\triangle 8 \rightarrow \triangle 7$-isomerase and $\Delta 14$-reductase by fenpropimorph, tridemorph and fenpropidin in cell-free enzyme systems from *Saccharomyces cerevisiae*', *Phytochemistry*, **26**, 663–668.

Baloch, R. I., Mercer, E. I., Wiggins, T. E., and Baldwin, B. C. (1984). 'Inhibition of ergosterol biosynthesis in *Saccharomyces cerevisiae* and *Ustilago maydis* by tridemorph, fenpropimorph and fenpropidin', *Phytochemistry*, **23**, 2219–2226.

Buchenauer, H. (1977). 'Mode of action of triadimefon in *Ustilago avenae*', *Pestic. Biochem. Physiol.*, **7**, 309–320.

Buchenauer, H. (1978). 'Inhibition of ergosterol biosynthesis by triadimenol in *Ustilago avenae*', *Pestic Sci.*, **9**, 507–512.

Burden, R. S. Carter, G. A., James, C. S., Clark, T., and Holoway, P. J. (1988). 'Selective effects of paropamocarb and prothiocarb on the fatty acid composition of some *Oomycetes*', *Proceedings of Brighton Crop Protection Conf.*, pp. 403–407.

Clemons, G. P. and Sisler, H. D. (1969). 'Formation of a fungitoxic derivative from Benlate', *Phytopathology*, **59**, 705–706.

Davidse, L. C. (1973). 'Antimitotic activity of methyl benzimidazol-2-ylcarbamate in *Aspergillus nidulans*', *Pestic. Biochem. Physiol.*, **3**, 317–325.

Davidse, L. C. and Flach, W. (1977). 'Differential binding of methyl benzimidazol-2-ylcarbamate to fungal tubulin as a mechanism of resistance to this antimitotic agent in mutant strains of *Aspergillus nidulans*', *J. Cell Biology*, **72**, 174–193.

Davidse, L. C. Hofman, A. E., and Velthuis, G. C. M. (1983). 'Specific interference of metalaxyl with endogenous RNA polymerase in isolated nuclei from *Phytophthora megasperma* f.sp. *medicaginis*', *Experiment. Mycol.*, **7**, 344–361.

De Waard, M. A. (1974). 'Mechanism of action of the organophosphorus fungicide pyrazophos', *Med. Landbouwhogesch. Wageningen*, **74**, 1–97.

Gasztonyi, M. and Josepovits, G. (1979). 'The activation of triadimefon and its role in the selectivity of fungicide action', *Pestic. Sci.*, **10**, 57–65.

Guo, Z., Miyoshi, H., Komyoji, T., Haga, T., and Fujita, T. (1991). 'Uncoupling activity of newly developed fungicide, fluazinam', *Biochim. Biophysica Acta*, **1056**, 89–92.

Halos, P. M. and Huisman, O. C. (1976). 'Inhibition of respiration in *Pythium* spp. by ethazole', *Phytopathology*, **66**, 158–164.

Hammerschlag, R. S. and Sisler, H. D. (1973). 'Benomyl and methyl 2-benzimidazolecarbamate: Biochemical, cytological and chemical aspects of toxicity to *Ustilago maydis* and *Saccharomyces cerevisiae*', *Pestic. Biochem. Physiol.*, **3**, 42–54.

Hayashi, K., Watanabe, M., Tanaka, T., and Uesugi, Y. (1996). 'Cyanide-insensitive respiration of phytopathogenic fungi demonstrated by antifungal joint action of respiration inhibitors', *J. Pesticide Sci.*, **21**, 399–403.

Hollomon, D. W. and Chamberlain, K. (1981). 'Hydroxypyrimidine fungicides inhibit adenosine deaminase in barley powdery mildew', *Pestic. Biochem. Physiol.*, **16**, 158–169.

Kato, T. and Kawase, Y. (1976). 'Selective inhibition of the demethylation at C-14 in ergosterol biosynthesis by the fungicide, Denmert (S-1358)', *Agr. Biol. Chem.*, **40**, 2379–2388.

Kato, T., Shoami, M. and Kawase, Y. (1980). 'Comparison of tridemorph with buthiobate in antifungal mode of action', *J. Pesticide Sci.*, **5**, 69–79.

Kato, T., Tanaka, S., Ueda, M., and Kawase, Y. (1974). 'Effects of the fungicide, S-1358, on general metabolism and lipid biosynthesis in *Monilinia fructigena*', *Agr. Biol. Chem.*, **38**, 2377–2384.

Kato, T., Tanaka, S., Ueda, M., and Kawase, Y. (1975). 'Inhibition of sterol biosynthesis in *Monilinia fructigena* by the fungicide, S-1358', *Agr. Biol. Chem.*, **39**, 169–174.

Kerkenaar, A., Uchiyama, M., and Versluis, G. G. (1981). 'Specific effects of tridemorph on sterol biosynthesis in *Ustilago maydis*', *Pestic. Biochem. Physiol.*, **16**, 97–104.

Kodama, O., Yamada, H., and Akatsuka, T. (1979). 'Kitazin P, inhibitor of phosphatidylcholine biosynthesis in *Pyricularia oryzae*', *Agr. Biol. Chem.*, **43**, 1719–1725.

Kodama. O., Yamashita, K., and Akatsuka, T. (1980). 'Edifenphos, inhibitor of phosphatidylcholine biosynthesis in *Pyricularia oryzae*', *Agr. Biol. Chem.*, **44**, 1015–1021.

Koenraadt, H. and Jones, A. L. (1992) 'The use of allele-specific oligonucleotide probes to characterize resistance to benomyl in field strains of *Venturia inaequalis*', *Phytopathology*, **82**, 1354–1358.

Koenraadt, H., Sommerville, S. C., and Jones, A. L. (1992). 'Characterization of mutations in the beta-tubulin gene of benomyl-resistant field strains of *Venturia inaequalis* and other plant pathogenic fungi', *Phytopathology*, **82**, 1348–1354.

Kurahashi, Y., Hattori, T., Kagabu, S., and Pontzen, R., (1996). 'Mode of action of the novel rice blast fungicide KTU 3616', *Pestic. Sci.*, **47**, 199–202.

Langcake, P., Kuhn, P. J., and Wade, M. (1983). 'The mode of action of systemic fungicides', in *Progress in Pesticide Biochemistry and Toxicology* (eds D. H. Hutson and T. R. Roberts). **Vol. 3**, p. 16, John Wiley, Chichester.

Leroux, P. and Gredt, M. (1979). 'Effects du barbane, du cholrobufame, du chlorprophame et du prophame sur diverses souches de *Botrytis cinerea* Pers et de *Penicillium expansum* Link sensibles ou résistantes au carbendazime et au thiabendazole', *C. R. Acad. Sc. Paris*, **t. 289 Série D**, 691–693.

Leroux, P., Lanen, C., and Fritz, R. (1992). 'Similarities in the antifungal activities of fenpiclonil, iprodione and tolclofos-methyl against *Botrytis cinerea* and *Fusarium nivale*', *Pestic. Sci.*, **36**, 255–261.

Loeffler, R. S. T., Butters, J. A., and Hollomon, D. W. (1992). 'The sterol composition of powdery mildews', *Phytochemistry*, **31**, 1561–1563.

Lyr, H. (1995). 'Aromatic hydrocarbon fungicides and their mechanism of action', in *Modern Selective Fungicides*, 2nd edition (ed H. Lyr), pp. 79–98, Gustav Fischer.

Masner, P., Muster, P., and Schmid, J. (1994). 'Possible methionine biosynthesis inhibition by pyridinamine fungicides', *Pestic. Sci.*, **42**, 163–166.

Milling, R. J. and Richardson, C. J. (1995). 'Mode of action of anilinopyrimidine fungicide pyrimethanil. 2. Effects on enzyme secretion in *Botrytis cinerea*', *Pestic. Sci.*, **45**, 43–48.

Mitani, S., Nakano, K., Matsuo, N., and Komyoji, T. (1995). 'Biological properties of fungitoxic propargyl *N*-(6-ethyl-5-iodo-2-pyridyl)carbamate', *J. Pesticide Sci.*, **20**, 153–160.

Miura, I., Kamakura, T., Maeno, S., Hayashi, S., and Yamaguchi, I. (1994). 'Inhibition of enzyme secretion in plant pathogens by mepanipyrim, a novel fungicide', *Pestic. Biochem. Physiol.*, **48**, 222–228.

Mizutani, A., Yukioka, H., Tamura, H., Miki, N., Masuko, M., and Takeda, R. (1995). 'Respiratory characteristics in *Pyricularia oryzae* exposed to a novel alkoxyiminoacetamide fungicide', *Phytopathology*, **85**, 306–311.

Nakamura, S. and Kato, T. (1984). 'Mode of action of tolclofos-methyl in *Ustilago maydis*', *J. Pesticide Sci.*, **9**, 725–730.

Nakanishi, T. and Sisler, H. D. (1983). 'Mode of action of hymexazol in *Pythium aphanidermatum*', *J. Pesticide Sci.*, **8**, 173–181.

Okuno, T., Furusawa, I., Matsuura, K., and Shishiyama, J. (1989). 'Mode of action of ferimzone, a novel systemic fungicide for rice diseases: Biological properties against *Pyricularia oryzae* in vitro', *Phytopathology*, **79**, 827–832.

Orth, A. B., Rzhetskaya, M., Pell, E. J., and Tien, M. (1995). 'A serine (threonine) protein kinase confers fungicide resistance in the phytopathogenic fungus *Ustilago maydis*', *Appl. Environ. Microbiol.*, **61**, 2341–2345.

Papavizas, G. C., O'Neill, N. R., and Lewis, J. A. (1978). 'Fungistatic activity of propyl *N*-(γ-dimethylaminopropyl)carbamate on *Pythium* spp. and its reversal by sterols', *Phytopathology*, **68**, 1667–1671.

Pontzen, R., Poppe, B., and Berg, D. (1990). 'Mode of action of sterol biosynthesis inhibitors in obligate parasites', *Pestic. Sci.*, **30**, 357–360.

Ragsdale, N. N. (1975). 'Specific effects of triarimol on sterol biosynthesis in *Ustilago maydis*', *Biochimica Biophysica Acta*, **380**, 81–96.

Ragsdale, N. N., and Sisler, H. D. (1973). 'Mode of action of triarimol in *Ustilago maydis*', *Pestic. Biochem. Physiol.*, **3**, 20–29.

Sekizawa, Y. and Watanabe, T. (1981). 'Mode of action of probenazole against rice blast', *J. Pesticide Sci.*, **6**, 247–255.

Selling, H. A., Vonk, J. W., and Kaars Sijpesteijn, A. (1970). 'Transformation of the systemic fungicide methyl thiophanate into 2-benzimidazolecarbamic acid methyl ester', *Chemistry and Industry*, **51**, 1625–1626.

Sherald, J. L. and Sisler, H. D. (1975). 'Antifungal mode of action of triforine', *Pestic Biochem. Physiol.*, **5**, 477–488.

Sherald, J. L., Ragsdale, N. N., and Sisler, H. D. (1973). 'Similarities between the systemic fungicide triforine and triarimol', *Pestic. Sci.*, **4**, 719–727.

Sugiura, H., Hayashi, K., Tanaka, T., Takenaka, M., and Uesugi, Y. (1993). 'Mutual antagonism between sterol demethylation inhibitors and phosphorthiolate fungicides on *Pyricularia oryzae* and the implication for their mode of action', *Pestic. Sci.*, **39**, 193–198.

Suzuki, K., Kato, T., Takahashi, J., and Kamoshita, K. (1984). 'Mode of action of methyl *N*-(3,5-dichlorophenyl)carbamate in the benzimidazole-resistant isolate of *Botrytis cinerea*', *J. Pesticide Sci.*, **9**, 497–501.

Takahashi, J., Nakamura, S., Noguchi, H., Kato, T., and Kamoshita, K. (1988). 'Fungicidal activity of *N*-phenylcarbamates against benzimidazole-resistant fungi', *J. Pesticide Sci.*, **13**, 63–69.

Tokousbalides, M. C., and Sisler, H. D. (1979). 'Site of inhibition by tricyclazole in the melanin biosynthetic pathway of *Venturia dahliae*', *Pestic. Biochem. Physiol*, **11**, 64–73.

Tuyl, J. M. van (1977). 'Genetics of fungal resistance of systemic fungicides', *Mededelingen Landbouwhogeschool, Wageningen*, **77-2**, 1–136.

Uesugi, Y. and Takenaka, M. (1992). 'The mechanism of action of phosphorothiolate fungicides', *Proceedings of the 10th International Symposium on Systemic Fungicides and Antifungal Compounds*, pp. 159–164.

Ueyama, I., Araki, Y., Kurogochi, S., Yoneyama, K., and Yamaguchi, I. (1990). 'Mode of action of phenylurea fungicide pencycuron in *Rhizoctonia solani*', *Pestic. Sci.*, **30**, 363–366.

Vincent, P. G. and Sisler, H. D. (1968). 'Mechanism of action of 2,4,5,6-tetrachloroisophthalonitrile', *Physiol. Plantarum*, **21**, 1249–1264.

Vonk, J. W. and Kaars Sijpesteijn, A. (1971). 'Methyl benzimidazol-2-ylcarbamate, fungitoxic principle of methyl thiophanate', *Pestic. Sci.*, **2**, 160–164.

White, G. A., Thorn, G. D., and Georgopoulos, S. G. (1978). 'Oxathiin carboxamides highly active against succinic dehydrogenase complexes from carboxin-selected mutants of *Ustilago maydis* and *Aspergillus nidulans*', *Pestic. Biochem. Physiol.*, **9**, 165–182.

Woloshuk, C. P. and Sisler, H. D. (1982). 'Tricyclazole, pyroquilon, tetrachlorophthalide, PCBA, coumarin and related compounds inhibit melanization and epidermal penetration by *Pyricularia oryzae*', *J. Pesticide Sci.*, **7**, 161–166.

Yoshida, M., Moriya, S., and Uesugi, Y. (1984). 'Observation of transmethylation from methionine into choline in the intact mycelia of *Pyricularia oryzae* by ^{13}C NMR under the influence of fungicides', *J. Pesticide Sci.*, **9**, 703–708.

Ziogas, B. N., and Davidse, L. C. (1987). 'Studies on the mechanism of action of cymoxanil in *Phytophthora infestans*', *Pestic. Biochem. Physiol.*, **29**, 89–96.

3 Natural Product-Derived Fungicides as Exemplified by the Antibiotics

I. YAMAGUCHI
RIKEN, Wako, Saitama, Japan

INTRODUCTION

While pesticides are indispensable to assure the high agricultural productivity to meet the ever growing population in the world, there is much concern about their side effects on mammals and wildlife, and about their potential as sources of environmental pollution. Some older conventional pesticides such as DDT and HCH remain in the natural environment for a long period and they are reportedly hazardous to mammals through their biological concentration in the food chain. This points to the need for selective agrochemicals with higher degradability in nature.

Fungicidal Activity. Edited by D. H. Hutson and J. Miyamoto

Pesticides of microbial origin, so-called agricultural antibiotics, are generally specific to target organisms and are considered to be easily decomposable in the environment. The antibiotics first introduced in agriculture were those originally developed for medicinal purposes, e.g. streptomycin and oxytetracycline. Most other medicinal antibiotics, however, did not meet agricultural field requirements, such as low phytotoxicity, high durability to sunlight, and low manufacturing cost. Therefore, many attempts were made to find microbial products for the prime purpose of plant protection and eventually some antibiotics were developed for practical use in agriculture. Recently world-wide interest in them appears to have been renewed since strobilurin and pyrrolnitrin were found to be good lead compounds for excellent synthetic fungicides (for example see Chapter 5). In this article, the natural product-derived fungicides as exemplified by the antibiotics are reviewed, and recent advances in applications of molecular biology in relation to microbial products will be also discussed.

FUNGICIDES OF MICROBIAL ORIGIN

Through an extensive screening of microbial products for the prime purpose of rice blast control, an epoch-making substance named blasticidin S was discovered (Takeuchi *et al.*, 1958). It exhibited significant efficacy in controlling

Table 1 Microbial products used in agriculture

Substances	Effective against
Antifungal	
Blasticidin S [a]	Rice blast
Kasugamycin [a]	Rice blast
Polyoxins [a]	Rice sheath blight, Fungal diseases of fruit trees
Validamycin A [a]	Rice sheath blight, Fungal diseases of vegetables
Mildiomycin [a]	Powdery mildew of rose and crape myrtle
Antibacterial	
Streptomycin and Dihydrostreptomycin	Bacterial diseases of fruit trees and vegetables
Oxytetracyclin	Bacterial diseases of fruit trees and vegetables
Insecticidal	
Avermectin-B	Acarina and Nematode
Milbemectins [a]	Mites of fruit trees and tea plants
Polynactins [a]	Mites of fruit trees and tea plants
Herbicidal	
Bilanafos [a]	Weeds in orchard and mulberry fields
Plant growth regulatory	
Gibberellin [a]	Parthenocarpy of grapes and timely flowering

[a] Substances discovered in Japan.

blast disease, the most serious and damaging of all the diseases of rice in temperate and humid climates. The success of blasticidin S in practical use inspired further research for new pesticides of microbial origin, leading to the development of excellent antifungal substances such as kasugamycin (Umezawa *et al.*, 1965), polyoxins (Isono *et al.*, 1965, 1967), and validamycin A (Iwasa *et al.*, 1971b,c). Table 1 shows the pesticides of microbial origin practically used in agriculture. Here characteristics of major fungicides of microbial origin and promising microbial products for future development as fungicides are outlined.

ANTIBIOTICS USED AS FUNGICIDES IN PRACTICE

Blasticidin S

Blasticidin S produced by *Streptomyces griseochromogenes* has been used for the control of rice blast disease caused by *Pyricularia oryzae* (teleomorph: *Magnaporthe grisea*). It was isolated from the culture filtrates of *S. griseochromogenes* by Takeuchi *et al.* in 1958, and its potent curative effect on rice blast was found by Misato *et al.* (1959). Since the benzylaminobenzene sulfonate of blasticidin S proved to be the least phytotoxic to the rice plant without reduction of antifungal activity against *P. oryzae* (Asakawa *et al.*, 1963), the salt formulation has been industrially produced for agricultural use.

Blasticidin S is obtained as white needle crystals; it is highly soluble in water, but mostly insoluble in organic solvents, stable in solution at pH 5–7, but unstable in alkaline conditions. The chemical structure of blasticidin S has been elucidated as shown in Figure 1 (Otake *et al.*, 1966; Yonehara and Otake, 1966). The molecule of blasticidin S comprises a novel nucleoside designated cytosinine and a β-amino acid named blastidic acid (*N*-methyl-β-arginine). By using the producing organism and ^{14}C-labeled precursors, Seto *et al.* (1966, 1968) showed that the pyrimidine ring and the 2,3-unsaturated sugar moiety of the antibiotic were derived from cytosine and glucose, respectively; arginine served as the precursor for the blastidic acid skelton, and the *N*-methyl group of blastidic acid came from methionine.

Blasticidin S exhibits a wide range of biological activities; besides its significant inhibitory effects on the growth of *P. oryzae*, it also showed inhibition of bacterial and fungal cells as well as antitumor and antiviral activities (Tanaka *et al.*, 1961; Hirai and Shimomura, 1965). The acute oral toxicity to mice is rather high (LD_{50} 53.3 mg/kg), but it can be used for blast control in the paddy field because toxicity to fish is low (LD_{50} to carp >40 ppm).

Misato *et al.* (1959) found the effect of blasticidin S on rice blast to be due to its strong inhibitory action on mycelial growth of *P. oryzae*. The antibiotic markedly inhibits the incorporation of ^{14}C-labeled amino acids into protein in the cell-free system of *P. oryzae* (Huang *et al.*, 1964). Yamaguchi and Tanaka

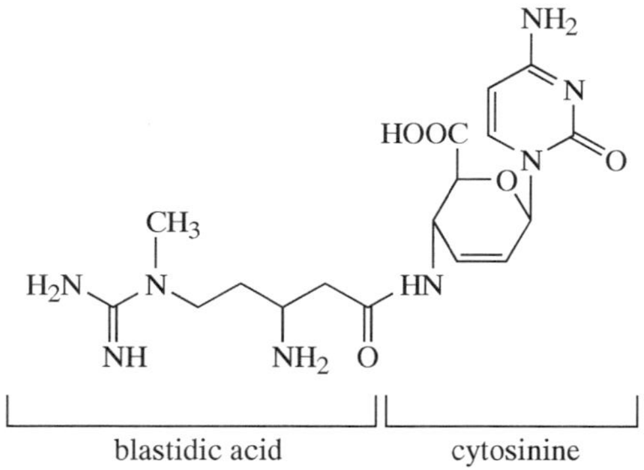

Blasticidin S

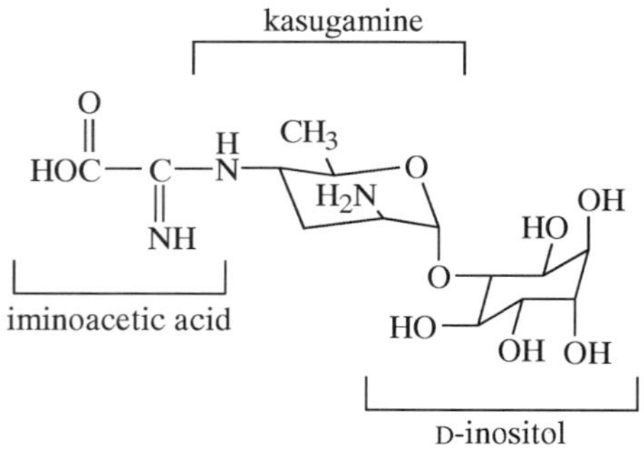

Kasugamycin

Figure 1 Chemical structures of blasticidin S and kasugamycin

(1966) reported that blasticidin S inhibited the incorporation of leucine and phenylalanine into polypeptides in the cell-free protein-synthesizing system from *Escherichia coli*, suggesting that blasticidin S acts in the step of peptide transfer from peptidyl-tRNA on the donor site with the incoming amino acyl-tRNA on the acceptor site. They also showed that blasticidin S markedly interfered with the puromycin reaction, in accordance with the findings by Kinoshita *et al.* (1970) and Yukioka *et al.* (1975) that the action of blasticidin S involves the formation of peptidyl-blasticidin S which remains bound to the ribosomes. Recently the fine structure of the peptidyl transferase center on 23 S-like rRNAs was deduced from the chemical structure of the blasticidin S-ribosome complex (Rodriguez-Fonsece *et al.*, 1995).

Efficacy of blasticidin S was evaluated in field trials for rice blast control and it was put into practical use in 1962. The effective spraying concentration of blasticidin S in the field is 10–30 ppm (1–3 g a.i./10 a). With regard to phytotoxicity, rice plants, water melon, and cucumber are resistant to blasticidin S, followed by grape, pear and peach; eggplant, tomato, potato and mulberry plants are rather susceptible; the tobacco plant is most susceptible to the antibiotic.

As for the mammalian toxicity, application of blasticidin S by dusting occasionally causes conjunctivitis if the dust accidentally comes in contact with the eyes. Many attempts to alleviate this toxic effect on mammals were performed, involving chemical modification and biological transformation of blasticidin S. Yonehara *et al.* (1968) discovered novel substances designated as detoxins produced by *Streptomyces caespitosus* var. *detoxicus*. Detoxins exhibited unique biological activities such as a selective antagonism against blasticidin S, negating the adverse effect of the antibiotic on mammals without a loss in the control efficacy on rice blast. Separation of components of the detoxin complex was done and their chemical structures were elucidated (Otake *et al.*, 1973; Kakinuma *et al.*, 1974). While the defects of blasticidin S were significantly reduced by the addition of detoxins, the complex was not introduced into practical use due to the difficulties in cost performance. However, a simpler method to alleviate eye irritation caused by blasticidin S was developed; the addition of calcium acetate (5%) to blasticidin S dust specifically reduced the eye trouble without reducing the antiblast effect. This improved dust is now in practical use.

The fate of blasticidin S in the environment was investigated using radioactive compounds prepared biosynthetically by the producing organism (Yamaguchi *et al.*, 1972). The hydrophilic antibiotic located mostly on the hydrophobic surface of the rice plant, and little was absorbed and translocated in the tissue. From the infected parts or wounded tissue, however, the antibiotic was incorporated and translocated to the apices. Blasticidin S located on the plant surface was decomposed by sunlight and gave rise to cytosine as a main degradation product. A considerable quantity of the blasticidin S sprayed fell to the ground and was bound onto the soil surface. Since remarkable generation of

^{14}C-carbon dioxide from the ^{14}C-blasticidin S-treated soil was observed and several microbes usually inhabiting the paddy field were found to reduce the biological activity of blasticidin S, it was suggested that, after application of blasticidin S to the crop field, the antibiotic was easily broken down in the environment. It was also shown that the residual amount of blasticidin S in rice grains to be less than detectable level.

Kasugamycin

Kasugamycin is an aminoglycoside antibiotic produced by *Streptomyces kasugaensis* (Umezawa *et al.*, 1965). It has been widely used for rice blast control with no phytotoxicity and very low toxicity to mammals and fish.

Kasugamycin is a water-soluble and basic substance, and its hydrochloride (practical form) is also insoluble in most organic solvents. The molecule consists of three moieties, i.e. D-inositol, kasugamine and an iminoacetic acid side chain as shown in Figure 1 (Suhara *et al.*, 1966, 1968; Ikekawa *et al.*, 1966); the kasugamine moiety is biosynthetically derived from glucose or mannose, and the other parts from myo-inositol and glycine (Fukagawa *et al.*, 1968a,b,c).

Kasugamycin inhibits the growth of *P. oryzae* specifically under acidic conditions (pH 5.0), but it is barely active in neutral media (Hamada *et al.*, 1965), suggesting that its effect on *P. oryzae* is expressed *in planta* under acidic conditions. The antibiotic also interferes with the growth of some bacteria including *Pseudomonas* species, but unlike fungicidal activity, it shows stronger inhibition against bacteria at pH 7 than at pH 5.

Kasugamycin inhibits protein synthesis in a cell-free system of *E. coli,* by interfering with the binding of aminoacyl-tRNA to the mRNA-30S ribosomal subunit complex without causing miscoding (Tanaka *et al.*, 1967). Thus it does not cause any cross-resistance with streptomycin or kanamycin. Kasugamycin sensitivity involves the 16S RNA of the 30S ribosomal subunit, and the resistant mutant of *E. coli* cannot methylate two adjacent adenine residues near the 3' end of the 16S RNA (Hesler *et al.*, 1971). The antibiotic has quite low acute and chronic toxicity to mice, rats, rabbits, dogs and monkeys. The oral LD_{50} for mice is 2 g/kg, and the TLm (median tolerance limit) to carp is more than 1000 ppm.

Rice blast disease is well controlled by kasugamycin at a concentration as low as 20 ppm (Ishiyama *et al.*, 1965). The antibiotic is mainly applied as a dust containing 0.3% of the active ingredient. When the rice seeds are coated with 2% kasugamycin wettable powder, the plants are protected from rice blast for a month in the fields.

The emergence of resistance to kasugamycin was recognized in a few years after its first application in 1965 and the development of resistant strains of *P. oryzae* in fields became a serious problem. Since then, mixtures of kasugamycin and chemicals with different modes of action such as fthalide have been used in practice. In laboratory experiments, kasugamycin-resistant strains of *P. oryzae* proved less infective and of lower fitness than the sensitive parent strains (Ohmori, 1967). In fact, once the application of kasugamycin was discontinued in the field, the population of the resistant strains declined quite rapidly.

Polyoxins

The polyoxins, a group of peptidylpyrimidine nucleoside antibiotics, produced by *Streptomyces cacaoi* var. *asoensis* (Isono *et al.*, 1965) are composed of 14 components (A–N). They are safely used for the control of many fungal diseases with no toxicity to livestock, fish and plants. Such excellent characteristics are due to the fact that polyoxins selectively inhibit the synthesis of cell-wall chitin in sensitive fungi (Hori *et al.*, 1974b). Polyoxins have been widely used for crop protection against such pathogenic fungi as *Alternaria alternata*, *Rhizoctonia solani* and *Cochliobolus miyabeanus* since 1967.

The chemical structures of polyoxins were elucidated by Isono *et al.*, (1965, 1967, 1969) and Isono and Suzuki (1968a,b) as shown in Figure 2. The identification of the nucleoside with a 5-aminohexuronic acid as its sugar group is the first example in nature. This prompted the total chemical synthesis of polyoxin J (R:CH_3) by Kuzuhara *et al.* (1973). The biosynthesis of polyoxins was also attractive since they contain a unique 5-substituted pyrimidine ring in their molecules; the 5-substituted pyrimidine of the polyoxins was biosynthesized from uracil and C-3 of serine by a new enzyme which resembles but differs from thymidylate synthetase (Isono and Suhadolnik, 1976).

Polyoxins inhibit the growth of various plant pathogenic fungi but are inactive against bacteria and yeasts. Among polyoxins, polyoxins B and L were specifically effective against pear black spot and apple cork spot caused by *Alternaria* spp. at 50–100 ppm (Eguchi *et al.*, 1968), while polyoxin D was most effective for rice sheath blight caused by *R. solani*. Thus the polyoxin complex has been used in practice in duplicate forms; B and L-rich fractions for *Alternaria* diseases and polyoxin D-rich fraction for rice sheath blight. Application of polyoxins could be made at any growth stage of rice plants without causing phytotoxicity even at 800 ppm application. Foliar sprays of 200 ppm of polyoxins have produced no phytotoxicity on all the other crops tested (Sasaki *et al.*, 1968). Oral administration at 15 g/kg and injection at 800 mg/kg to mice did not cause any adverse effect, nor were they toxic to fish during 72 hours exposure at 10 ppm.

With regard to the mechanism of fungicidal action of polyoxins, a specific physiological phenomenon was observed in *Alternaria* spp; polyoxins caused a marked abnormal swelling on germ tubes of spores and hyphal tips of the pathogen, and this made the pathogen non-infectious (Eguchi *et al.*, 1968). In a cell-free system of *Neurospora crassa*, polyoxin D was proved to inhibit the incorporation of *N*-acetylglucosamine (GlcNAc) into chitin in a competitive manner between UDP-GlcNAc and polyoxin D (Endo and Misato, 1969). The relation between polyoxin structure and inhibitory activity on chitin synthase

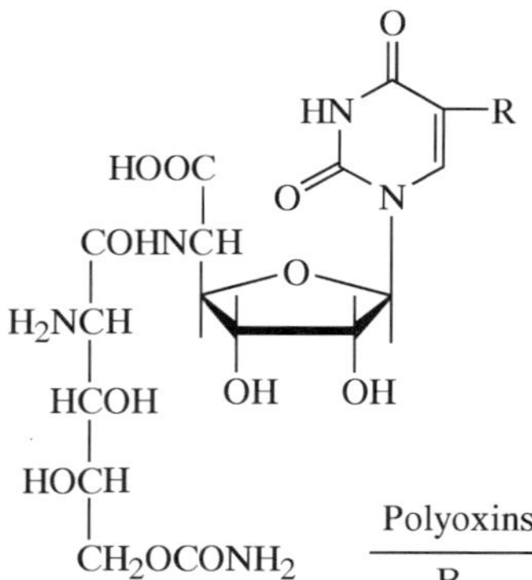

Polyoxins	R
B	CH_2OH
D	COOH
L	H

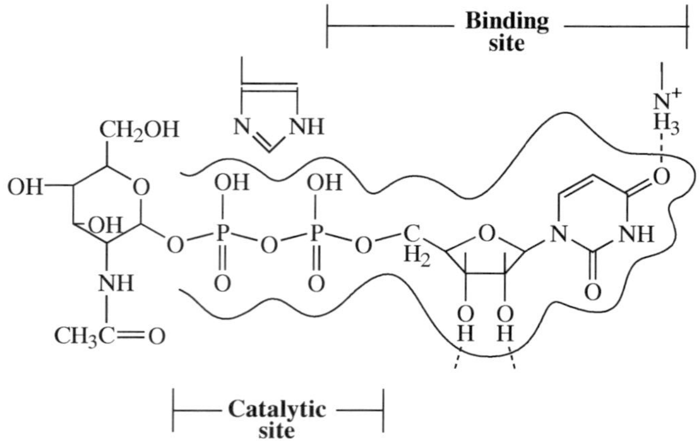

Figure 2 Chemical structures of major components of polyoxins and binding features of UDP-*N*-acetylglucosamine on chitin synthase

was clarified by kinetic analysis; the pyrimidine nucleoside moiety of the antibiotics was shown to fit into the binding site of the enzyme protein, and the carbamoylpolyoxamic acid moiety of polyoxins stabilizes the polyoxin-enzyme complex (Hori *et al.*, 1974b; see Figure 2). Therefore, the excellent characteristics of polyoxins can be explained by the fact that the antibiotics inhibit cell wall synthesis of sensitive fungi but have no adverse effects on other organisms that have no chitinous cell walls.

Polyoxin resistant strains of *A. alternata* were recognized in pear orchards in Japan after several years of intensive use of the antibiotics. Since the inhibition in mycelial growth of *R. solani* by polyoxins was antagonized by glycylglycine, glycyl-L-alanine, glycyl-DL-valine and DL-alanylglycine, the resistance was suggested to be caused by a lowered permeability of the antibiotic through the cell membrane into the site of chitin synthesis (Hori *et al.*, 1974a, 1977). Polyoxins may be incorporated into the fungal cells through the dipeptide uptake channels.

Validamycins

Validamycin A was isolated from the culture filtrate of *Streptomyces hygroscopicus* var. *limoneus*, which produced five additional components designated as validamycins B to F, together with validoxylamine A (Iwasa *et al.*, 1971b,c; Horii *et al.*, 1972). It has been used for the control of *Rhizoctonia* diseases without phytotoxicity and with very low toxicity to mammals, birds, fish and insects (Matsuura, 1983).

The chemical structure of validamycin A was determined to have two kinds of new hydroxymethyl-branched cyclitols in its molecule (Suami *et al.*, 1980), giving a unique example in the field of pseudosugar chemistry (Figure 3). Validamycins A, C, D, E and F contain validoxylamine as a common moiety in their molecules, but they differ from each other in the configuration of the anomeric centre of the glucoside, the position of the glucosidic linkage, or the number of D-glucose molecules. On the other hand, validamycin B contains validoxylamine B in its molecule, which yielded hydroxyvalidamine instead of validamine derived from other validamycins (Horii *et al.*, 1971). The studies on microbial transformation of validamycins showed the conversion of validamycin C to A by *Endomycosis* spp. and *Candida* spp. (Kameda *et al.*, 1975), which is interesting because validamycin A is about 1000 times more active than validamycin C against *R. solani* in the 'dendroid-test' established by Iwasa *et al.* (1971a).

Validamycin A is specifically effective against plant diseases caused by *Rhizoctonia* spp., such as sheath blight of rice, web blight, bud rot, damping-off, seed decay, root rot, and black scurf of several crops (Wakae and Matsuura, 1975). When it was applied to rice, sufficient control of sheath blight was achieved by a spray solution of 30 ppm. Thus validamycin A has been commercially used for the control of the disease since 1973. In China, its

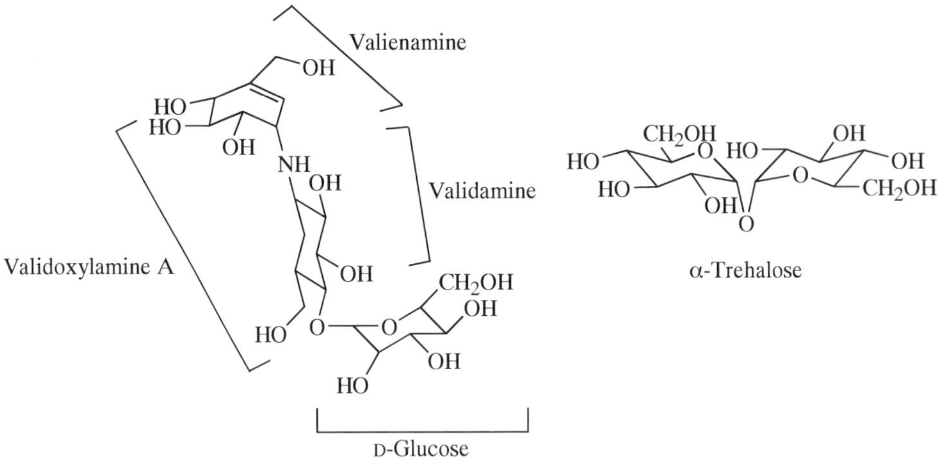

Figure 3 Chemical structures of validamycin A and α-trehalose

production has recently reached as much as 5000 t a.i. per year (Y. C. Shen, personal communication).

Although validamycin A did not significantly suppress the growth of *R. solani* on a nutritionally rich medium, it caused extensive branching of hyphae and the cessation of colony development on a water agar, i.e. *under nutrientless conditions* (Nioh and Mizushima, 1974). It was also observed that validamycin A had no need of continual contact with the pathogen for the effective control of fungal growth (Wakae and Matsuura, 1975). This is important under field conditions because *R. solani* is one of the typical fungi which grows rapidly by transporting nutrients from the basal part to hyphal tips through long stretches of hyphae. This process provides the tips with relatively poor nutrition. As for the mode of action of validamycin A, it was shown that validamycin A exerted potent inhibitory activity against trehalase in *R. solani* AG-1, in a competitive manner between validoxylamine A (the possible active form of validamycin A) and the substrate, trehalose (Shigemoto *et al.*, 1989; Figure 3). Since trehalose is a storage carbohydrate in some fungi, trehalase is suggested to play an essential role for the digestion of trehalose to D-glucose and for its transportation to the hyphal tips.

Except for the significant activity of validamycin A on *R. solani*, its antimicrobial activity was not detected on many fungi and bacteria, nor was disturbance of microflora on rice plants and other field crops tested (Wakae and Matsuura, 1975). No phytotoxicity was observed for over 150 species of plants sprayed with validamycin A even at a concentration of 1000 ppm. Furthermore, toxicity to mammals is markedly low; in oral administration of validamycin A at a dose of 10 g/kg to mice and rats, or in subcutaneous and intravenous administration at the does of 2 g/kg to mice, no adverse effects were observed on any of the animals tested for seven days. Oral subacute toxicity for four months in beagle dogs indicated no significant abnormalities of the morphological and biochemical parameters accompanying the daily administration of 200 mg/kg of validamycin A. The LD_{50} for killifish is greater than 1000 ppm.

Validamycins are susceptible to microbial degradation and their addition to the soil resulted in quick loss of biological activity via the action of soil microbes. Microbial degradation of validamycin A by *Pseudomonas denitrificans* gave rise to D-glucose and validoxylamine A, which was further decomposed into valienamine and validamine (Kameda *et al.*, 1975; see Figure 3). Orally administered validamycin A is readily decomposed up to CO_2 by enteric bacteria in animals, and when injected intravenously it is rapidly excreted in the urine. No cross-resistance with medicinal aminoglycoside antibiotics, such as dihydrostreptomycin or kanamycin, has been reported in human pathogens.

Validamycin A has been used to protect rice sheath blight mainly in formulations of 3–5% solution or 0.3% dust. Residues in rice grains and straws were less than the detectable limit by gas chromatography. Thus, validamycin A is considered to be one of the ideal microbial product-derived fungicides particularly with respect to safety and environmental pollution.

Mildiomycin

Mildiomycin was isolated from the culture filtrate of *Streptoverticillium rimofaciens* B-98891 (Harada and Kishi, 1978; Harada *et al.*, 1978). It is a water-soluble and basic antibiotic which belongs to the nucleoside antibiotic class. Mildiomycin is specifically active against the pathogens which cause powdery mildews, but less active on bacteria, most fungi and yeasts, and thus it has been used for the control of powdery mildew on rose, spindle tree and Indian lilac.

The toxicity of mildiomycin is very low; the LD_{50} in rats and mice is 500–1000 mg/kg by intravenous and subcutaneous injections, and 2.5–5.0 g/kg by oral administration. At a concentration of 1 g/ml mildiomycin, no irritation to the cornea and skin of rabbits was recognized for 10 days, and its toxicity to killifish was not observed at a concentration of 20 ppm for seven days.

Protein synthesis was remarkably inhibited by mildiomycin at concentrations down to 0.02 mM; it markedly interfered with the incorporation of ^{14}C-phenylalanine into polypeptides in the cell-free system of *E. coli* (Om *et al.*, 1984). The mammalian cell-free system from rabbit reticulocytes, however, proved to be much less sensitive to mildiomycin in the synthesis of polypeptides than in the system from *E. coli*.

Streptomycins and Oxytetracycline

Streptomycin and dihydrostreptomycin are well known aminoglycoside antibiotics which have been used for the control of many bacterial diseases on plants such as apple- and pear-fire blights, wild fire of tobacco, and bacterial leaf blight of rice. A mixture of streptomycin and oxytetracycline is also highly effective for the control bacterial canker of peach, citrus canker, soft rot of vegetables, and various other bacterial diseases.

Streptomycin-resistant strains are distributed in a wide range of plant pathogenic bacteria such as *Xanthomonas oryzae, X. citri, Pseudomonas syringae* pv. tabaci and *P. lachrymans.* In agricultural use, the alternate or combined applications of streptomycin with other chemicals having different mechanisms of action was recommended in order to reduce the development of resistant strains in the field. Since it occasionally causes chemical injuries to vegetables if it is sprayed at high concentrations, a mixture of low concentration of streptomycin sulphate and iron chloride or citrate is used to reduce the phytotoxicity.

Streptomycin is known to inhibt protein synthesis in bacterial cells by binding to the 30S ribosomal subunit, and it causes misreading of the genetic codes in protein synthesis (Likover and Kurland, 1967). The mutants of *E. coli* highly resistant to streptomycin were shown to involve modification of the P10 protein of the bacterial ribosome 30S subunit.

Oxytetracycline produced by *Streptomyces rimosus* is active against bacteria, rickettsiae and protozoa, and it is effective in the control of some plant bacterial diseases caused by *Pseudomonas*, *Xanthomonas*, and *Erwinia* species. It also shows protective effects on plants against diseases caused by mycoplasma-like organisms (Ishii *et al.*, 1967). Oxytetracycline is easily taken up by plant leaves from stomata, and rapidly translocated to plant tissues. In the agricultural field, oxytetracycline has been used for the control of bacterial diseases mostly by mixing with streptomycin to avoid the development of resistant strains.

Tetracycline is a potent inhibitor of bacterial protein biosynthesis with less activity on mammalian cells. It binds to the 30S and 50S bacterial ribosomal subunits, and inhibits the binding of aminoacyl-tRNA, the termination factors RF1, and RF2 to the A site of bacterial ribosome.

Bilanafos

Bilanafos (bialaphos) was found in screening tests for microbial products effective against rice sheath blight (Tachibana *et al.*, 1982). It also showed a strong phytotoxic activity against a wide range of weeds, and bilanafos was developed as a herbicide in practice. The application rates are 1 to 3 kg/ha depending upon the growth stage of the weeds and whether they are annuals or perennials. While bilanafos exerts potent herbicidal activity by foliage application, it shows very weak phytotoxic activity towards vegetables when it is present in soils due to its remarkable biodegradability by soil microorganisms. Bilanafos has been used for presowing treatments and in applications to orchard and mulberry fields. Interestingly, the combined application of bilanafos with fertilizers such as urea or ammonium sulphate enhances the herbicidal activity. This may be because bilanafos inhibits the glutamine synthase in weeds, which results in a deficiency of glutamine in the plant cells as well as abnormal accumulation of ammonia, acting as an uncoupler of photophosphorylation.

PROMISING MICROBIAL PRODUCTS

Although the discovery of new antibiotics is becoming more and more difficult, there are still promising microbial products discovered when novel screening systems are developed or new sources for producing organisms are investigated. Further, high throughput screening for microbial products known to inhibit target enzymes such as those in melanin biosynthesis, ergosteral biosynthesis, chitin synthase and chitinase may be possible. In addition, the receptor proteins for elicitors and the related enzymes in the systemic acquired resistance in host plants would be good targets for future screening of microbial products. Promising fungicidal products at present are listed below.

Rice Blast

Miharamycin A and B are novel nucleoside antibiotics produced by *Streptomyces miharaensis* SF-489 (Seto *et al.*, 1983a). They exert strong controlling activity against rice blast disease, though they sometimes cause phytotoxic effects on host plants. Nitropeptin, a dipeptide antibiotic possessing a nitro group, produced by *S. xanthochromogenus* 6257-MC_1 also shows anti-blast activity at less than 200 ppm with low mammalian toxicity (Ohba *et al.*, 1987). Nitropeptin exhibits inhibitory activity on *E. coli* K-12 on a synthetic medium, but its activity is antagonized by supplementation of L-glutamine, suggesting that it is an antimetabolite of L-glutamine. As its chemical structure is unique and rather simple, nitropeptin can be a good lead compound for synthetic anti-blast chemicals (Figure 4).

Rice Sheath Blight

Dapiramicin A (Figure 4) is a nucleoside antibiotic produced by *Micromonospora* sp. SF-1917 (Seto *et al.*, 1983b) which is effective on rice sheath blight in a green-house test. Its mammalian toxicity is low. A macrolide antibiotic, Notonesomycin A, is the main component of the antifungal substances produced by *S. aminophilus* subsp. *notonesogenes* 647 AV_1 (Sasaki *et al.*, 1986). It is as effective as validamycin against the sheath blight in a green-house test, but its acute toxicity is high (intraperitoneal LD_{50} in mice; 0.25 to 0.50 mg/kg). Benanomicins produced by *Actinomadura* sp. MH193-16F4 show wide antifungal and antiviral activities (Kondo *et al.*, 1991). Benanomicin B can inhibit the growth of the sheath blight pathogen (*Pellicularia sasakii* = *R. solani*) at 12.5 μg/ml *in vitro* and its mammalian toxicity is low.

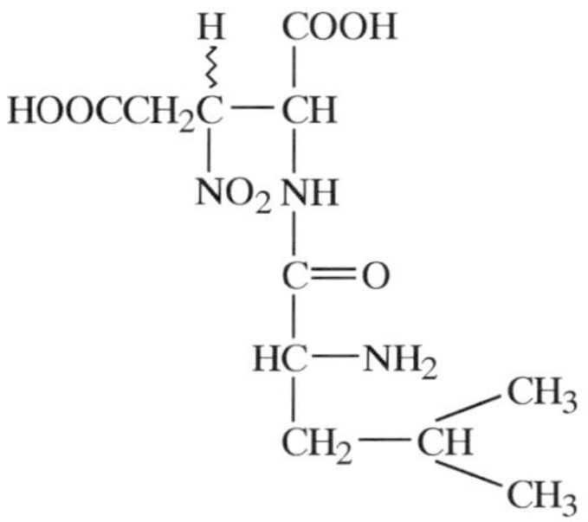

Nitropeptin

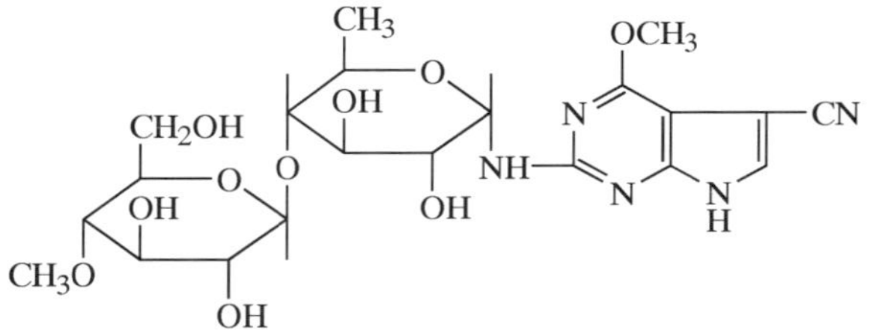

Dapiramicin A

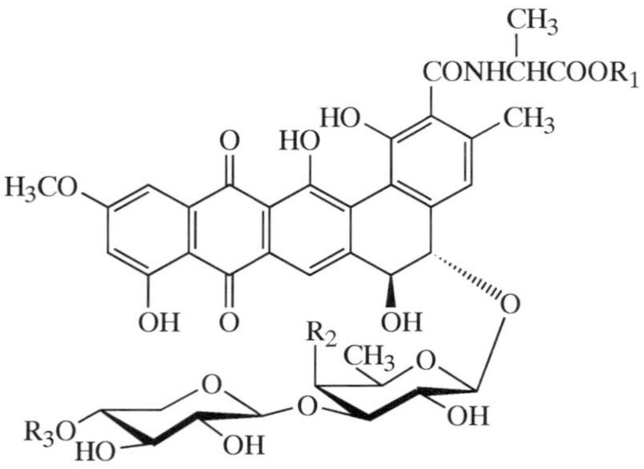

Benanomicin A $R_1 = H$, $R_2 = OH$, $R_3 = H$
Benanomicin B $R_1 = H$, $R_2 = NH_2$, $R_3 = H$

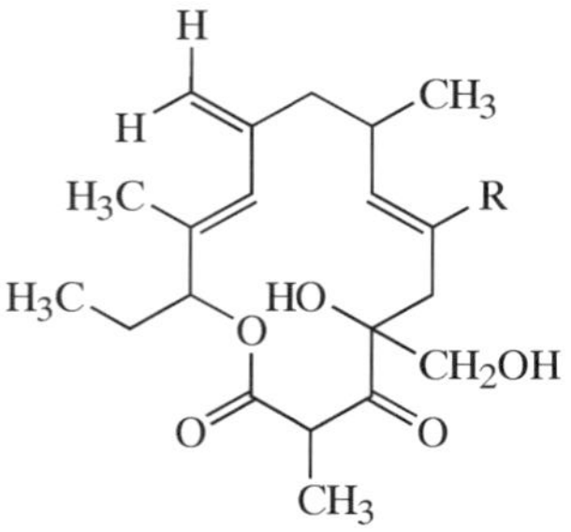

Rustmicin $R = OCH_3$
Neorustmicin A $R = CH_3$

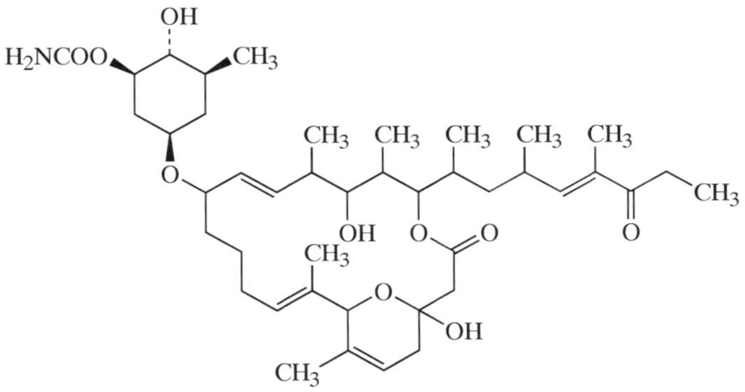

Irumamycin

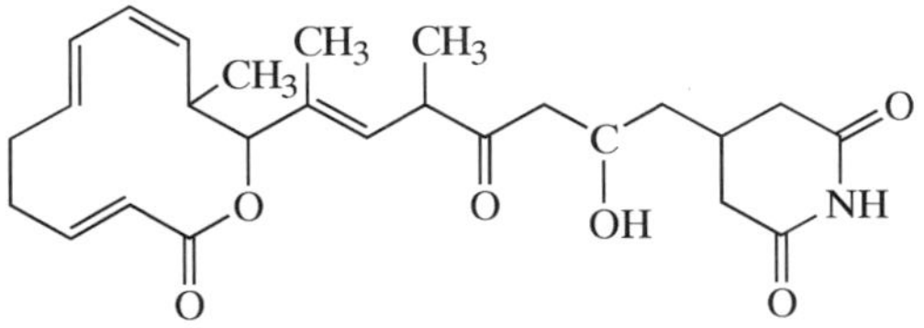

Lactimidomycin

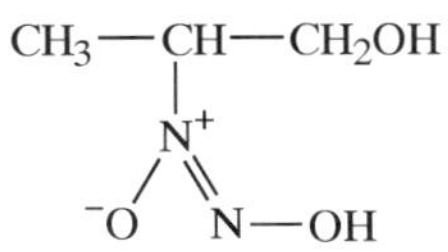

Propanosine

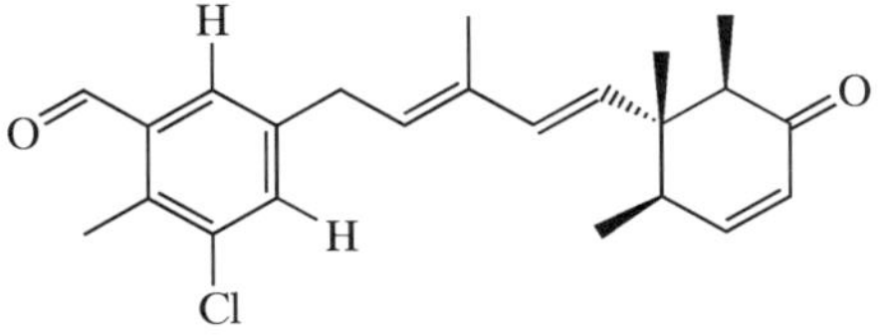

Ilicicolin E

Figure 4 Chemical structures of promising antibiotics as fungicides and leads

Wheat Stem Rust

Stem rust caused by *Puccinia graminis* f. sp. *tritici* is a serious wheat disease observed in many areas of the world. Through the *in vitro* screening of microbial products using uredospores of the pathogen, rustmicin and neorustmicins were found to be effective against the disease (Takatsu *et al.*, 1985; Nakayama *et al.*, 1986). Rustmicin and neorustmicin A (Figure 4), which belong to the 14-membered macroloide antibiotics, can protect wheat leaves from rust infection at concentrations as low as 4 ppm, but their toxicity has yet to be evaluated due to the low productivity in fermentation.

Gray Mold

Botrytis cinerea causes a troublesome gray mold disease with a wide host range, and the emergence of its resistant strains to benzimidazoles and dicarboximide fungicides has become a serious problem. Irumamycin (Figure 4) is a 20-membered macrolide antibiotic effective against gray mold, rice blast and pear black spot (Omura *et al.*, 1982). It is stable to sunlight, and its mammalian toxicity and phytotoxicity are low. Albopeptin (Isono *et al.*, 1986), tautomycin (Cheng *et al.*, 1987), tautomycetin (Cheng *et al.*, 1989) and RS-22 (Ubukata *et al.*, 1995) were also found to be effective on gray mold. These antibiotics were isolated from *Streptomyces* spp. of Chinese origin, in collaboration with Japanese scientists, providing a good example of new sources of organisms affording new antibiotics. Other microbial products such as deoxymulundocandin (Mukhopadhyay *et al.*, 1992), illudins (Lee *et al.*, 1996), malolactomycins (Tanaka *et al.*, 1997), and oxysporidinone (Breinholt *et al.*, 1997) were recently reported to have inhibitory activities on *B. cinerea*, but further study on their *in vivo* efficacy against gray mold as well as their toxicity on mammals and host plants are needed.

Other Diseases

Soilborne diseases such as *Fusarium* vascular wilts and clubroots are difficult to control using conventional synthetic chemicals. Lactimidomycin (Figure 4) (Sugawara *et al.*, 1992) and pradimicin A (Oki *et al.*, 1992) were reported to be active against *Fusarium moniliforme* (MIC: 0.8 and 3.1μg/ml, respectively). Their mammalian toxicities are rather low but durability to sunlight and phytotoxicity are not known. With regard to clubroot disease, *Phoma glomerata* and its product, epoxydon, were recently found to be effective against diseases of cruicifers (Arie *et al.*, 1998). Since epoxydon has an anti-auxin activity, clubroot suppression may be related to the anti-IAA activity of epoxydon.

Apple canker caused by *Valsa ceratosperma* is also a difficult disease to control with synthetic chemicals. Propanosine (Figure 4), a new antibiotic selectively active against the pathogen, was found by Abe *et al.* (1983). As the chemical structure of propanosine is simple, its derivatives can be synthesized for further study *in vivo*.

Ilicicolins were reported to be fungicidal to the oomycetes such as *Phytophthora infestans* and *Plasmopara viticola* (Hayakawa *et al.*, 1971), and they appear to be useful leads for further synthetic study.

Pathogenesis-Related enzymes

Cell wall chitin and glycan synthases are good targets for the development of microbial products as new fungicides because they do not exist in mammalian cells. Satomi *et al.* (1982) found neopeptins by *in vitro* assays using proteoheteroglycan and β-1,3-glucan synthase from *Saccharomyces cerevisiae*. Neopeptins exerted control efficacy against powdery mildew on cucumber at 100 ppm in pot tests with no phytotoxicity.

A novel mushroom metabolite, 3,5-dichloro-4-methoxybenzyl alcohol, was isolated from submerged cultures of *Stropharia* spp. as an inhibitor against chitin synthase prepared from *Coprinus cinereus* (Pfefferle *et al.*, 1990). The synthetic analogues showed the necessity of chlorine substituents for enzyme inhibition, and hexachlorophene was found to strongly inhibit chitin synthase.

Nishimoto *et al.*, (1991) have reported allosamidins as yeast chitinase inhibitors but the effects against phytopathogenic fungi *in planta* have not been widely investigated.

Germinating conidium of *Magnaporthe grisea* (anamorph: *P. oryzae*), the rice blast fungus, differentiates an infection structure called appressorium upon host surface signalling. The formation of appressoium is a prerequisite for the successful infection by the pathogen. From cultures of a *Gliocladium roseum* strain, glisoprenins were isolated (Tomoda *et al.*, 1992) and they proved to interfere with hydrophobic signalling necessary for appressorium formation (Thines *et al.*, 1997). Fusarin C and its derivatives were isolated from *Nectria coccinca* (Eilbert *et al.*, 1997) and they showed inhibition of fungal melanin biosynthesis which is inevitable for the appressoria of *M. grisea* to mature and obtain pathogenicity.

Some metabolites from plant pathogens have been reported to be self-growth inhibitors; macrophorins produced by *Botryosphaeria berengeriana* (the pathogen of *Macrophoma* fruit rot of apple) inhibit the mycelial growth of the pathogen itself (Sassa and Nukina, 1984), and a benomyl-resistant strain of cherry rot fungus *Monilinia fructicola* produces chloromonilicin as a self-growth inhibitor (Sassa *et al.*, 1985). *Puccinia coronata* (oat crown rust fungus), *P. graminis* (wheat stem rust fungus), and *P. oryzae* produce methyl *cis*-3,4-dimethoxycinnamate, methyl *cis*-ferulate, and pyriculol-related substances (Kono *et al.*, 1991), respectively. These microbial products specifically inhibit the spore germination of their producers.

APPLICATIONS OF MOLECULAR BIOLOGY

BLASTICIDIN S DEAMINASE GENES AS NOVEL SELECTIVE MARKERS IN MOLECULAR BIOLOGY

Blasticidin S is a microbial fungicide with a potent control efficacy on rice blast disease, as described earlier (Misato *et al.*, 1959). It occasionally exhibits adverse phytotoxic effects on some sensitive plants such as tobacco and eggplant due to its inhibitory activity on protein synthesis.

In the course of studies on the fate of blasticidin S in the environment, *Aspergillus terreus* S-712 isolated from a paddy soil was found to markedly decrease the activity of blasticidin S and thus to be highly resistant to the antibiotic (Yamaguchi *et al.*, 1972). The inactivation proved to be based on the deamination of the cytosine moiety in blasticidin S, and a new aminohydrolase named blasticidin S deaminase (BSD; EC 3.5.4.23) was purified from the fungus (Yamaguchi *et al.*, 1975, 1985, 1986). Later, another deaminase which also confers blasticidin S resistance was discovered in a strain of *Bacillus cereus* K55-S1 (Endo *et al.*, 1987, 1988). This enzyme, named BSR, catalyzed the same hydrolytic deamination with high substrate specificity towards blasticidin S and several of its analogues. Both BSD and BSR catalyzed no hydrolytic deamination of cytosine, cytidine, deoxycytidine, CMP, dCMP, or other purine nucleosides.

The responsible genes, *bsr* and *bsd*, have been cloned (Kamakura *et al.*, 1987; Kimura *et al.*, 1994a), their nucleotide sequences determined (Kobayashi *et al.*, 1991; Kimura *et al.*, 1994b), and the deduced amino acid sequences were compared each other. The *bsd* and *bsr* genes were predicted to code for polypeptides of 130 and 140 amino acid residues, respectively, with calculated molecular weights of 13 468 and 15 560 daltons. No overall similarity was detected between the deduced peptide sequences of these blasticidin S deaminases except the local regions corresponding to residues 85–95 of BSD and 97–107 of BSR (Figure 5). This short span of contiguous amino acids was also found in cytidine deaminase, which has no hydrolytic activity on blasticidin S. Recent crystallographic studies of cytidine deaminase from *E. coli* show that the region belongs to a part of a novel zinc binding motif and the 'TVHA sequence' (Bett *et al.*, 1994; Yang et al., 1992). The cytosine nucleoside/nucleotide deaminase families retain the striking similarity of local region for the coordination of catalytic zinc, while other parts of the deaminases are highly divergent. When BSD was analysed for zinc content by inductively coupled plasma emission spectrometry, one zinc per deaminase subunit was detected as in the case of cytidine deaminase.

Using the T7 transcription/expression plasmid, *bsd* was overexpressed in *E.coli* BL21 and purified to homogeneity by affinity chromatography. The native BSD was shown to be a tetrameric enzyme with a molecular weight of about 50 kDa (Kimura, M., unpublished). A comparison of nucleotide sequ-

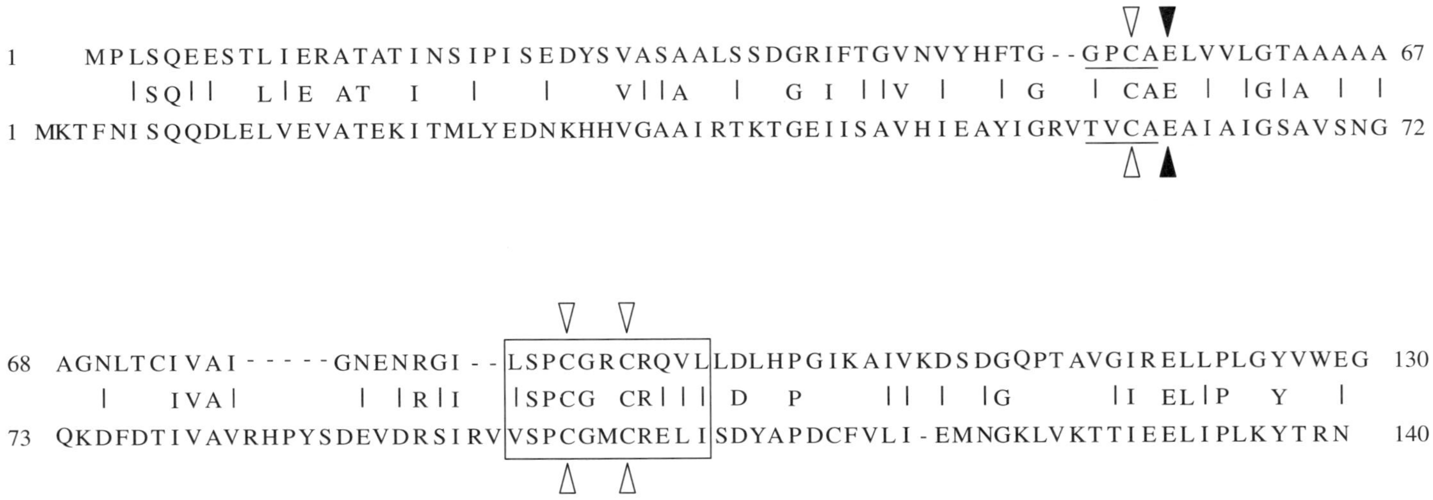

Figure 5 Comparison of the amino acid sequences of BS deaminases from *A. terreus* (BSD; upper) and *B. cereus* (BSR; lower). The identical or similar amino acids are shown in the middle. A local region showing extensive similarity is enclosed by a frame. Sequences corresponding to the ‘TVHA sequence’ are underlined. A catalytic zinc atom is coordinated to three cysteine residues marked by empty arrows. The active site glutamic acid residue is indicated by a filled arrow

ences showed that there exists no similarity between *bsd* and *bsr*, and a large difference was observed in preference for some codon use; the *bsr* gene originated from *B. cereus* has a 74.3% of codon use either A or U at the third position. Such extreme biased usage of codons was particularly evident for Ile, Tyr, Gln, Asn, Cys and Arg. In contrast, *bsd* does not possess such a bias in codon use, and as *bsd* is the inactivating gene of eukaryotic origin, it is expected to be more suitable for expression in eukaryotes. Overall, the availability of different genes, *bsd* and *bsr*, with the same function would be of great advantage for heterologous expression of blasticidin S deaminase activity in various organism, e.g. to be used as selectable marker genes.

In transfection of mammalian cells, blasticidin S served as a good selective reagent; HeLa cells transfected with *bsr* were isolated more easily and at less cost than the conventinal G418-*neo* transfection system (Izumi *et al.*, 1991). In other cell lines such as FM3A cells, however, substantially no blasticidin S resistant cells were obtained by the use of *bsr*. This low efficiency was greatly improved by the use of *bsd* instead for *bsr*, to the level comparable to that with *neo* (Kimura *et al.*, 1994b). The results suggest that efficient transfection of mammalian cells could be achieved by the choice of appropritate blasticidin S resistance gene depending on the recipient cells to be transfected.

Blasticidin S also proved to be useful for an efficient transformation of *P. oryzae*, one of the model phytopathogenic fungi recently drawing attention world-wide. While no transformants were obtained with *bsr*, transformation with *bsd* yielded many putative blasticidin S resistant colonies on the selective agar plate. Then a new method which makes use of the intrinsic property of regenerating fungal protoplasts to loosely adhere to dishes coated with collagen in liquid media was developed. By using this selection method, stable transformants of *P. oryzae* were obtained with the transformation efficiency of 190 transformants/μg DNA, indicating that the selection system with blasticidin S is much more efficient than the conventional system with hygromycin B in transformation of *P. oryzae* (Kimura *et al.*, 1995).

When blasticidin S was tested as a selective reagent in transformation of plants, *Nicotiana tabacum* and *Arabidopsis thaliana* transformed by *bsr* and *bsd* showed high levels of blasticidin S resistant phenotype (Kamakura *et al.*, 1990; Tamura *et al.*, 1995). No difference was observed in either growth or morphology between the transformed and the untransformed plants. However, when the leaves of wild-type and transgenic tobacco were inoculated with *Botrytis cinerea* and sprayed with 100 ppm of blasticidin S, the wild-type tobacco plants were severely damaged but the transgenic plants were well protected by blasticidin S from the development of gray mold disease with negligible phytotoxicity. The result suggests that blasticidin S may be applicable to control diseases in genetically engineered plants.

The discovery of blasticidin S deaminases and the cloning of the genes revealed a new aspect of the fungicide as a useful research reagent in molecular biology. Use of blasticidin S has several advantages over other selective reagents as follows; (a) conversion of blasticidin S to the deaminohydroxy-derivative can be spectrophotometrically determined, so that the deaminase activity is easily determined, (b) blasticidin S is sufficiently stable in the usual medium, but it is inactivated in alkaline solution, (c) blasticidin S has been registered as an agricultural chemical, its safety in the environment was already evaluated. Thus blasticidin S is a promising selective reagent for efficent introduction and expression of foreign DNA in molecular biology.

DISEASE RESISTANCE IN TRANSGENIC PLANTS ENGINEERED FOR DETOXIFICATION OF PATHOGENESIS-RELATED TOXINS

Advances in genetic engineering of plants have allowed the development of transgenic plants with agronomically important traits such as herbicide, insect and viral resistances. Such strategies could be applied to achieve plant protection against bacterial and fungal diseases. In particular, detoxification of pathogenesis-related toxins has an advantage that greater resistance can be exerted with lower levels of gene expression of detoxifying enzyme. Tabtoxin, a phytotoxic dipeptide produced by *Pseudomonas syringae* pv. *tabaci*, is composed of tabtoxinine β-lactam and either serine or threonine (Stewart, 1971). Several lines of evidence suggest that this toxin causes the chlorotic symptom associated with the disease wildfire caused by the pathogen. It is suggested that *in planta* some peptidases convert tabtoxin to tabtoxinine β-lactam, which inhibits the target enzyme glutamine synthetase (Sinden and Durbin, 1968). This inhibition results in the abnormal accumulation of ammonium in tobacco cells which causes the characteristic chlorosis. The bilanafos resistance gene which encodes an acetyltransferase originally functions in the biosynthesis of bilanafos in *S. hygroscopicus* (Kumada *et al.*, 1988). This example of 'self-resistance' suggests the existence of the tabtoxin resistance genes in the wildfire pathogen *P. syringae* pv. *tabaci*.

Bcl I-digested genomic DNA of *P. syringae* pv. *tabaci* MAFF 03-01075 was ligated to the *Bam* HI site of pUC13 and transformed into *E. coli* DH1 sensitive to tabtoxin. Consequently the recombinant plasmid pARK10 including a 2 kb insert of genomic DNA was obtained and the responsible gene *ttr* specific for the inactivation of tabtoxin was cloned (Anzai *et al.*, 1989). To test whether this gene could confer resistance to wildfire disease in tobacco, the chimeric *ttr* gene inserted between the cauliflower mosaic virus 35S promoter and the nopaline synthase polyadenylation signal on the binary vector pBI121 was introduced into tobacco by the leaf disk infection method. When the resulting transgenic tobacco plants were inoculated with *P. syringae* pv. *tabaci*, none of the transformed plants showed any chlorotic halo typical of wildfire disease. This indicates that the transgenic tobacco plants expressing the *ttr* gene have become resistant not only to tabtoxin but also to infection by *P. syringae* pv. *tabaci*. The results demonstrate a successful approach to obtain disease resistant plants by detoxification of the pathogenic toxins which play an important role in pathogenesis. Such a strategy could be applied to other diseases caused by pathogenesis-related toxins.

CONCLUSIONS AND OUTLOOK

Since pesticides of microbial origin (biochemical pesticides) are synthesized biologically, they are generally specific for target organisms and are likely to be inherently biodegradable. Among the pesticides listed in Table 1, polyoxins and validamycins are quite safe fungicides; they are not only non-phytotoxic, but also non-toxic to humans, livestock and wildlife. Such excellent characteristics are due to their modes of action; polyoxins selectively inhibit the synthesis of fungal cell-wall chitin, which does not exist in mammalian cells, and validamycin A only deteriorates the normal mycelial growth of the pathogen on plants. Because of such moderate activity of validamycin A, no occurrence of resistance has been reported despite the largest use among the pesticides of microbial origin.

Further, components of avermectins (Dybas, 1983), insecticidal microbial products, were successfully transformed into significantly more active derivatives by simple chemical modification of the original structures. This finding leaves considerable scope for the prospect of novel microbial products. In addition, new microbial products with novel characteristics, e.g. strobilurin, were found against plant pathogens resistant to conventional synthetic chemicals. The following considerations might be important for the future development of microbial products as fungicidal agents.

1. Establishment of screening directions. The developments of some new antibiotics were achieved shortly after establishing a new screening project, e.g. polyoxins against rice sheath blight.
2. Establishment of new screening test methods. Introduction of a novel screening test method carries with it the possibility for the development of new effective antibiotics. Kasugamycin and validamycins were discovered by adopting new assay systems.
3. Avoidance of pathogens' resistance to antibiotics. Kasugamycin and polyoxins have highly selective toxicity to pathogenic fungi. When pathogens resistant to these antibiotics emerged, the alternate or combined application of chemicals with different action mechanisms was effective. Such microbial products which can exert a negative cross resistance with conventional chemicals may be also possible.
4. The use of antibiotics as leads for new pesticides. Modifications of existing antibiotics provided new potent chemicals. The best examples of these would be semi-synthetic avermectin and methoxyacrylates. More attention should be focused on finding the molecular relationship between chemical structure and biological activity.

In recent years, standards for the disposal of industrial wastes have become more and more strict and consequently the manufacture of synthetic chemicals becomes quite expensive if waste disposal facilities, as well as production plants, must be set up for each new product. Pesticides of microbial origin have an economic advantage in this respect, since a variety of substances can be manufactured using one set of equipment and facilities. Additionally, they are produced not from limited fossil resources but from renewable agricultural products through fermentation by microorganisms. From these aspects, pesticides of microbial origin may be more advantageous especially for developing countries.

However, as is true for every scientific technique, their use also has limitations. One disadvantage is the difficulty in their microanalysis, especially when they consist of many components. The second demerit is that, because of their highly specific mechanism of action, they are apt to suffer from the emergence of resistant strains. Further, the biggest limitation to the use of microbial products in agriculture would be a concern that their wide use might create resistant strains which could hinder medical treatment of humans by the antibiotics. However, this concern should not be limited to pesticides of microbial origin. There is no difference between microbial products and synthetic chemicals in this respect and the important point is whether or not any pesticides may induce cross resistance to medicines. Crop and pest biochemistry and molecular biology will make further rapid advances in the near future, and new biotechnology will be an important factor for plant protection research and strategy. For example, more efficient production of microbial products is becoming possible by modern gene engineering. Biorational approaches will also become feasible in the design of new pesticides from microbial products as leads by the extensive use of computer and data-processing procedures.

REFERENCES

Abe, Y., Kadokura, J., Shimazu, A., Seto, H., and Otake, N. (1983). 'Propanosine (K-76), a new antibiotic active against *Valsa ceratosperma*, the pathogen of apple canker disease', *Agric. Biol. Chem.*, **47**, 2703–2705.

Anzai, H., Yoneyama, K., and Yamaguchi, I. (1989). 'Transgenic tobacco resistant to a bacterial disease by the detoxification of a pathogenic toxin', *Mol. Gen. Genet.*, **219**, 492–494.

Arie, T., Kobayashi, Y., Okada, G., Kono, Y., and Yamaguchi, I. (1998). 'Epoxydon produced by *Phoma glomerata* controls soilborne clubroot disease of cruciferous plants', *Plant Pathol.*, submitted.

Asakawa, M., Misato, T. and Fukunaga, K. (1963). 'Studies on the prevention of the phytotoxicity of blasticidin S', *Pestic. and Technique.*, **8**, 24–29.

Bett, L., Xiang, S., Short, S-A., Wolfenden, R., and Carter, C.-W. Jr (1994). 'The 2.3 A crystal structure of an enzyme: transition-state analog complex', *J. Mol. Biol.*, **235**, 635–656.

Breinholt, J., Ludvigsen, S., Rassing, B. R., Rosendahl, C. N., Nielsen S. E., and Olsen, C. E. (1997). 'Oxysporidinone: a novel *N*-methyl-4-hydroxy 2-pyridone from *Fusarium oxysporum*', *J. Nat. Prod.*, **60**, 33–35.

Cheng, X.-C., Kihara, T., Kusakabe, H., Magae, J., Kobayashi, Y., Fant, R.-P., Ni, Z.-F., Shen, Y.-C., Ko, K., Yamaguchi, I. and Isono, K. (1987). 'A new antibiotic, tautomycin', *J. Antibiot.*, **40**, 907–909.

Cheng, X.-C., Kihara, T., Ying, X., Uramoato, M., Osada, H., Kusakabe, H., Wang B-N., Kobayashi, Y., Ko, K., Yamaguchi, I., Shen, Y.-C., and Isono, K. (1989). 'A new antibiotic, tautomycetin', *J. Antibiot.*, **42**, 141–144.

Dybas, R. A. (1983). 'Avermectins: their chemistry and pesticidal activities', in *Pesticide Chemistry; Human Welfare and the Environment, vol. 1* (eds Miyamoto, J. and Kearney, P. C.), pp. 83–90, Pergamon Press, Oxford.

Eguchi, J., Sasaki, S., Ohta, N., Akashiba, T., Tsuchiyama, T., and Suzuki, S. (1968). 'Studies on polyoxins, antifungal antibiotics. Mechanism of action on the diseases caused by *Alternaria* spp.', *Ann. Phytopathol. Soc. Japan*, **34**, 280–288.

Eilbert, F., Thines, E., Arendholz, W. R., Sterner, O., and Anke, H. (1997). 'Fusarin C, (7Z)-fusarin C and (5Z)-fusarin C; inhibitors of dihydroxynaphthalene-melanin biosynthesis from *Nectria coccinca*', *J. Antibiot.*, **50**, 443–445.

Endo, T., Furuta, K., Kaneko, A., Katsui, T., Kobayashi, K., Azuma, A., Watanabe, A., and Shimadzu, A. (1987). 'Inactivation of blasticidin S by *Bacillus cereus*. I. Inactivation mechanisms', *J. Antibiot.*, **40**, 1791–1793.

Endo, T., Kobayashi, K., Nakayama, N., Tanaka, T., Kamakura, T., and Yamaguchi, I. (1988). 'Inactivation of blasticidin S by *Bacillus cereus*. II. Isolation and characterization of a plasmid, pBSR8, from *Bacillus cereus*', *J. Antibiot.*, **41**, 271–273.

Endo, A. and Misato, T. (1969). 'Polyoxin D, a competitive inhibitor of UDP-*N*-acetylglucosamine. Chitin *N*-acetylglucosaminyl-transferase in *Neurospora crassa*', *Biochem. Biophys. Res. Comm.*, **37**, 718–722.

Fukagawa, Y., Sawa, T., Takeuchi, T., and Umezawa, H. (1968a). 'Studies on biosynthesis of kasugamycin. I. Biosynthesis of kasugamycin and the kasugamine moiety', *J. Antibiot.*, **21**, 50–54.

Fukagawa, Y., Sawa, T., Takeuchi, T., and Umezawa, H. (1968b). 'Studies on biosynthesis of kasugamycin. III. Biosynthesis of the D-inositol' *J. Antibiot.*, **21**, 185–188.

Fukagawa, Y., Sawa, T., Takeuchi, T., and Umezawa, H. (1968c). 'Studies on biosynthesis of kasugamycin. V. Biosynthesis of the amidine group', *J. Antibiot.*, **21**, 410–421.

Hamada, M., Hashimoto, T., Takahashi, S., Yoneyama, M., Miyake, T., Takeuchi, Y., Okami, Y., and Umezawa, H. (1965). 'Antimicrobial activity of kasugamycin', *J. Antibiot.*, **Ser. A. 18**, 104–106.

Harada, S. and Kishi, T. (1978). 'Isolation and characterization of mildiomycin, a new nucleoside antibiotic', *J. Antibiot.*, **31**, 519–524.

Harada, S., Mizuta, E., and Kishi, T. (1978). 'Structure of mildiomycin, a new antifungal nucleoside antibiotic', *J. Am. Chem. Soc.*, **100**, 4895–4897.

Hayakawa, S., Minato, H., and Katagiri, K (1971). 'The ilicicolins, antibiotics from *Cylindrocladium ilicicola*' *J. Antibiot.*, **24**, 653-654.

Helser, T. L., Davies, J. E., and Dahlberg, J. E. (1971). 'Change in methylation of 16S ribosomal RNA associated with mutation to kasugamycin resistance in *Escherichia coli.*', *Nature*, **233**, 12–14.

Hirai, T. and Shimomura, T. (1965). 'Blasticidin S, an effective antibiotic against plant virusmultiplication', *Phytopathol.*, **55**, 291–295.

Hori, M., Eguchi, J., Kakiki, K., and Misato, T. (1974a). 'Studies on the mode of action of polyoxins. VI. Effect of polyoxin B on chitin synthesis in polyoxin-sensitive and resistant strains of *Alternaria kikuchiana*', *J. Antibiot.*, **27**, 260–266.

Hori, M., Kakiki, K., and Misato, T. (1974b). 'Studies on the mode of action of polyoxin. Part IV. Further study on the relation of polyoxin structure to chitin synthetase inhibition', *Agric. Biol. Chem.*, **38**, 691–698.
Hori, M., Kakiki, K., and Misato, T. (1977). 'Antagonistic effect of dipeptides on the uptake of polyoxin A by *Alternaria kikuchiana*', *J. Pestic. Sci.*, **2**, 139–149.
Horii, S., Iwasa, T., and Kameda Y. (1971). 'Studies on validamycins, new antibotics. V. Degradation studies', *J. Antibiot.*, **24**, 57–58.
Horii, S., Iwasa, T., and Kawahara, K. (1972). 'Studies on validamycins, new antibiotics. VIII. Validamycins C, D, E and F', *J. Antibiot.*, **25**, 48–53.
Huang, K. T., Misato, T., and Asuyama, H. (1964). 'Effect of blasticidin S on protein synthesis of *Piricularia oryzae*', *J. Antibiot.*, **Ser. A. 17**, 65–70.
Ikekawa, T., Umezawa, H., and Iitaka, Y. (1966). 'The structure of kasugamycin hydrobromide by X-ray crystallographic analysis', *J. Antibiot.*, **Ser. A. 19**, 49–50.
Ishii, T., Doi, Y., Yora, K., and Asuyama, H. (1967). 'Suppressive effects of antibiotics of tetracycline group on symptom development of mulberry dwarf disease', *Ann. Phytopatol. Soc. Japan*, **33**, 267–275.
Ishiyama, T., Hara, I., Matsuoka, M., Sato, K., Shimada, S., Izawa, R., Hashimoto, T., Hamada, M., Okami, Y., Takeuchi, T., and Umezawa, H. (1965). 'Studies on the preventive effect of kasugamycin on rice blast', *J. Antibiot.*, **Ser. A. 18**, 115–119.
Isono, K., Asahi, K., and Suzuki, S. (1969). 'Studies on polyoxins, antifungal antibiotics. XIII. The structures of polyoxins', *J. Am. Chem. Soc.*, **91**, 7490–7505.
Isono, K., Kobinata, K., Oikawa, H., Kusakabe, H., Uramoto, M., Ko, K., Misato, T., Tai, S-H., Ni, C-T., and Shen, Y-C. (1986). 'New antibiotics, albopeptins A and B', *Agric. Biol. Chem.*, **50**, 2163–2165.
Isono, K., Nagatsu, J., Kawashima, Y., and Suzuki, S. (1965). 'Studies on polyoxins, antifungal antibiotics. Part I. Isolation and characterization of polyoxins A and B', *Agric. Biol. Chem.*, **29**, 848–854.
Isono, K., Nagatsu, J., Kobinata, K., Sasaki, K., and Suzuki, S. (1967). 'Studies on polyoxins, antifungal antibiotics. Part V. Isolation and characterization of polyoxins C, D, E, F, G, H and I', *Agric. Biol. Chem.*, **31**, 190–199.
Isono, K. and Suhadolnik, R. H. (1976). 'The biosynthesis of natural and unnatural polyoxins by *Streptomyces cacaoi*', *Arch. Biochem. Biophy.*, **173**, 141–153.
Isono, K. and Suzuki, S. (1968a). 'The structures of polyoxins D, E, F, G, H, I, J, K and L', *Agric. Biol. Chem.*, **32**, 1193–1197.
Isono, K. and Suzuki, S. (1968b). 'The structure of polyoxin C', *Tetrahedron Lett.*, **2**, 203–208.
Iwasa, T., Higashide, E., and Shibata, M. (1971a). 'Studies on validamycins, new antibiotics. III. Bioassay methods for the determination of validamycin', *J. Antibiot.*, **24**, 114–118.
Iwasa, T., Higashide, E., Yamamoto, H., and Shiba, M. (1971b). 'Studies on validamycins, new antibiotics. II. Production and biological properties of VM-A and B', *J. Antibiot.*, **24**, 107–113.
Iwasa, T., Kameda, Y., Asai, M., Horii, S., and Mizuno, K. (1971c). 'Studies on validamycins, new antibiotics. IV. Isolation and characterization of VM-A and B', *J. Antibiot.*, **24**, 119–123.
Izumi, M., Miyazawa, H., Kamakura, T., Yamaguchi, I., Endo, T., and Hanaoka, F. (1991). 'Blasticidin S resistance gene (*bsr*): a novel selectable marker for mammalian cells', *Exp. Cell Res.*, **197**, 229–233.
Kakinuma, K., Otake, N., and Yonehara, H. (1974). 'The structure of detoxin D_1', *Agr. Biol. Chem.*, **38**, 2529–2538.

Kamakura, T., Yoneyama, K., and Yamaguchi, I. (1990). 'Expression of the blasticidin S deaminase gene (*bsr*) in tobacco: fungicide tolerance and a new selective marker for transgenic plants', *Mol. Gen. Genet.*, **223**, 332–334.

Kamakura, T., Kobayashi, K., Tanaka, T., Yamaguchi, I., and Endo, T. (1987). 'Cloning and expression of a new structural gene for blasticidin S deaminase, a nucleoside aminohydrolase', *Agric. Biol. Chem.*, **51**, 3165–3168.

Kameda, Y., Horii, S., and Yamoto, T. (1975). 'Microbial transformation of validamycins', *J. Antibiot.*, **28**, 298–306.

Kimura, M., Izawa, K., Yoneyama, K., Arie, T., Kamakura, T., and Yamaguchi, I. (1995). 'A novel transformation system for *Pyricularia oryzae*: Adhesion of regenerating fungal protoplasts to collagen-coated dishes', *Biosci. Biotech. Biochem.*, **59**, 1177–1180.

Kimura, M., Kamakura, T., Tao, Q-Z., Kaneko, I., and Yamaguchi, I. (1994a). 'Cloning of the blasticidin S deaminase gene (*BSD*) from *Aspergillus terreus* and its use as a selectable marker for *Schizosaccharomyces pombe* and *Pyricularia oryzae*', *Mol. Gen. Genet.*, **242**, 121–129.

Kimura, M., Takatsuki, A., and Yamaguchi, I. (1994b). 'Blasticidin S deaminase gene from *Aspergillus terreus* (*BSD*): A new drug resistance gene for transfection of mammalian cells', *Biochim. Biophys. Acta*, **1219**, 653–659.

Kinoshita, T., Tanaka, N., and Umezawa, H. (1970), 'Binding of blasticidin S to ribosomes', *J. Antibiot.*, **23**, 288–290.

Kobayashi, K., Kamakura, T., Tanaka, T., Yamaguchi, I., and Endo, T. (1991). 'Nucleotide sequence of the *bsr* gene and *N*-terminal amino acid sequence of the blasticidin S deaminase from blasticidin S resistant *Escherichia coli* TK121', *Agric. Biol. Chem.*, **55**, 3155–3157.

Kondo, S., Gomi, S., Ikeda, D., Hamada, M., Takaeuchi, T., Iwai, H., Seki, J., and Hoshino, H. (1991). 'Antifungal and antiviral activities of benanomicins and their analogues', *J. Antibiot.*, **44**, 1228–1236.

Kono, Y., Sekido, S., Yamaguchi, I., Kondo, H., Suzuki, Y., Neto, G. C., Sakurai, A., and Yaegashi, H. (1991). 'Structures of two novel pyriculol-related compounds and identification of naturally produced epipyriculol from *Pyricularia oryzae*', *Agric. Biol. Chem.*, **55**, 2785–2791.

Kumada, Y., Anzai, H., Takano, E., Murakami, T., Hara, O., Itoh, R., Imai, S., Satoh, A., and Nakagawa, K. (1988). 'The bialaphos resistance gene (*bsr*) plays a role in both self-defense and bialahos production in *Streptomyces hygroscopicus*', *J. Antibiot.*, **41**, 1838-1845.

Kuzuhara, H., Ohrui, H., and Emoto, S. (1973). 'Total synthesis of polyoxin J', *Tetrahedron Lett.*, **50**, 5055–5058.

Lee, I. K., Jeong, C. Y., Cho, S. M., Yun, B. S., Kim, Y. S., Yu, S. H., Koshino, H., and Yoo, I. D. (1996), 'Illudins C2 and C3, new illudin C derivatives from *Coprinus atramentarius* ASI 20013', *J. Antibiot.*, **49**, 821–822.

Likover, T. E., and Kurland, C. G. (1967). 'The contribution of DNA to translation errors induced by streptomycin *in vitro*', *Proc. Natl. Acad. Sci. U.S.A.*, **58**, 2385–2392.

Matsuura, K. (1983). 'Characteristics of validamycin A in controlling *Rhizoctonia* diseases', in *Pesticide Chemistry; Human Welfare and the Environment, vol. 2* (eds Miyamoto, J. and Kearney, P. C.), pp. 301–308, Pergamon Press, Oxford.

Misato, T., Ishii, I., Asakawa, M., Okimoto, Y., and Fukunaga, K. (1959). 'Antibiotics as protectant fungicides against rice blast. II. The therapeutic action of blasticidin S', *Ann. Phytopathol. Soc. Japan.*, **24**, 302–306.

Mukhopadhyay, T., Roy, K., Bhat, R. G., Sawant, S. N., Blumbach, J., Ganguli, B. N., Fehlhaber, H. W., and Kogler, H. (1992). 'Dexymulundocandin— a new echinocandin type antifungal antibiotic', *J. Antibiot.*, **45**, 618–623.

Nakayama, H., Hanamura, T., Abe, Y., Shimazu, A., Furihata, K., Ikeda, K., Seto, H., and Otake, N. (1986). 'Structures of neorustmicins B, C and D; new congeners of rustmicin and neorustmicin A', *J. Antibiot.*, **39**, 1016–1020.

Nioh, T. and Mizushima, S. (1974). 'Effect of validamycin on the general growth and morphology of *Pellicularia sasakii*', *J. Gene. Applied Microbiol.*, **20**, 373–383.

Nishimoto, Y., Sakuda, S., Takayama, S., and Yamada, Y. (1991). 'Isolation and characterization of new allosamidins', *J. Antibiot.*, **44**, 716–722.

Ohba, K., Nakayama, H., Shimazu, A., Seto, H., and Otake, N. (1987). 'Nitropeptin, a new dipeptide antibiotic possessing a nitro group', *J. Antibiot.*, **40**, 709–713.

Ohmori, K. (1967). 'Studies on characters of *Piricularia oryzae* made resistant to kasugamycin', *J. Antibiot.*, **Ser. A. 20**, 109–114.

Oki, T., Kakushima. M., Hirano, M., Takahashi, A., Ohta, A., Masuyoshi, S., Hatori, M., and Kamei, H. (1992). '*In vitro* and *in vivo* antifugal activities of BMS-181184', *J. Antibiot.*, **45**, 1512–1517.

Om, Y., Yamaguchi, I., and Misato, T. (1984). 'Inhibition of protein synthesis by mildiomycin, an anti-mildew substance', *J. Pesticide Sci.*, **9**, 317–323.

Omura, S., Nakagawa, A., and Tanaka, Y. (1982). 'Structure of a new antifungal antibiotic, irumamycin', *J. Org. Chem.*, **47**, 5413–5415.

Otake, N., Kakinuma, K., and Yonehara, H. (1973). 'Separation of detoxin complex and characterization of two active principles, detoxin C and D', *Agr. Biol. Chem.*, **37**, 777–780.

Otake, N., Takeuchi, S., Endo, T., and Yonehara, H. (1966). 'Chemical studies on blasticidin S. III. The structure of blasticidin S', *Agric. Biol. Chem.*, **30**, 132–141.

Pfefferle, W., Anke, H., Bross, M., and Steglich, W. (1990). 'Inhibition of solubilized chitin synthase by chlorinated aromatic compounds isolated from mushroom cultures', *Agric. Biol. Chem.*, **54**, 1381–1384.

Rodriguez-Fonseca, C., Amils, R., and Garrett, R. A. (1995). 'Fine structure of the peptidyl transferase centre on 23 S-like rRNAs deduced from chemical probing of antibiotic-ribosome complexes', *J. Mol. Biol.*, **247**, 224–235.

Sasaki, S., Ohta, N., Eguchi, J., Furukawa, Y., Akashiba, T., Tsuchiyama, T., and Suzuki, S. (1968). 'Studies on polyoxins, antifungal antibiotics', *Phytopathol. Soc. Japan.*, **34**, 272–279.

Sasaki, T., Furihata, K., Shimazu, A., Seto, H., Iwata, M., Watanabe, T., and Otake, N. (1986). 'A novel macrolide antibiotic, notonesomycin A', *J. Antibiot.*, **39**, 502–509.

Sassa, T. and Nukina, M. (1984). 'Macrophorin D, a new self-growth inhibitor of the causal fungus of *Macrophoma* fruit rot of apple', *Agric. Biol. Chem.*, **48**, 1923–1925.

Sassa, T., Kachi, H., Nukina, M., and Suzuki, Y. (1985). 'Chloromonilicin, a new antifungal metabolite produced by *Monilinia fructicola*', *J. Antibiot.*, **38**, 439–441.

Satomi, T., Kusakabe, H., Nakamura, G., Nishio, T., Uramoto, M., and Isono, K. (1982). 'Neopeptins A and B, new antifungal antibiotics', *Agric. Biol. Chem.*, **46**, 2621–2623.

Seto, H., Koyama, M., Ogino, H., Tsuruoka, T., Inouye, S., and Otake, N. (1983a). 'The structures of novel nucleoside antibiotics, miharamycin A and miharamycin B', *Tetrahedron Lett.*, **24**, 1805–1808.

Seto, H., Sasaki, S., Otake, N., Koyama, M., Kodama, Y., Ogino, H., Nishizawa, Y., Tsuruoka, T., and Inouye, S. (1983b). 'The structure of a novel nucleoside antibiotic, dapiramicin A', *Tetrahedron Lett.*, **24**, 495–498.

Seto, H., Otake, N., and Yonehara, H. (1968). 'Studies on the biosynthesis of blasticidin S. Part II. Leucylblasticidin S, a metabolic intermediate of blasticidin S biosynthesis', *Agric. Biol. Chem.*, **32**, 1299–1305.

Seto, H., Yamaguchi, I., Otake, N., and Yonehara, H. (1966). ''Biogenesis of blasticidin S', *Tetrahedron Lett.*, 3793–3799.

Shigemoto, R., Okuno, T., and Matsuura, K. (1989). 'Effect of validamycin A on the activity of trehalase of *Rhizoctonia solani* and several sclerotial fungi', *Ann. Phytopathol. Soc. Japan*, **55**, 238–241.

Sinden, S. L., and Durbin, R. D. (1968). 'Glutamine synthetase inhibition: possible mode of action of wildfire toxin from *Pseudomonus tabaci*', *Nature*, **219**, 378–380.

Stewart, W. (1971). 'Isolation and proof of structure of wild fire toxin', *Nature*, **229**, 174–178.

Suami, T., Ogawa, S., and Chida, N. (1980). 'The revised structure of validamycin A', *J. Antibiot.*, **33**, 98–99.

Sugawara, K., Nishiyama, Y., Toda, S., Komiyama, N., Hatori, M., Moriyama, T., Sawada, Y., Kamei, H., Konishi, M., and Oki, T. (1992). 'Lactimidomicin, a new glutarimide group antibiotic', *J. Antibiot.*, **45**, 1433–1441.

Suhara, Y., Maeda, K., and Umezawa, H. (1966). 'Chemical Studies on kasugamycin. V. The structure of kasugamycin', *Tetrahedron Lett.*, **12**, 1239–1244.

Suhara, Y., Sasaki, F., Maeda, K., Umezawa, H., and Ohno, M. (1968). 'The total synthesis of kasugamycin', *J. Am. Chem. Soc.*, **90**, 6559–6560.

Tachibana, K., Watanabe, T., Sekizawa, Y., Konnai, M., and Takematsu, T. (1982). 'Finding of the herbicidal activity and the mode of action of biolaphos, L-2-amino-4-[(hydroxy) (methyl) phosphinoyl] butyrylalanyl-alanine', *Abstr. 5th Int. Congr. Pestic. Chem.*, IVa-19.

Takeuchi, S., Hirayama, K., Ueda, K., Sasaki, H., and Yonehara, H. (1958). 'Blasticidin S, a new antibiotic', *J. Antibiot.*, **Ser. A. 11**, 1–5.

Takatsu, T., Nakayama, H., Shimazu, A., Furihata, K., Ikeda, K., Furihata, K., Seto, H., and Otake, N. (1985). 'Rustmicin, a new macrolide antibiotic active against wheat stem rust fungus', *J. Antibiot.*, **38**, 1806–1809.

Tamura, K., Kimura, M., and Yamaguchi, I. (1995). 'Blasticidin S deaminase gene (*BSD*): a new selection marker gene for transformation of *Arabidopsis thaliana* and *Nicotiana tabacum*', *Biosci. Biotech. Biochem.*, **59**, 2336–2338.

Tanaka, N., Matsukawa, H., and Umezawa, H. (1967). 'Structural basis of kanamycin for miscoding activity', *Biochem. Biophys. Res. Comm.*, **26**, 544–549.

Tanaka, N., Sakagami, Y., Yamaki, H., and Umezawa, H. (1961). 'Activity of cytomycin and blasticidin S against transplantable animal tumors', *J. Antibiot.*, **Ser. A. 14**, 123–126.

Tanaka, Y., Yoshida, H., Enomoto. Y., Shiomi, K., Shinose, M., Takahashi, Y., and Omura, S. (1997). 'Malolactomycins C and D, new 40-membered macrolides active against *Botrytis*', *J. Antibiot.*, **50**, 194–200.

Thines, E., Eilbert, F., Sterner, O., and Anke, H. (1997). 'Glisoprenin A, an inhibitor of the signal transduction pathway leading to appressorium formation in germinating conidia of *Magnaporthe grisea* on hydrophobic surfaces,' *FEMS Microbiol. Lett.*, **151**, 219–224.

Tomoda, H., Huang, X.-H., Nishida, H., Masuma, R., Kim, Y. K., and Omura, S. (1992). 'Glisoprenins, new inhibitors of acyl-Co A: Cholesterol acyltransferase produced by *Gliocladium* sp. FO-1523', *J. Antibiot.*, **45**, 1202–1206.

Ubukata, M., Shiraishi, N., Kobinata, K., Kudo, T., Yamaguchi, I., Osada, H., Shen, Y-C., and Isono, K. (1995). 'RS-22A, B and C: New macrolide antibiotics from *Streptomyces violaceusniger*', *J. Antibiot.*, **48**, 289–292.

Umezawa, H., Okami, Y., Hashimoto, T., Suhara, Y., Hamada, M., and Takeuchi, T. (1965). 'A new antibiotic, kasugamycin', *J. Antibiot.*, **Ser. A. 18**, 101–103.

Wakae, O., and Matsuura, K. (1975). 'Characteristics of validamycin as a fungicide for *Rhizoctonia* disease control', *Rev. Plant Protec. Res.*, **8**, 81–92.

Yamaguchi, H. and Tanaka, N. (1966). 'Inhibition of protein synthesis by blasticidin S. II. Studies on the site of action in *E. coli* polypeptide synthesizing systems', *J. Biochem.*, **60**, 632–642.

Yamaguchi, I. and Misato, T. (1985). 'Active center and mode of reaction of blasticidin S deaminase', *Agric. Biol. Chem.*, **49**, 3355–3361.
Yamaguchi, I., Seto, H., and Misato, T. (1986). 'Substrate binding by blasticidin S deaminase, an aminohydrolase for novel 4-aminopyrimidine nucleosides', *Pestic. Biochem. Physiol.*, **25**, 54–62.
Yamaguchi, I., Shibata, H., Seto, H., and Misato, T. (1975). 'Isolation and purification of blasticidin S deaminase from *Aspergillus terreus*', *J. Antibiot.*, **28**, 7–14.
Yamaguchi, I., Takagi, K., and Misato, T. (1972). 'The sites for degradation of blasticidin S', *Agric. Biol. Chem.*, **36**, 1719–1727.
Yang, C., Carlow, D., Wolfenden, R., and Short, S-A. (1992). 'Cloning and nucleotide sequence of the *Escherichia coli* cytidine (*cdd*) deaminase gene', *Biochemistry*, **31**, 4168–4174.
Yonehara, H. and Otake, N. (1966). 'Absolute configuration of blasticidin S', *Tetrahedron Lett.*, 3785–3791.
Yonehara, H., Seto, H., Aizawa, S., Hidaka, T., Shimazu, A., and Otake, N. (1968). 'The detoxin complex, selective antagonists of blasticidin S', *J. Antibiot.*, **21**, 369–370 .
Yukioka, M., Hatayama, T., and Morisawa, S. (1975). 'Affinity labelling of the ribonucleic acid component adjacent to the peptidyl recognition center of peptidyl transferase in *Escherichia coli* ribosomes', *Biochim. Biophys. Acta.*, **390**, 192–208.

4 Fungicide Resistance

S. J. KENDALL and D. W. HOLLOMON
Long Ashton Research Station, Bristol, UK

INTRODUCTION

Prior to 1970, nearly all fungicides used for the control of plant pathogens were multisite inhibitors with protectant activity. Despite their widespread use, resistance to these compounds was a rare event and only occurred many years after introduction. Where resistance did occur, the mechanism was likely to be non-specific such as detoxification of the fungicide, reduction in uptake, efflux of the fungicide from the pathogen or immobilization on fungal spore surfaces. With the introduction of the site-specific fungicides, the incidence of fungicide resistance problems greatly increased and the time interval between commercial introduction and the emergence of resistance was very much shorter than that seen with multisite fungicides. The mechanisms of resistance to site-specific fungicides can be attributed not only to detoxification, reduction in uptake and increased efflux of the fungicide from fungal cells, but also, to alternate metabolic pathways, an increase in the target enzyme or reduced affinity of the target site. In general, greater effectiveness and the eradicant action of the systemic fungicides may also have increased the risk of resistance through increasing selection pressure. So resistance may develop to any disease control agent, and in any fungal pathogen. In this chapter we review scientific and

Fungicidal Activity. Edited by D. H. Hutson and J. Miyamoto

technological aspects of resistance against a background of an increasing need to quantify resistance risk with greater precision. Definitions of resistance are explored in the context of epidemiology and factors which affect resistance build up. We expose the lack of information about resistance mechanisms in field strains of plant pathogens in contrast to information about laboratory mutants. The potential flowing from understanding resistance at a molecular level is discussed in terms of exploiting negative cross-resistance as a durable anti-resistance strategy, and developing more rapid methods for detecting and monitoring resistance. Detailed studies are beginning to make some sense of the effect of dose rate on selection for resistance, but this remains one of the unanswered questions affecting practical use of fungicides. More positively, molecular biology provides an opportunity to bring together for the first time fundamental studies on insecticide, herbicide and fungicide resistance, which hopefully will reveal common themes underlying the initial development of resistance, and suggest novel approaches to combating the problem.

EPIDEMIOLOGY OF RESISTANCE

Mutations of all kinds are continuously and spontaneously evolving and disappearing. Problems of resistance can arise only if the genetic mutation for resistance to the toxicant already exists or there is the potential for genetic mutation in the target population and if pleiotropic effects of the mutation do not constitute a major disadvantage to the organism.

QUALITATIVE RESISTANCE

Qualitative, one-step or discrete resistance is the result of a mutation on a single or, at most, a small number of major genes. This type of resistance is characterized by what appears to be a sudden shift in the population from totally sensitive to one where a very high proportion of individuals are totally resistant to the fungicide, at agricultural use levels at least. The resistant individuals in the population may only have an initial frequency of 1×10^{-8}, but they constitute a distinct subpopulation (Figure 1) which have a selective advantage over the sensitive wild-type strains after application of the fungicide. The selection of the resistant subpopulation proceeds largely unnoticed until the frequency of resistant individuals reaches detectable levels (10^{-2} or 10^{-1}, Brent, 1986). Once the resistant subpopulation predominates, what was once a highly effective treatment rapidly becomes ineffective invariably leading to total breakdown of control. The phenylamides and benzimidazoles typically show this type of resistance, although the mechanism of phenylamide-resistance has not yet been established. Furthermore, closer examination of benzimidazole-resistance in several plant pathogens reveals different levels of resistance (weakly resistant WR, moderately resistant MR, highly resistant HR and very

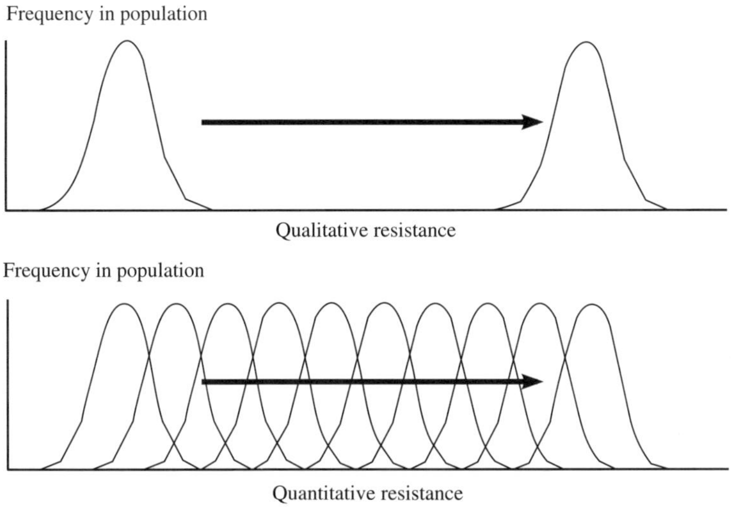

Figure 1 Selection effects on qualitative and quantitative resistance

highly resistant VHR; Davidse and Ishii, 1995). Except for WR strains, these different resistant phenotypes reflect different alleles of the target β-tubulin gene, involving substitution of different amino acids with subsequent weaker binding of the fungicide to its target (Hollomon *et al.*, 1997). Interestingly, many resistance mutations spread throughout β-tubulin can be generated in the laboratory but in field populations of several pathogens, resistance is confined to just a few tightly linked changes at amino acid codons 198 and 200 (Figure 2). The reasons for this apparent restriction in the range of resistance mutations under field conditions is not entirely clear, although it presumably reflects adverse effects of many mutations on fitness. It is perhaps surprising that these mutations are in highly conserved regions of the β-tubulin protein, which might well be assumed important for normal function. Indeed, Phe_{200} is critical for

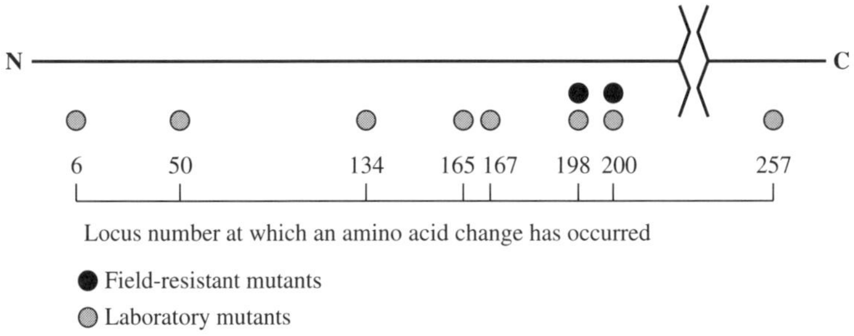

Figure 2 Benzimidazole resistance and the β-tubulin target

selectivity and action of benzimidazoles; when other amino acids replace Phe_{200}, fungicide activity is lost. A full understanding of how these different amino acid changes affect β-tubulin structure may soon emerge following the recent resolution of the structure of tubulin at 3.7 Å by electron crystallography (Nogales *et al.*, 1998). It is interesting, however, that a similar phenomenon of restricted mutations in field strains exists both in bacteria and insects where the mechanism of resistance has been identified as involving target site changes.

QUANTITATIVE RESISTANCE

Quantitative or multi-step resistance is the result of the mutation of several genes each contributing, to a greater or lesser degree, to the development of resistance (polygenic resistance). This type of resistance is observed in field populations as a continuous unimodal response; the sensitive wild-type population is gradually replaced with a more resistant population with a lower mean fungicide sensitivity (Figure 1; Hollomon, 1981; Brent, 1982). Quantitative resistance results in a slowly decreasing fungicide efficacy but, unlike qualitative resistance, rarely leads to total breakdown of disease control. The C-14 demethylase inhibitors (DMIs) and hydroxypyrimidines generally exhibit this type of resistance although, major-gene resistance has been recognized to the DMI, triadimenol, in laboratory resistant mutants of *Nectria haematococca* (Kalamarakis *et al.*, 1989) and in field strains of *Rhynchosporium secalis* (Kendall *et al.*, 1993), where a discontinuous sensitivity distribution has been observed.

In general, qualitative and quantitative resistance are characteristics of a given fungicide group. Practical resistance to benzimidazole and phenylamide fungicides is always qualitative, whereas DMI-resistance is almost always quantitative, reflecting many different resistance mechanisms. However, both types of resistance can operate within a single pathogen; for instance, *R. secalis* shows qualitative resistance to benzimidazoles (Kendall *et al.*, 1994) and triadimenol but a quantitative pattern is observed for all other DMIs.

NEGATIVE CROSS-RESISTANCE

Changes that reduce fungicide binding in resistant strains may increase the binding of other compounds, resulting in increased sensitivity to a second fungicide. This negative cross-resistance is not uncommon. The best practical example occurs between benzimidazoles and phenylcarbamates (Kato *et al.*, 1984) in many plant pathogens and the molecular mechanisms involved are gradually being uncovered. Replacing Glu_{198} (wild-type) of β-tubulin with a small amino acid, such as glycine or alanine, is responsible for this negative cross-resistance. Replacement with a non-polar amino acid such as lysine, or substitution of Phe_{200} with tyrosine, destroys the negative cross-resistance, strains carrying these β-tubulin alleles are resistant to both fungicide groups.

Attempts to use this negative cross-resistance as an anti-resistance strategy have met with a mixed response in a number of diseases. Doubly resistant strains of eyespot (*Pseudocercosporella herpotrichoides*) emerged very rapidly in fields following treatment with carbendazim (Leroux and Clerjeau, 1985), and in grey mould in vineyards in Champagne following carbendazim/diethofencarb treatments (Leroux and Moncombie, 1994), but elsewhere anti-resistance management has been successful and grey mould remains well controlled. Likewise, carbendazim/diethofencarb mixtures have not selected strains of *Rhynchosporium secalis* resistant to both fungicides, even though doubly resistant strains can be isolated, albeit at low frequencies, from barley crops before treatment (Hollomon and Kendall, unpublished observation 1996). Negative cross-resistance is also reported between carboxin and 4′-substituted carboxins (White *et al.*, 1978) and between fenpropimorph/fenarimol mixtures in *Penicillium italicum* (DeWaard and Van Nistelrooy, 1982), but these have not been evaluated as anti-resistance strategies.

FACTORS AFFECTING BUILD UP OF RESISTANCE

Data accumulated over the last 20 years have pinpointed factors influencing the build up of fungicide resistance. As a rule, a qualitative response is faster than a quantitative one, however, the response rates vary depending on the selection pressure exerted by the fungicide and the comparative fitness of sensitive and resistant strains. As mentioned earlier, the initial frequency of resistant mutants within the wild-type population prior to specific fungicide usage can be as little as 1×10^{-8}. Accumulation of resistant mutants in any population depends on their fitness in terms of growth rate, ability to infect the host tissue, reproductive potential and their ‘overwintering-oversummering’ capabilities. If differences in fitness between the sensitive population and the resistant mutants are large, without the selection pressure from fungicides, the resistant mutants may remain below detection levels.

The extent of the selection pressure exerted by the fungicide is governed by the difference in sensitivity between the wild-type population and the resistant mutants (resistance factor); the greater the difference, the greater the selection for resistance. A fungicide with high efficacy increases the selection pressure by killing or arresting the development of a high percentage of the wild-type population. This effectively negates competition for infection sites and therefore, any penalty for fitness the resistant mutants may have. Nevertheless, even low levels of resistance eventually, however, cause problems. The dodine resistance factor in apple scab (*Venturia inaequalis*) was no more than two to four-fold, but after 20 years’ continuous use dodine resistance became a practical problem in apple orchards (Gilpatrick and Blower, 1974).

The persistence, coverage and frequency of fungicide application influence the rate of development of resistance. With high persistency at the site of infection, good coverage and frequent usage, there is little escape for the sensitive population, and the resistant strains will increase in numbers. Indeed, inhibition of sensitive strains themselves can influence the spread of resistance depending on the mode of action of the fungicide. Benzimidazoles are frequently used in citrus packing and storage sheds for the control of fungal storage rots. Like many site-specific fungicides, they do not inhibit spore germination, but instead severely stunt germ-tube growth. Wild and Eckert (1982) reported that benzimidazole-sensitive strains of the storage rot, *Penicillium digitatum*, increased the infection frequency of resistant spores when a mixed population was inoculated into benomyl treated oranges. Dcspite the inhibition of germ-tube growth, the germinating conidia produced extracellular pectolytic enzymes which aided the infection of the few resistant spores in the population, leading to poor disease control under storage.

The spread of fungicide resistance is affected by the speed at which the pathogen reproduces. Resistance develops faster where pathogens produce several asexual generations and infection cycles within a season, and is slower with pathogens having a single reproductive cycle (Tables 1 and 2). Shorter

Table 1 Effects of apparent pathogen infection rate and latent period on standard selection time

	Standard selection time for latent period equal to: (days)		
Apparent infection rate per day (r)[a]	10	8	5
0.4	5.8	5.0	3.9
0.2	7.9	7.2	6.4
0.1	12.8	12.4	11.8
0.05	23.6	23.3	22.7

[a] van der Plank (1963).

Table 2 Predicted and observed selection-pressure duration required for resistant subpopulations of selected pathogens to cause practical resistance outbreak

			Selection-pressure duration in time units as indicated (d: days; y: years)	
Pathogen	Chemical	Standard selection time in days	Predicted	Observed
Erysiphe cichoracearum	dimethirimol	8.5–16.5	98–263 d	112–224 d
Phytophthora infestans	metalaxyl	3.7–3.8	51–70 d	1–2 seasons
Cercospora beticola	benomyl	9.5–14.3	130–263 d	140–200 d
Cereal smut	carboxin	158.5	5–7 y	13 y

Data adapted from Skylakakis (1982).

latent periods also favour the spread of resistance. Infection rates at the height of potato blight (*Phytophthora infestans*) and barley powdery mildew (*Erysiphe graminis*) epidemics are no more than 4–6 days, and metalaxyl and ethirimol resistance developed respectively, within two growing seasons. By contrast, at least 13 seasons elapsed before limited carboxin resistance was detected in barley loose smut (*Ustilago nuda*), and 8–10 years before resistance ended the usefulness of benzimidazole fungicides for eyespot control, a pathogen with 2–3 generations each year. These differences in standard selection times (Skylakakis, 1985) in field populations occurred despite the fact that resistant mutations can be generated equally rapidly in the laboratory, in all four pathogens.

Unless resistance is fully dominant, haploid pathogens are expected to have a higher frequency of resistance in the population than diploid pathogens, since the sexual phase provides an opportunity for rearrangement of dominant and recessive genes for fitness and fungicide resistance. Certainly DMI resistance has not developed so far in dikaryotic rust fungi (Bayles *et al.*, 1994) whilst it has done so in several ascomycete haploid pathogens (Hollomon, 1997). Fungal pathogens that produce multinucleate asexual spores can produce heterokaryotic mycelia containing both fungicide-sensitive and resistant nuclei. These heterokaryons are capable of growth in either the presence or absence of the fungicide and, thus, enable the pathogen to maintain resistant and sensitive nuclei in the population and adapt to any selection pressure within their environment (Summers *et al.*, 1984).

The dispersal and intermixing of fungal spores affects the development and spread of resistance. Movement and ingress of sensitive individuals into a selected population is critical when a fitness penalty accompanies resistance and may be especially so in greenhouse and polythene tunnel environments, where populations are otherwise contained. Resistance to the hydroxypyrimidine fungicide, dimethirimol, became a serious, practical problem in the control of cucurbit powdery mildew (*Sphaerotheca fuliginea*) in greenhouses in Northern Europe within two years of its introduction, yet hydroxypyrimidines still perform well in field-grown cucurbits after more than 25 years. This is despite the detection of a few dimethirimol resistant strains in Spanish field crops (Brent, personal communication). Where pathogen dispersal is very localized and there is little intermixing of populations, resistance is concentrated in specific areas, developing unchecked by outside influences.

PREDICTION OF RESISTANCE

MONITORING

The use of benzimidazole fungicides heralded the beginning of fungicide resistance problems. Because of their curative and eradicant activity, specificity,

Table 3 Detecting rare mutations in cereal powdery mildew populations

Mutant frequency	Sample size required[a] (No. of single spore isolates)	Sample area required[a] (ha d^{-1})
10^{-3}	3×10^3	0.2
10^{-4}	3×10^4	2.1
10^{-5}	3×10^5	23.1
10^{-6}	3×10^6	230.8
10^{-7}	3×10^7	2307.0
10^{-8}	3×10^8	23077.0

[a] Sample size and sample area required each day to detect a unique mutation occurring at a particular frequency.

broad spectrum of activity and high efficacy at low concentrations, they had a great advantage over traditional protectant fungicides. As a result, they were widely used and often as the only disease control agent. Although their site-specific mode of action was described soon after they were introduced (Davidse, 1973), the practical significance of this work and its association with resistance problems was not fully appreciated during the developmental phase of these fungicides and monitoring to establish baseline sensitivity levels was limited. However, in order to detect major-gene resistance at the early stages of the build up of resistance it is necessary to test an inordinate number of fungal strains. Table 3 illustrates the scale of the problem, even for a moderate powdery mildew epidemic on barley, when as many as 2000 loci ha^{-1} have been calculated to mutate each day (Wolfe, 1984). Conventional monitoring techniques based on bioassay are labour-intensive, time-consuming, slow and, at best, the detection range is 0.1–1% of the population. Therefore, at these resistance levels, it is unlikely that evasive action is possible. In contrast, the monitoring of polygenic resistance, may well detect the gradual sensitivity shift of the population and allow for successful evasive action.

During the period 1985–95 no new fungicide groups were introduced, yet chemistry provided several novel DMI fungicides, which helped overcome resistance problems associated with early DMIs, such as triadimenol, triadimefon, and fenarimol. Although some level of cross-resistance exists between DMIs, resistance factors can differ significantly between them (Kendall, 1986; Kendall *et al.*, 1993). This suggests that different resistance mechanisms may be selected by different DMIs, creating the possibility that certain DMI mixtures could provide durable anti-resistance strategies, providing that monitoring methods were available to identify which mechanisms were present in pathogen populations.

The chemist has done much in recent years to overcome existing resistance problems, and this looks set to continue with the introduction of several new fungicide groups. These include anilinopyrimidines, strobilurins and quinolines, as well as compounds activating systemic-acquired resistance (SAR), e.g.

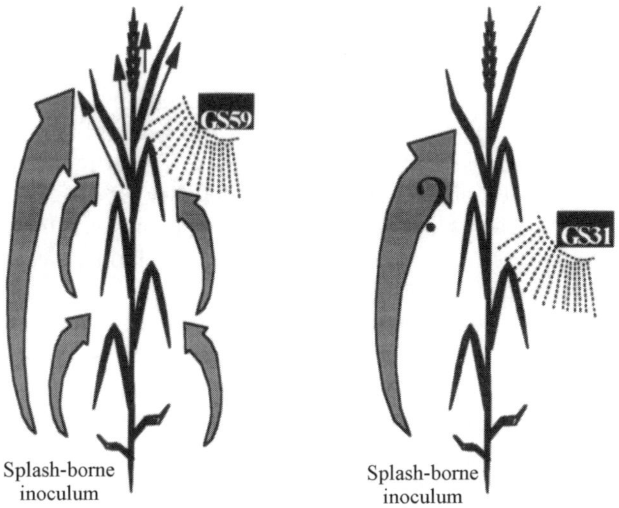

Figure 3 Improved persistence adds new options for cereal disease control. Curative activity of DMI fungicides requires treatment at GS59 for *Septoria tritici* control; improved mobility and persistence of strobilurins allows earlier application between GS31–39

CGA245704 (Ruess *et al.*, 1995). Not only do these groups have novel modes of action, but they have a greater persistence of action and this introduces a new concept of 'systemic protectant' activity which, if fully exploited, could change timing and approaches to disease control (Figure 3). In some cases, normal systemic activity is combined with redistribution within the crop as vapour, protecting the whole of the crop as it grows. This improved persistence of action may allow a single early application to healthy plants to give season-long protection, but this will extend the period of selection and increase any resistance risk. Indeed, with the exception of the SAR compound, benzthiadiazole (CGA 2457047), resistant strains have been identified in several fungi to these new fungicide groups, albeit in some cases only in mutants generated in the laboratory (di Rago *et al.*, 1989; Förster *et al.*, 1995; Hollomon, *et al.*, 1996; Ziogas *et al.*, 1997). These new fungicides will be used in mixtures with existing, mainly systemic fungicides, largely to improve disease control spectra, and to add curative activity to a protectant one. This is, however, unlikely to be an effective anti-resistance strategy to preserve the life of these new products, since their persistence will ensure that they exert a selection pressure on the population well beyond that of the mixture partner. Management of this resistance risk will require suitable monitoring techniques in order to alert early problems of resistance.

NEW DEVELOPMENTS IN MONITORING TECHNOLOGY

Assessment of resistance risk is not only a key element in developing effective, and practical, anti-resistance strategies, but it is now important in the registration of new fungicides. At present, resistance-risk assessment is imprecise, largely because current methods fail to detect resistance alleles at low enough frequencies, and because of lack of precision to measure any fitness penalty. Available bioassay methods detect only phenotypes rather than known resistance mutations; they are resource-intensive and seldom detect resistance at frequencies less than 1 in 1000. This is not useful for risk assessment studies and may be too late to implement anti-resistance strategies. New technologies must be explored if these real limitations are to be overcome.

Resistance to benzimidazole fungicides is well understood and provides a good model system to develop new concepts and approaches that might detect resistance alleles at frequencies below 1×10^{-3}. Point mutations conferring benzimidazole-resistance in several fungi have been identified (Hollomon *et al.*, 1997), and the highly conserved nature of the β-tubulin target allows easy amplification by PCR (Polymerase Chain Reaction) of DNA fragments containing the expected mutations. Detection of these mutations has been achieved in a number of fungi using short (15 mer) allele specific oligonucleotide (ASOs) probes and appropriate stringency through various temperatures, which discriminate between matched and mismatched probes (Koenraadt and Jones, 1992; Wheeler *et al.*, 1995; Yarden and Katan, 1993). These dot-blot assays, carried out on membranes, used a DNA template from well-characterized pure cultures, but the technology has been extended to look at the frequency of resistant alleles in lesions collected from field crops, without prior isolation of the pathogen. Although these assays gave similar results to those obtained by conventional bioassay, direct PCR always gave higher resistance frequencies (Table 4), highlighting the inadequacies of bioassay where isolation and culture on fungicide-free media may allow loss of resistant mutations, certainly where heterokaryons exist. Even so, these studies only examined a few hundred lesions and not the thousands needed to detect rare mutations at low enough frequencies to be of use in resistance-risk assessments.

Table 4 Detection of benzimidazole resistance in *Rhynchosporium secalis*. Comparison of PCR diagnostic test with a standard bioassay

Test method	Total tested[a]	% benzimidazole resistant[b]
Bioassay	75	2.7
PCR-diagnostic	57	8.7

[a] Bioassay isolates were obtained from surface-sterilized lesions according to the method described in Kendall *et al.* (1993). PCR-diagnostic was performed without isolation of the pathogen but simply using template DNA obtained by boiling lesions in water. In both cases, lesions were from the same untreated plot.

[b] Isolates grow normally at $1 \mu g\, ml^{-1}$ carbendazim.

Other PCR-based molecular methods detect benzimidazole resistance but all rely on electrophoretic separations on agarose gels to identify base-pair changes. Alteration of Glu_{198} (GAG) to alanine (GCG) creates a unique *ThaI* (CGCG) restriction site which has been used to identify highly benzimidazole-resistant phenotypes in *Venturia nashicola* (Ishii *et al.*, 1997). ASOs can be used not only as hybridization probes but also as PCR primers. Under the right stringency conditions, and especially where detection involves an A-T base pair rather than a G-C pair, having the potentially mismatched base-pair at the 3′-end of a primer will produce a PCR product only when the sequences match exactly (Martin *et al.*, 1993). Targeted PCR products can also be separated by capillary electrophoresis and where single DNA strands migrated in a denaturing gel (Single Strand Conformation Polymorphism, SSCP), single base substitutions in the β-tubulin gene of *V. nashicola* were identified through changes in migration time as a result of the different conformation of each DNA strand (Figure 4, Ishii *et al.*, 1997). To achieve the necessary resolution electrophoretic conditions must be well controlled and identification is achieved only through comparison with known standards. It is not possible to predict what effect a mutation will have on strand conformation but SSCP offers the advantage that analysis can be automated, and high-throughput of PCR products is possible.

Most molecular diagnostic methodologies evaluated so far generally involve isolation of the pathogen prior to testing, and even with pathogens that are readily grown in culture, it is only feasible to detect mutations occurring at

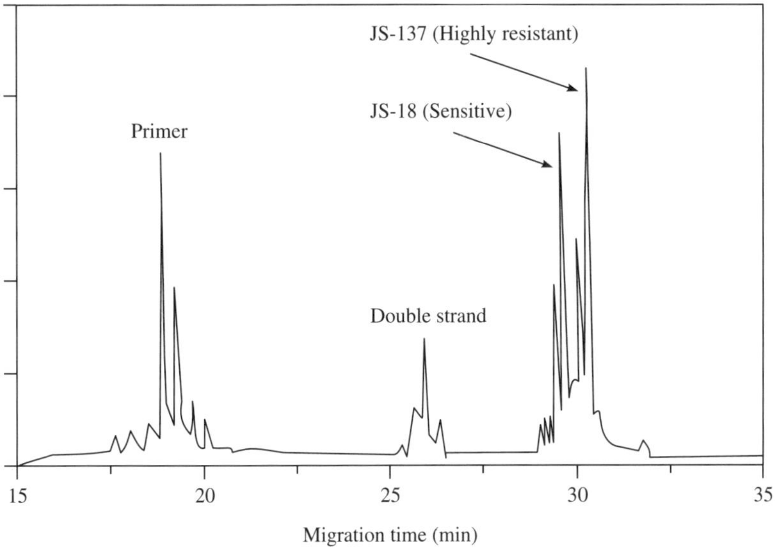

Figure 4 Detection of one base change in the β-tubulin gene *Venturia nashicola* by SSCP analysis

frequencies between 1×10^{-2} and 1×10^{-3}. This is clearly not low enough to give useful early warning of any build up of resistance; detection of mutations at frequencies of 1×10^{-6} or below, would be more meaningful. This might seem a daunting task, but medical diagnostic screening programmes for both rare genetic disorders, and antibiotic resistances such as Multiple Resistant Staphylococcus (MRS), are already moving towards these detection levels. Ligation assays exist in kit form which allow automated detection of 19 point mutations which influence the risk of heart disease through hypercholesterolemia (Baron *et al.*, 1996). Using a pool of ASO tags as PCR primers, and probing a matrix of 90 of these ASO tags on a solid support with the PCR product produced from a template derived from a population of cells, allows identification of a single point mutation in this matrix (Holden, 1997). 'TAQMAN' chemistry (Perkin Elmer Corporation, Seer Green, UK) offers the advantage of detecting point mutations in one PCR reaction avoiding any additional capture or probing step. Whilst this methodology is ideal for testing many samples rapidly, at present it is only suited to situations where detection of one mutation is needed, rather than identifying the several mutations that might exist in any pathogen population.

In addition to developments in molecular technologies for detection of fungicide resistance mutations, improvements have been made in the throughput of conventional bioassays by adapting tests to microtitre plate technologies (Pijls *et al.*, 1994). Much of this has evolved in response to demands for high-throughput screening systems, and is largely confined to pathogens that grow in liquid culture in a yeast-like manner, since this allows automatic measurement of growth in microtitre plate readers. Fortunately, this includes the important cereal pathogens *S. tritici* and *R. secalis*, but it is still difficult to examine sample sizes of more than 1000. In general, sensitivity values for existing systemic fungicides obtained with *in vitro* assays correlate well with *in vivo* tests on treated plants. However, for the novel strobilurin fungicides, *in vitro* tests exposed a range of variation in sensitivity (ED_{95} values) in *V. inaequalis* and *S. tritici* which was wider for *in vitro* assays than for *in vivo* tests (Heaney and Lorenz, 1997). Unfortunately, none of these bioassay developments are applicable to sensitivity monitoring in important obligate pathogens, such as downy and powdery mildews, emphasizing the potential benefits if PCR diagnostics can be applied to these diseases.

SAMPLE SIZE

New technology may eventually provide monitoring with a tool to identify rare resistance mutations before they generate a practical problem, but at present monitoring is used to establish base-line sensitivity distributions and to evaluate the performance of anti-resistance strategies. Experience from several cases of practical resistance suggests that a decline in the mean sensitivity of a pathogen population of around tenfold will clearly be noticed by growers through poor

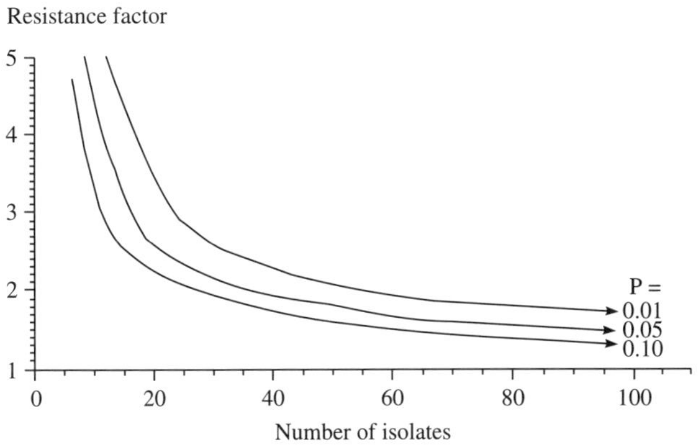

Figure 5 Sample size required to detect differences between mean ED_{50} sensitivity values of *Venturia inaequalis* populations to flusilazole. Resistance factors represent the ratio of mean ED_{50} values for two populations. Sample sizes are shown for three significance levels. Redrawn from data in Smith *et al.* (1991)

fungicide performance. Shifts of one-hundredfold or more will be reflected in little or no control, even at treatment rates above those normally recommended. Establishing base-line sensitivity distributions before the introduction of a new fungicide provides a benchmark, which can be used to confirm any subsequent decline in sensitivity. Sample sizes needed to detect shifts in sensitivity of practical importance reliably are surprisingly small (Smith *et al.*, 1991). Ten to twenty individuals were sufficient to identify a tenfold shift in sensitivity in *V. inaequalis* (apple scab) to flusilazole (Figure 5) although this sample size will be dependent on pathogen/fungicide combinations, and the precision with which each sensitivity value is obtained. Where accurate *in vitro* assays are available alongside sample sizes of 100 or more, quite small shifts in sensitivity can be detected; these may be of little practical significance. This is illustrated from results of monitoring *V. inaequalis* populations from three orchards in New York State for sensitivity to flusilazole, which is used in the region for apple scab control (Figure 6). Significant shifts in mean sensitivity were detected but these were of the order of only 2–3-fold and were not related at all to the previous history of DMI use in that orchard.

ANTI-RESISTANCE STRATEGIES

Anti-resistance strategies evolved in the early 1980s around a series of modelling studies (Kable and Jeffery, 1980; Levy and Cohen, 1983; Skylakakis, 1985) and included inputs of a number of pathogen- and fungicide-related parameters. These models were all based on selection of a major gene for resistance rather than a polygenically controlled change; their outcomes were

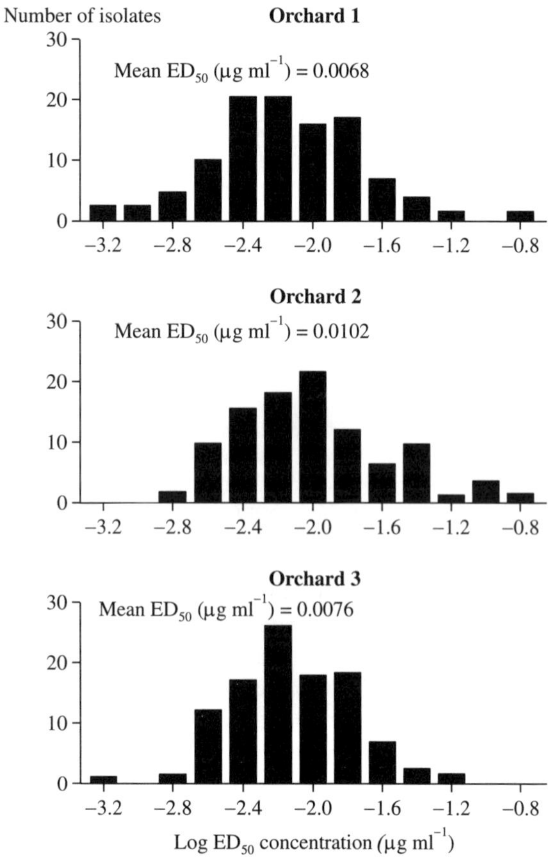

Figure 6 Frequency distribution of ED_{50} sensitivity values of *Venturia inaequalis* to flusilazole in populations from three apple orchards. Orchard 1 was well isolated from other orchards and had never been treated with fungicides. Orchard 2 was abandoned, and had never been treated with a DMI fungicide, but was surrounded by managed orchards. Orchard 3 was an intensively managed research orchard that had been treated with various DMI fungicides for 12 years prior to sampling. 100 isolates were tested from each orchard. The mean sensitivity of the population from Orchard 2 is significantly different ($P = 0.05$) from that of the other orchard populations. Data from Smith *et al.* (1991)

surprisingly similar. They suggested that resistance was best kept in check using mixtures (or alternations) of fungicides with different modes of action. If possible, at least one mixture partner should be a multisite inhibitor. This simple concept was easily grasped by growers, and was attractive to manufacturers who could manage anti-resistance strategies as pre-packed mixtures. Unlike insecticide-resistance, cross-resistance patterns generally follow mode of action with fungicides, presumably reflecting target-site alterations rather than uptake or detoxification changes. Mixtures with partners selecting different mechanisms of resistance would have provided a sounder scientific strategy but, in the early 1980s, little information was available on resistance mechanisms on which to base the choice of mixture partners. Mixtures have not always met the criteria laid down by modelling and in reality commercial fungicide mixtures have included partners with the same mode of action. Benzimidazole/phenylcarbamate (diethofencarb) mixtures have been successfully used in some regions as an anti-resistance strategy to control grey mould (*Botrytis cinerea*) in a range of crops. Both fungicide groups have the same β-tubulin target site, but their mechanisms of resistance are sufficiently well understood to exploit the negative cross-resistance that occurs. In theory, at least, this should provide the perfect anti-resistance strategy with each mixture partner always having a greater effect on isolates resistant to the other 'at risk' fungicide than to wild-type isolates. Unfortunately, as explained earlier, isolates of *B. cinerea* resistant to both fungicides may also occur, and in some situations their frequency can increase so that the mixture strategy is no longer effective.

Between the period 1980–95 when very few new fungicide groups were introduced, anti-resistance strategies, especially in cereal diseases, relied very much on morpholines backed up by the DMI and hydroxypyrimidine groups. None of these fungicides was a multisite inhibitor, yet mixtures have shown evidence of effectively combating resistance (Heaney *et al.*, 1988; Brent *et al.*, 1989). A number of different resistance mechanisms are now being identified for DMI fungicides (Joseph-Horne and Hollomon, 1997), and where these are reflected by differences in resistance levels for any particular DMI, anti-resistance strategies based on DMI mixtures may well be durable. Certainly it is the chemist and biochemist rather than the biologist who has contributed most in recent years towards the evolution of long-lasting anti-resistance strategies. Unfortunately, with emphasis shifting towards mixtures based on knowledge of resistance mechanisms, and back-up by rapid diagnostic methods to identify each mechanism, the concepts become complex, and more difficult for growers to grasp.

DOSE RATE

Despite recommendations by manufacturers to use 'label' rates, growers often reduce these in order to lower the cost of disease control inputs. For control of several diseases, including powdery mildews, this is often effective since 'label' dose rates are often higher than needed, reflecting a requirement to provide sufficient persistence to control other, less sensitive, diseases. Where resistance is under the control of several genes, each with an additive effect, reducing dose rates may allow the least sensitive individuals to survive. As a result, resistance alleles would accumulate in increasingly resistant strains, eventually leading to an erosion of performance. A full rate, however, would kill all individuals and reduce this risk of resistance. By contrast, where resistance is controlled by a single major gene, often giving effective immunity to the fungicide, dose rate will have little effect on this selection process in the longer term. However, lowering the dose rate will allow some escape of sensitive strains and initially slow down the rate of change within the population.

The impact of dose rate on resistance is certainly the most common question asked by growers, but despite many years of study, there is still no unifying concept emerging in answer to this question. Some field experiments support the view that low dose rates encourage resistance (Kuck, 1994; Schulz, 1994; Steva, 1994; Förster *et al.*, 1994; Engels *et al.*, 1996), whilst others arrived at the opposite conclusion suggesting that increasing dose rates favour selection for resistance (Hunter *et al.*, 1984; Porras *et al.*, 1990; Peever *et al.*, 1994). Still other experiments suggest that dose rate, at least within agriculturally accepted terms, is selectively neutral (Pijls and Shaw, 1997; Burnett and Zziwa, 1997). Unfortunately, many of these studies were carried out with cereal powdery mildews, which are very mobile wind-dispersed pathogens, and where it is not easy to identify any contribution to sensitivity changes coming from outside the trial site. These outside populations will generally be selected to some extent by the same fungicides being tested in trials. Furthermore, many experiments relate not simply to reduced dose rates, but compensate for this through increasing the number of doses. Thus it is not easy to distinguish between the effect of reducing dose rate from increasing the number of applications.

Perhaps the most thorough study of this question was recently reported by Pijls and Shaw (1997), which examined the effect of dose rate on selection for flutriafol-resistance in the wheat pathogen, *S. tritici*. This fungus offers opportunities to overcome some of the difficulties associated with powdery mildew experiments. *S. tritici* is an extremely variable pathogen with the *in vitro* EC_{50} of isolates to flutriafol varying by 40-fold, whilst *in vivo* the least sensitive isolates were unaffected by the recommended flutriafol dose rate (Hollomon and Mapstone, unpublished observation, 1989). This unimodal sensitivity distribution suggests polygenic control of resistance, rather than the involvement of a major gene one. Epidemics begin from wind-borne ascospores (Shaw and Royle, 1989) so that populations established each year are genetically well mixed and independent of any fungicide used the previous year at the trial site. Subsequent development of the epidemic within the crop is clonal and restricted to dispersal by rain-splash, so that inter-plot interference is minimal and selection is confined throughout to one population. Unlike powdery mildews, *S. tritici* can be grown *in vitro* and fortunately is amenable to bioassay using microtitre plate methodology. This allows thousands of isolates to be tested rather than the 100 or so examined in other dose rate experiments. Despite this large sample size, and three years of field experiments, significant decreases in sensitivity were observed after spraying with a water control, as well as with fungicide. Allowing for these changes in sensitivity in plots treated with water only, no dose of fungicide produced any effect on resistance; indeed mixing flutriafol with the multisite inhibitor chlorothalonil, had no effect on selection either.

The dilemma of these results is that, even at the lowest flutriafol dose (1/4 field rate) significant disease control was achieved. This implies that the least sensitive individuals were removed from the population which, bearing in mind the range of flutriafol sensitivities known to exist in a *S. tritici* population, should cause a shift in sensitivity detectable by an experimental protocol designed to identify any change in mean sensitivity (ED_{50}) greater than 9% of the starting value. Certainly, in cereal powdery mildew, the level of control was related to the extent of any shift in fungicide sensitivity (Schulz, 1994). The answer to this problem may reside in the extremely rapid mobility of flutriafol, which accumulates in leaf tips leaving the rest of the leaf unprotected. Since DMIs are fungistatic rather than fungicidal, and the curative action of flutriafol is not great, resumption of growth and sporulation in these unprotected areas would reduce any effect of selection. Although a major experimental input, it would be worth while repeating these experiments with a much less mobile DMI, such as tebuconazole, in order to evaluate any influence of fungicide mobility on dose rate effects. In any event, these results highlight that particular properties of each fungicide/pathogen combination are likely to overlay any underlying principles governing the effect of dose rate on selection.

IMPACT ON DISEASE CONTROL PRACTICES

So far, we have discussed the scientific and technical issues raised by resistance. These have undoubtedly been a factor in the search for novel modes of action, and the knowledge has increased awareness of the need to use novel chemistry carefully through more innovative ways to control diseases. But resistance is a complex issue requiring more than just scientific understanding of the problem. Social, regulatory and legal issues also play a part (Urech *et al.*, 1997). The responsibility placed on the agrochemical industry to protect novel chemistry has required new approaches to marketing, and increased communication and cooperation between companies, academia, extension services and growers. These were the driving forces behind the setting up by industry of the Fungicide Resistance Action Committee (FRAC), and more recently local groups with both private and public participation. In some regions, most notably the European Union, government has responded with a registration requirement for information on resistance risk and anti-resistance management strategies. (Rotteveel *et al.*, 1997). Data needed to fulfil these registration requirements are not yet fully worked out but these should be simple, applicable to all product types, easy to implement in member states, and rapid so as not to delay market introduction. Overall, resistance has had a positive impact on disease control measures, and if managed sensibly it plays a part in good agricultural practice, ensuring sustainability.

CONCLUSIONS

Entomologists had been grappling with insecticide-resistance for many years before the arrival of serious practical fungicide-resistance caught plant pathologists somewhat by surprise. Site-specific fungicides had exposed the power of natural selection on pathogen populations so large that even rare resistance mutations occurred quite frequently. Realization of the problem led to a great deal of research, both practical and theoretical, which helped define factors which influenced the spread of resistance and identify strategies to combat it. There is no doubt that this research, much of it summarized by Brent (1995), has helped manage resistance so that disasters have largely been avoided, and chemical control of plant diseases has remained effective. Input of new chemistry in achieving this objective has been important. This innovation is continuing with the arrival of a new generation of systemic protectant fungicides and the possibility of effective control through chemical manipulation of disease resistance. This will undoubtedly produce new challenges for resistance management and will focus research sharply on to the assessment of resistance-risk associated with this novel and persistent chemistry.

Resistance-risk assessment remains an inexact science but molecular biology has widened horizons and, coupled with improvements in biochemistry, should provide a better understanding of resistance mechanisms which was the factor largely missing from earlier attempts to assess risk and devise durable anti-resistance strategies. Negative cross-resistance is not an uncommon phenomenon which potentially offers a durable way to manage resistance, provided that allelic changes involving resistance can be monitored quickly, and with sufficient precision. Harnessing developments in diagnostic technologies towards more effective monitoring will be important in managing resistance, because the utility of traditional bioassay is largely limited by sample size and cost. Certainly, monitoring is the key to many questions which will inevitably surround the use of these new generation fungicides. Is resistance the cause of poor disease control, and are anti-resistance strategies working? If resistance alleles can be identified at low enough frequencies, any fitness penalty associated with resistance can be measured in the real world of field populations, something which has 'dogged' resistance-risk evaluation to date.

Overall, the future for managing resistance is a positive one. New chemical synthesis techniques, linked with high-throughput screening, should ensure that novel chemistry continues to flow into disease control strategies. The behaviour of resistance ultimately depends on the particular organism/pesticide combination, and this has kept the paths of fungicide, insecticide, herbicide and antibiotic research largely apart over the past 25 years. But molecular biology has begun to identify common research objectives and unifying concepts which are likely to emerge from understanding resistance at the level of a small inhibitor molecule interacting with its target. For instance, similarities in the molecular basis or pyrethroid-, benzimidazole-, and several antibiotic resistances should stimulate research into how these resistances are generated; this could well identify a common factor and a sounder foundation for managing resistance.

ACKNOWLEDGEMENTS

We thank numerous colleagues in the agrochemical industry and elsewhere who have helped shape the thinking on which this article is based. We are particularly grateful to Dr. Keith Brent for critical reading of the manuscript and his helpful suggestions. Financial support from the Home Grown Cereals Authority, MAFF and the Agrochemical Chemical Industry is gratefully acknowledged. IACR receives grant-aided support from the Biotechnology and Biological Sciences Research Council of the United Kingdom.

REFERENCES

Baron, H., Fung, S., Atakan, A., Bähring, S., Luft, F. C., and Schuster, H. (1996). 'Oligonucleotide ligation assay (OLA) for the diagnosis of familial hypercholesterolemia', *Nature-Biotech.*, **14**, 1279–1282.

Bayles, R. A., Stigwood, P. L., and Barnard, E. G. (1994). 'Sensitivity to morpholine fungicides in yellow rust of wheat (*Puccinia striiformis*)', BCPC Monograph No. 60, *Fungicide Resistance*, pp. 309–321.

Brent, K. J. (1982). 'Case study 4. Powdery mildew of barley and cucumber', in *Fungicide Resistance in Crop Protection* (eds J. Dekker and S. G. Georgopoulos), The Pudoc, Wageningen, The Netherlands, pp. 219–230.

Brent, K. J. (1986). 'Detection and monitoring of resistant forms: an overview', in *Pesticide Resistance Strategies and Tactics for Management*, NRC Board on Agriculture, National Academy Press, Washington, DC, pp. 298–312.

Brent, K. J. (1995). 'Fungicide Resistance in Crop Pathogens: How can it be managed?', FRAC Monograph NO. 1. GIFAP, Brussels.

Brent, K. J., Carter, G. A., Hollomon, D. W., Hunter, T., Locke, T., and Poven, M. (1989). 'Factors affecting build-up of fungicide resistance in powdery mildew in spring barley', *Neth. J. Pl. Path.*, **95**, 31–41.

Burnett, F. J. and Zziwa, M. C. N. (1997). 'Effect of fungicide dose rate on the sensitivity of *Erysiphe graminis* f. sp. *tritici*', *Pestic. Sci.*, **51**, 335–340.

Davidse, L. C. (1973). 'Antimycotic activity of methyl benzimidazole-2-yl carbamate (MCB) in *Aspergillus nidulans*', *Pestic. Biochem. Physiol.*, **3**, 317–325.

Davidse, L. C., and Ishii, H. (1995). 'Biochemical and molecular aspects of the mechanism of action of benzimidazoles, *N*-phenylcarbamates and *N*-phenylformamidoximes and the mechanism of resistance to these compounds in fungi', in *Modern Selective Fungicides*, 2nd edition (ed. H. Lyr) Gustav Fischer Verlag, Berlin, pp. 305–322.

De Waard, M. A., and Van Nistelrooy, J. G. M. (1982). 'Toxicity of fenpropimorph to fenarimol-resistant isolates of *Penicillium italicum*', *Neth. J. Pl. Path.*, **88**, 231–236.

di Rago, J- P., Coppee, J- Y., and Colson, A- M. (1989). 'Molecular basis for resistance to myxothiazol, mucidin (Strobilurin A) and stigmatellin', *J. Biol. Chem.*, **264**, 14543–14548.

Engels, A. J. G., Mantel, B. C., and de Waard, M. A. (1996). 'Effect of split applications of fenpropimorph-containing fungicides on sensitivity of *Erysiphe graminis* f. sp. *tritici*', *Plant Path.*, **45**, 636–643.

Förster, B., Chavaillaz, O., Steden, C., Radtke, W. Käsbohrer, M., and Kühl, A. (1994). 'Influence of split application of fenpropimorph mixtures on disease control and on the sensitivity of *Erysiphe graminis* f. sp. *tritici*', in *Fungicide Resistance*, (eds Heaney, S., Slawson, D., Holloman, D. W., Smith, M., Russell, P. E. and Parry D. W.), Farnham, UK, BCPC Monograph No. 60, pp. 331–337.

Förster, B., Heye, U., Pillonel, C., and Staub, T. (1995). 'Methods to determine the sensitivity to cyprodinil; a new broad-spectrum fungicide, in *Botrytis cinerea*', in *Modern fungicides and Antifungal Compounds* (eds Lyr, H., Russell, P. E. and Sisler, H. D.), Intercept, Andover, UK, pp. 365–378.

Gilpatrick, J. D. and Blowers, D. R. (1974). 'Ascospore tolerance to Dodine in relation to orchard control of Apple scab', *Phytopath.*, **64**, 649–652.

Gisi, U. and Hermann, D. (1994). 'Sensitivity behaviour to *Septoria tritici* population on wheat to cyproconazole', in *Fungicide Resistance* (eds Heaney, S., Slawson, D., Hollomon, D. W., Smith, M., Russell, P. E. and Parry, D. W.), Farnham, UK, BCPC Monograph No. 60, pp. 11–18.

Heaney, S. P. and Lorenz, G. (1997). 'Accurate measurement of sensitivity to strobilurin fungicides *in vitro* or *in vivo*', *Resistance 97*, Institute of Arable Crops Research, Harpenden, Abstract.

Heaney, S. P., Martin, T. J., and Smith, J. M. (1988). 'Practical approaches to managing anti-resistance strategies with DMI fungicides', *Brighton Crop Prot. Conf.—Pests and Diseases—1988*, Farnham, UK, BCPC, pp. 1097–1106.

Holden, D. (1997). Asilomar Fungal Genetics Conference, USA, Abstract.

Hollomon, D. W. (1981). 'Genetic control of ethirimol resistance in a natural population of *Erysiphe graminis* f. sp. *hordei*', *Phytopath.*, **71**, 536–540.

Hollomon, D. W. (1997). 'Fungicide resistance in cereal pathogens 1991–96. *Rhynchosporium secalis* in barley; *Erysiphe graminis* on wheat and barley; *Septoria tritici* on wheat', Home Grown Cereals Authority, Project Report 143E, London.

Hollomon, D. W., Butters, J. A., and Barker, H. (1997). 'Tubulins: a target for antifungal agents', in *Anti-infectives. Recent Advances in Chemistry and Structure-Activity Relationships* (eds P. H. Bentley and P. J. O'Hanlon), Royal Society of Chemistry, Cambridge, pp. 152–162.

Hollomon, D. W., Wheeler, I., Dixon, K., Longhurst, C., and Skylakakis, G. (1996). 'Resistance profiling of the new powdery mildew fungicide Qunoxyfen (DE-795), in cereals', *Brighton Crop Prot. Conf.—Pests and Diseases—1996*, Farnham, UK, pp. 701–706.

Hunter, T., Brent, K. J., and Carter, G. A. (1984). 'Effects of fungicide regimes on sensitivity and control of barley powdery mildew', *Brit. Crop Prot. Conf.—Pests and Disease—1984*, Farnham, UK, BCPC, pp. 471–476.

Ishii, H., Kamahori, M., Hollomon, D. W., and Narusaka, Y. (1997). 'DNA-based approaches for diagnosis of benzimidazole resistance in *Venturia nashicola* in scab fungus of Japanese pears', *Resistance 97*, Institute of Arable Crops Research, Harpenden, Abstract.

Joseph-Horne, T. and Hollomon, D. W. (1997). 'Molecular mechanisms of azole resistance in fungi', FEMS Microbiol, *Letters*, **149**, 141–149.

Kable, P. F. and Jeffery, H. (1980). 'Selection for tolerance in organisms exposed to sprays of biocide mixtures: a theoretical model', *Phytopath.*, **70**, 8–12.

Kalamarakis, A. E., Demopoulos, V. P., Ziogas, B. N., and Georgopoulos, S. G. (1989). 'Resistance to fenarimol in *Nectria haematococca* var. *cucurbitae*', *Neth. J. Pl. Path.*, **95** (supplement 1), 109–120.

Kato, T., Suzuki, K., Takahashi, J., and Kamoshita, K. (1984). 'Negatively correlated cross-resistance between benzimidazole fungicides and methyl N-(3,5-dichlorophenyl) carbamate', *Journal of Pestic. Sci.*, **9**, 489–495.

Kendall, S. J. (1986). 'Cross-resistance of triadimenol-resistant fungal isolates to other C-14 demethylation inhibitor fungicides', *Proc. Brit. Crop Prot. Conf.—Pests and Diseases*, pp. 539–546.

Kendall, S. J., Hollomon, D. W., Cooke, L. R., and Jones, D. R. (1993). 'Changes in sensitivity to DMI fungicides in *Rhynchosporium secalis*', *Crop Protect.*, **12**, 357–362.

Kendall, S. J., Hollomon, D. W., Ishii, H., and Heaney, S. P. (1994). 'Characterisation of benzimidazole-resistant strains of *Rhynchosporium secalis', Pestic. Sci.*, **40**, 175–181.

Koenraadt, H. and Jones, A. L. (1992). 'The use of allele-specific oligonucleotide probes to characterise resistance to benomyl in field strains of *Venturia inaequalis*', *Phytopath.*, **82**, 1354–1358.

Kuck, K. H. (1994). 'Evaluation of anti-resistance strategies', in *Fungicide Resistance* (eds Heaney, S., Slawson, D., Hollomon, D. W., Smith, M., Russell, P. E. and Parry, D. W.), Farnham, UK, BCPC Monograph No. 60, pp. 43–46.

Leroux, P. and Clerjeau, M. (1985). 'Resistance of *Botrytis cinerea*. Pers. and *Plasmopara viticola* to fungicides in French vineyards', *Crop Protect.*, **4**, 137–160.

Leroux, P. and Moncombie, D. (1994). 'Resistance of *Botrytis cinerea* to dicarboximides, benzimidazoles and phenylcarbamates in the Champagne vineyards', in *Fungicide Resistance* (eds Heaney, S., Slawson, D., Hollomon, D. W., Smith, M., Russell, P. E., and Parry, D. W.), British Crop Protection Council, Farnham, Surrey, pp. 267–270.

Levy, Y. and Cohen, Y. (1983). 'Build up of a pathogen subpopulation resistant to a systemic fungicide under various control strategies', *Phytopath.*, **73**, 1475–1480.

Martin, L. A., Fox, R. T. V., and Baldwin, B. C. (1993). 'Rapid methods for detection of MBC-resistance in fungi', in *Proc. 10th Symp. on Systemic Fungicides and Antifungal Compounds* (eds Lyr, H. and Polter, C.), Eugen Ulmer, Stuttgart, pp. 209–218.

Nogales, E., Wolf, S. G., and Downing, K. H. (1998). 'Structure of the $\alpha\beta$-tubulin dimer by electron crystallography', *Nature*, **391**, 199–203.

Peever, T. L., Brants, A., Bergstrom, G. C., and Milgroom, M. G. (1994). 'Selection for decreased sensitivity to propiconazole in experimental field populations of *Stagnospora nodorum* (syn. *Septoria nodorum*)', *Can. J. Pl. Path.*, **16**, 109–117.

Pijls, C. F. N., Shaw, M. W., and Parker, A. (1994). 'A rapid test to evaluate *in vitro* sensitivity of *Septoria tritici* to flutriafol, using a microtitre plate reader', *Plant Path.*, **43**, 726–730.

Pijls, C. F. N. and Shaw, M. W. (1997). 'Weak selection by field sprays for flutriafol resistance in *Septoria tritici*', *Plant Path.*, **46**, 247–264.

Porras, L., Gisi, U., and Staehle-Csech, U. (1990). 'Selection dynamics in triazole treated populations of *Erysiphe graminis* on barley', *Brighton Crop Prot. Conf.—Pests and Diseases—1990*, Farnham, UK, BCPC, pp. 1163–1168.

Rotteveel, A. J. W., de Goeij, J. W. F. M., and van Gemerden, A. F. (1997). 'Towards construction of a resistance risk evaluation scheme', *Pestic. Sci.*, **51**, 407–411.

Ruess, W., Kunz, W., Staub, T., Müller, K., Poppinger, N., Speich, J., and Ahl Goy, P. (1995). *XIII International Plant Protection Congress, The Hague, Netherlands*, Abstract 424.

Schulz, U. (1994). 'Evaluating anti-resistance strategies of control of *Erysiphe graminis* f. sp. *tritici*', in *Fungicide Resistance* (eds Heaney, S., Slawson, D., Hollomon, D. W., Smith, M., Russell, P. E., and Parry, D. W.), Farnham, UK, BCPC Monograph 60, pp. 55–58.

Shaw, M. W. and Royle, D. J. (1989). 'Airborne inoculum as a major source of *Septoria tritici (Mycosphaerella graminicola)* infections in winter wheat crops in the UK', *Plant Path.*, **83**, 35–43.

Skylakakis, G. (1982). 'The development and use of models describing outbreaks of resistance to fungicides', *Crop Protect.*, **1**, 249–262.

Skylakakis, G. (1985). 'Two different processes for the selection of fungicide-resistant sub-populations', *EPPO Bulletin*, **15**, 519–525.

Smith, F. D., Parker, D. M., and Köller, W. (1991). 'Sensitivity distribution of *Venturia inaequalis* to the sterol demethylation inhibitor flusilazole: Baseline sensitivity and implications for resistance monitoring'. *Phytopath.*, **81**, 392–396.

Steva, H. (1994). 'Evaluating anti-resistance strategies for control of *Uncinula necator*', in *Fungicide Resistance* (eds Heaney, S., Slawson, D., Hollomon, D. W., Smith, M., Russell, P. E. and Parry, D. W.) Farnham, UK, BCPC Monograph No. 60, pp. 59–66.

Summers, R. W., Heaney, S. P., and Grindle, M. (1984). 'Studies of a dicarboximide resistant heterokaryon of *Botrytis cinerea*', *Proc. Brit. Crop Prot. Conf.—Pests and Diseases*, pp. 453–458.

Urech, P. A., Staub, T., and Voss, G. (1997). 'Resistance as a concomitant of modern crop protection', *Pestic. Sci.*, **51**, 227–234.

van der Plank, J. E. (1963). *Plant Diseases: Epidemics and Control*, Academic Press, New York.

Wheeler, I., Kendall, S. J., Butters, J. A., Hollomon, D. W., and Hall, L. (1995). 'Using allele-specific oligonucleotide probes to characterise benzimidazole resistance in *Rhynchosporium secalis*', *Pestic. Sci.*, **43**, 201–209.

White, G. A., Thorn, G. D., and Georgopoulos, S. G. (1978). 'Oxathiin carboxamides highly active against carboxin-resistant succinic hydrogenase complexes from carboxin-selected mutants of *Ustilago maydis* and *Aspergillus nidulans*', *Pestic. Biochem. Physiol.*, **9**, 165–182.

Wild, B. L. and Eckert, J. W. (1982). 'Synergy between a benzimidazole-sensitive and benzimidazole-resistant isolates of *Penicillium digitatum*', *Phytopath.*, **72**, 1329–1332.

Wolfe, M. S. (1984). 'Trying to understand and control powdery mildew', *Plant Path.*, **33**, 451–466.

Yarden, O. and Katan, T. (1993). 'Mutations leading to substitution at amino acid 198 and 200 of beta-tubulin that correlate with benomyl resistance phenotypes of field strains of *Botrytis cinerea*', *Phytopath.*, **83**, 1478–1483.

Ziogas, B. N., Baldwin, B. C., and Young, J. E. (1997). 'Alternative respiration: a biochemical mechanism of resistance to azoxystrobin (ICIA 5504) in *Septoria tritici*', *Pestic. Sci.*, **50**, 28–34.

5 The Strobilurin Fungicides

JOHN M. CLOUGH AND CHRISTOPHER R. A. GODFREY
Zeneca Agrochemicals, Bracknell, UK

INTRODUCTION

The widespread use of fungicides in agriculture is a relatively recent phenomenon, and most of the major developments have taken place during the last 40 years. In days gone by, farmers often ignored or did not recognize the effect that fungal pathogens had on the yield and quality of their crops. Nowadays, however, these losses are unacceptable, and farmers rely on the use of chemicals to control fungal diseases. As a consequence, commercial fungicides have become an important component of the total agrochemical business, with world-wide sales in 1996 of about \$5.9 billion, equivalent to 18.9% of the total agrochemical market (Wood Mackenzie, 1997a).

A large number of fungicides are already available to the farmer; the most recent edition of *The Pesticide Manual* (Tomlin, 1994) contains 158 different

Fungicidal Activity. Edited by D. H. Hutson and J. Miyamoto fungicidal active ingredients in current use. Nevertheless, further industrial research aimed at the discovery and development of new compounds is extremely intensive, and this is due to a number of important factors. Firstly, the development of fungicides with novel modes of action remains an important strategy in the search for ways to overcome problems associated with resistance to established products. Secondly, it is becoming increasingly desirable (some would say essential) to replace certain existing products with compounds of lower toxicity to non-target species and acceptable levels of persistence in the environment. Finally, in an increasingly competitive world, agrochemical companies are forced to look for new compounds which show marketable technical advantages over their own and their competitors' products.

Traditionally, the discovery of new fungicide leads has been achieved in several ways, including the exploitation of loop-holes in competitors' patents ('me-too chemistry'), biorational design, and the screening of collections of compounds, including natural products. Fungicidal natural products, which can be obtained from a wide variety of sources, including plants, bacteria and (as will be seen later) even fungi, are a particularly attractive source of new leads due to their structural diversity (Godfrey, 1995; Copping, 1996).

In principle, fungicidal natural products can either be used as fungicides in their own right, or may be exploited as leads for the design of other novel synthetic materials. In the former approach, purified natural products constitute the active ingredient of a formulated mixture, or are used in mixture with a synthetic material. However, the use of natural products *per se* as fungicides has not been particularly successful for a number of reasons. Firstly, natural products possessing marketable levels of activity against a broad spectrum of commercially important diseases have proved to be very hard to find. Furthermore, they are often inherently unstable (for example, to sunlight) and consequently are not sufficiently persistent in the field to deliver a useful effect. In addition, some lack selectivity of action and this can manifest itself in the form of toxicity to plants or mammals. Finally, many natural products derived from fermentation broths are present in low concentrations and are difficult to purify on a large scale. Some of these limitations can be overcome by making semisynthetic derivatives, but this inevitably adds to the overall cost. The products which have been exploited commercially can be classified conveniently into two distinct groups, the aminoglycosides (e.g., streptomycin, validamycin A and kasugamycin), and the nucleosides (e.g., polyoxins B and D and blasticidin S). All of these compounds are manufactured in Japan by large-scale fermentation of various species of *Streptomyces*, but together they command less than 1% of the total fungicide market (Godfrey, 1995).

For the reasons above, the agrochemical industry has largely focused on the second approach: the design of novel, fully synthetic compounds from a consideration of the structure of appropriate natural product leads. These synthetic compounds ideally possess optimized biological, physical and environmental properties and are often simpler in structure than their naturally occurring

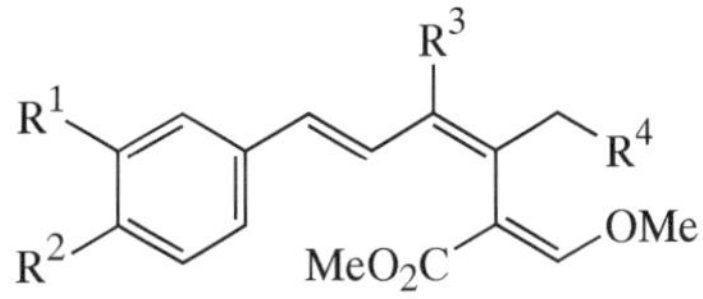

The Strobilurins

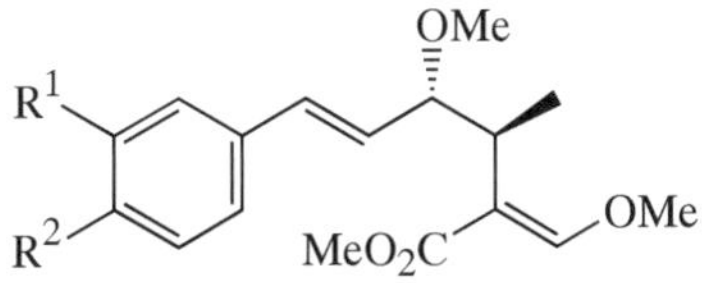

The Oudemansins

R^1, R^2 = H or a wide range of substituents
R^3 = H or OMe
R^4 = H or OH

Figure 1 The strobilurins and oudemansins

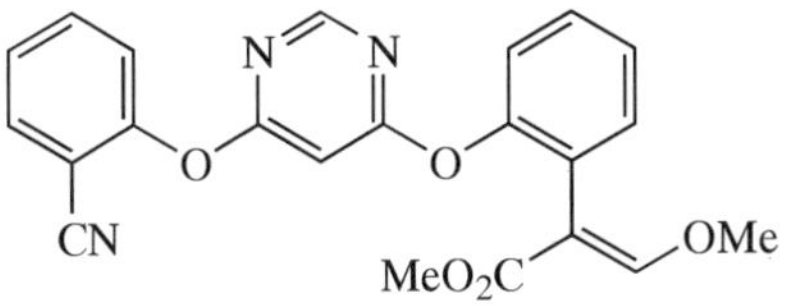

Azoxystrobin (ICIA5504)

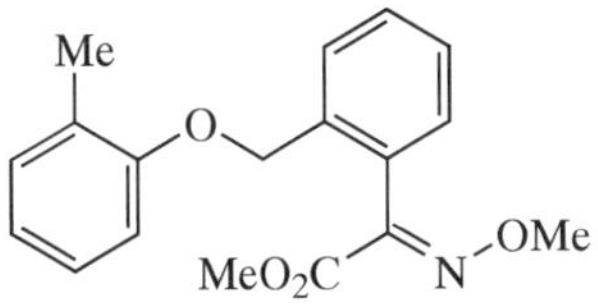

Kresoxim-Methyl (BAS 490 F)

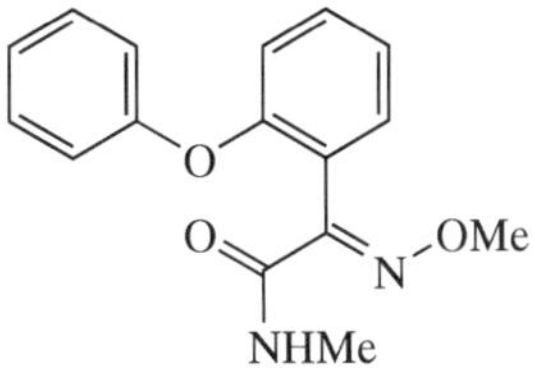

SSF-126

Figure 2 Strobilurin analogues of commercial interest

progenitors. The classic example of a family of agrochemicals discovered in this way are the insecticidal pyrethroids. In the fungicide area, various natural products have been studied as possible starting points for synthetic agrochemicals, including hadacidin, thiolutin, griseofulvin, pisiferic acid and pyrrolnitrin, and the last of these has led to the commercial product fenpiclonil (Godfrey, 1995). The subject of this chapter is a new class of fungicides, the 'strobilurins', which have been discovered as a result of work on a family of natural derivatives of β-methoxyacrylic acid: the strobilurins, oudemansins (Figure 1) and myxothiazols.

The three strobilurins of current commercial interest are azoxystrobin (ICIA5504) from Zeneca Agrochemicals†, kresoxim-methyl (BAS 490 F) from BASF and SSF-126 from Shionogi (Figure 2). Azoxystrobin was first registered

†Zeneca Agrochemicals was ICI Agrochemicals until mid-1993.

in April 1996 for use on cereals in Germany, and has received further registrations on 13 crops in 20 countries to 4 June 1997. These include cereals and vines in Europe, bananas in Central America, and vines, tomatoes, peaches, peanuts, pecans and turfgrass in the USA. It is being (or will be) sold by Zeneca Limited under its trademarks 'Amistar', 'Quadris', 'Abound', 'Heritage', 'Bankit' and 'Ortiva'. Recently a key registration in the USA has been granted under the Environmental Protection Agency's Reduced Risk Pesticide Scheme which provides a fast-track registration review for products with desirable toxicological and environmental profiles (Wood Mackenzie, 1997b). Kresoxim-methyl, first registered in March 1996 for use on cereals in Germany, is being (or will be) sold by BASF Aktiengesellschaft under its trademarks 'Brio' and 'Mentor' (combination products with fenpropimorph), 'Allegro' and 'Juwel' (combination products with epoxiconazole), and 'Candit' and 'Stroby WG' (Agrow, 1997). SSF-126 is not yet on sale.

THE NATURAL PRODUCTS

INTRODUCTION

The novel agricultural fungicides which are the subject of this chapter were inspired by a family of natural products which are all derivatives of β-methoxyacrylic acid: the strobilurins, oudemansins and myxothiazols. The strobilurins and oudemansins are produced by a wide variety of fungi, mainly Basidiomycetes, but several strobilurins are also produced by the Ascomycete *Bolinea lutea.* Most of these fungi grow on decaying wood. By contrast, the myxothiazols are produced by various myxobacteria (so-called 'gliding' bacteria). A total of 16 strobilurins, four oudemansins and 34 myxothiazols are now known. In addition, two compounds related to the myxothiazols have recently been isolated from another myxobacterium, and these have been given the names melithiazol A and melithiazol B.

The strobilurins and oudemansins have a variety of potentially useful biological activities and, indeed, it was these properties which first enabled them to be detected in and then isolated from fermentation broths. They have activity against a wide range of filamentous fungi and yeasts, but do not affect bacteria. Myxothiazol A is also highly active *in vitro* against many fungi and yeasts and, in addition, it controls the growth of certain Gram-positive bacteria. None of the myxothiazols has higher activity than myxothiazol A. The melithiazols are both fungicidal *in vitro.*

Musílek and his co-workers in Prague discovered the first of the strobilurins, strobilurin A, during the 1960s and, in fact, this was the earliest of all the fungicidal β-methoxyacrylates to be isolated. Anke and Steglich and their research groups in German universities have been responsible for most of the subsequent work on the strobilurins, as well as the leading work on the oudemansins. The isolation and characterization of the myxothiazols and melithiazols were carried out at the Gesellschaft für Biotechnologische Forschung (GBF) in Germany.

The key data for the natural products are summarized in this chapter in Figures 3 to 6 and Table 1. However, further details about most of the natural products of this class (all those reported to the end of 1992) can be found in a comprehensive review of this area (Clough, 1993).

THE STROBILURINS

Sixteen strobilurins have been isolated and characterized (Figure 3, Table 1). Each comprises an (*E*)-methyl β-methoxyacrylate group linked at the α-position to a (substituted) phenylpentadienyl unit. The first member of this family to be discovered, originally named mucidin but now generally known as strobilurin A, has the simplest structure of the group, and all the other strobilurins may be regarded as derivatives of this compound. In most cases, derivatization is only on the benzene ring, with substitution ranging from a simple hydroxyl group (strobilurin F-1) to complex spiroketal functionality (strobilurin E). However, more recently, two strobilurins with a hydroxyl substituent at the vinyl methyl position have been discovered, namely the hydroxystrobilurins A and D. Furthermore, four strobilurins with a methoxyl group at the 9-position, the 9-methoxystrobilurins A, E, K and L, were described for the first time as recently as 1995 and 1996.

THE OUDEMANSINS

Just four oudemansins are known (Figure 4, Table 1). Formally these are derivatives of the corresponding strobilurins in which the elements of methanol have been added across the central olefinic bond of the triene system. The oudemansins therefore have two contiguous chiral centres. Oudemansins A, B and X have the (9*S*,10*S*)-configuration, and it is likely that the same applies to oudemansin L, but this has not been confirmed.

THE MYXOTHIAZOLS

Myxothiazol A (Figure 5), originally known simply as myxothiazol, is the best known of the myxothiazols. However, a total of 34 myxothiazols have been isolated, and structures have been established for 25 of them (Table 1). Instead of the 'α-linked' (*E*)-methyl β-methoxyacrylate group found in each of the strobilurins and oudemansins, myxothiazol A contains an (*E*)-β-methoxyacrylamide group which is linked to the rest of the molecule at the β-position. The backbone to which this acrylamide group is attached bears some resemblance to that of the oudemansins. However, instead of the (substituted) benzene ring of the latter, myxothiazol A has an interesting bisthiazole moiety, substituted with

Table 1 Natural derivatives of β-methoxyacrylic acid

Compound	Source(s) [a]	Year first described	Leading reference(s) [b]
Strobilurin A = mucidin	*Oudemansiella mucida*, *Strobilurus tenacellus*, *Bolinea lutea*, *Xerula melanotricha*, *Mycena galopoda*	1965	Musílek, 1965; Vondráček *et al.*, 1967; Musílek *et al.*, 1969; Vondráček *et al.*, 1974; Anke *et al.*, 1977; Schramm *et al.*, 1978; Bäuerle and Anke, 1980; Anke *et al.*, 1983b; Fredenhagen *et al.*, 1990a. Strobilurin A and mucidin are identical: Sedmera *et al.*, 1981; von Jagow *et al.*, 1986
Strobilurin B	*Strobilurus tenacellus*, *Bolinea lutea*, *Xerula longipes*, *Xerula melanotricha*	1977	Anke *et al.*, 1977; Schramm *et al.*, 1978; Bäuerle and Anke, 1980; Anke *et al.*, 1983b; Fredenhagen *et al.*, 1990a
Strobilurin C	*Xerula longipes*	1983	Anke *et al.*, 1983b
Strobilurin D	*Cyphellopsis anomala*	1986	Schwalge, 1986; Weber *et al.*, 1990a
Strobilurin E	*Crepidotus fulvotomentosus*	1988	Daum *et al.*, 1988; Weber *et al.*, 1990b
[c]Strobilurin F-1	*Cyphellopsis anomala*	1990	Weber *et al.*, 1990a
[c]Strobilurin F-2	*Bolinea lutea*	1990	Fredenhagen *et al.*, 1990a and 1990b
Strobilurin G	*Bolinea lutea*	1990	Fredenhagen *et al.*, 1990a and 1990b
Strobilurin H	*Bolinea lutea*	1990	Fredenhagen *et al.*, 1990a and 1990b
[d]Strobilurin X	*Oudemansiella mucida*	1983	Vondráček *et al.*, 1983
Hydroxystrobilurin A	*Pterula* sp. 82168	1995	Engler *et al.*, 1995
Hydroxystrobilurin D	*Mycena sanguinolenta*	1983	Anke *et al.*, 1983b; Backens *et al.*, 1988
9-Methoxystrobilurin A	*Favolaschia* sp. 87129	1995	Zapf *et al.*, 1995
9-Methoxystrobilurin E	*Favolaschia pustulosa*	1996	Wood *et al.*, 1996
9-Methoxystrobilurin K	*Favolaschia* sp. 87129	1995	Zapf *et al.*, 1995
9-Methoxystrobilurin L	*Favolaschia pustulosa*	1996	Wood *et al.*, 1996
Oudemansin A (=oudemansin)	*Oudemansiella mucida*, *Mycena polygramma*, *Xerula melanotricha*	1979	Anke *et al.*, 1979; Bäuerle and Anke, 1980; Anke *et al.*, 1983b

Oudemansin B	*Xerula melanotricha*	1983	Anke *et al.*, 1983a and 1983b
Oudemansin X	*Oudemansiella radicata*	1990	Anke *et al.*, 1990
Oudemansin L	*Favolaschia pustulosa*	1996	Wood *et al.*, 1996
Myxothiazol A (= myxothiazol)	*Myxococcus fulvus*, *Stigmatella aurantiaca*, *Angiococcus disciformis*	1978	Augustiniak *et al.*, 1978; Gerth *et al.*, 1980; Trowitzsch *et al.*, 1980; Trowitzsch *et al.*, 1981; Kohl *et al.*, 1985; Reichenbach *et al.*, 1988
Myxothiazols B to I inclusive and K to Y inclusive (there is no myxothiazol J), plus 9 further uncharacterized myxothiazols	*Myxococcus fulvus*	1986	Bedorf *et al.*, 1986; Clough, 1993
Myxothiazol A, methyl ester	*Myxococcus fulvus*	1994	Höfle *et al.*, 1994a
Melithiazol A	*Melittangium lichenicola*	1994	Höfle *et al.*, 1994a and 1994b
Melithiazol B	*Melittangium lichenicola*	1994	Höfle *et al.*, 1994a and 1994b

[a] The column headed 'Source(s)' does not necessarily list all known sources. Strobilurin A, for example, has been isolated from at least 21 different fungi (Anke, 1995).
[b] Unless otherwise stated, the references cited refer to the isolation and characterization of the natural product.
[c] Two different strobilurins have been termed strobilurin F, the result of overlapping publications. In this chapter, the terms F-1, F-2 are used as suggested in a review (Clough, 1993).
[d] The strobilurin named strobilurin X in the table was not named when it was first described, but the term strobilurin X was coined later.

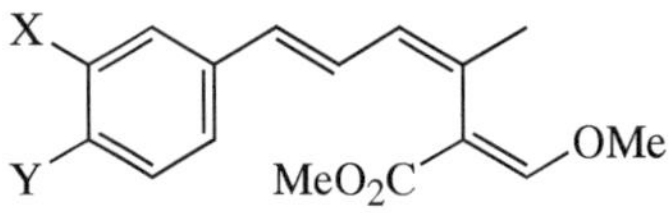

X = Y = H, Strobilurin A (Mucidin)
X = OMe, Y = Cl, Strobilurin B
X = OH, Y = H, Strobilurin F-1
X = OMe, Y = H, Strobilurin H
X = H, Y = OMe, Strobilurin X

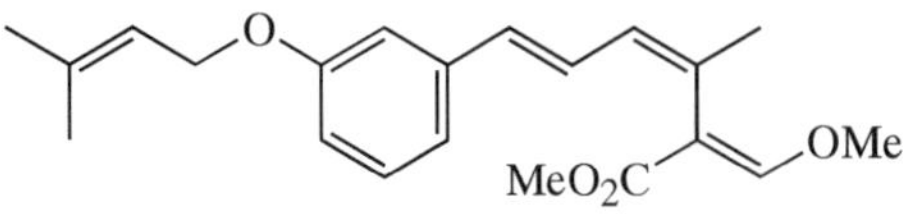

Strobilurin C

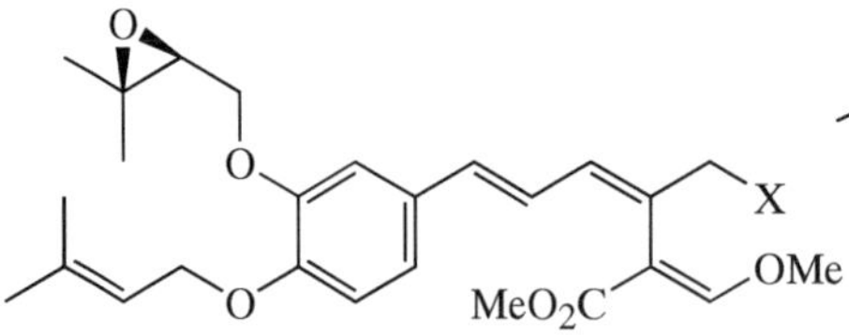

X = H, Strobilurin D
X = OH, Hydroxystrobilurin D

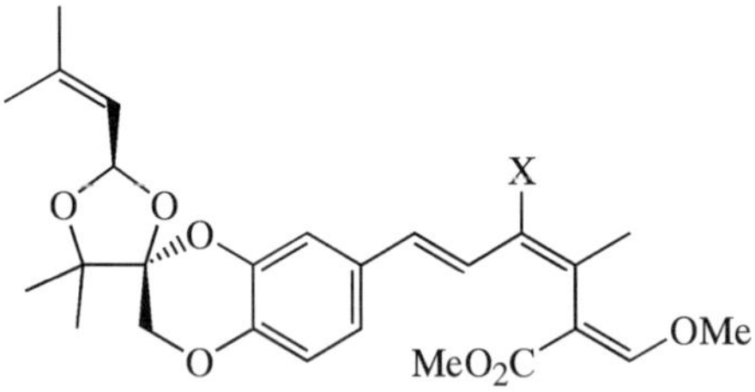

X = H, Strobilurin E
X = OMe, 9-Methoxystrobilurin E

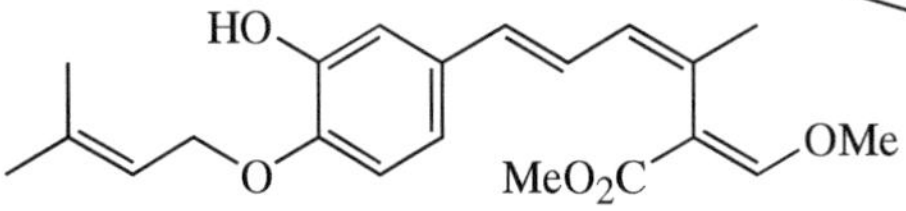

Strobilurin F-2

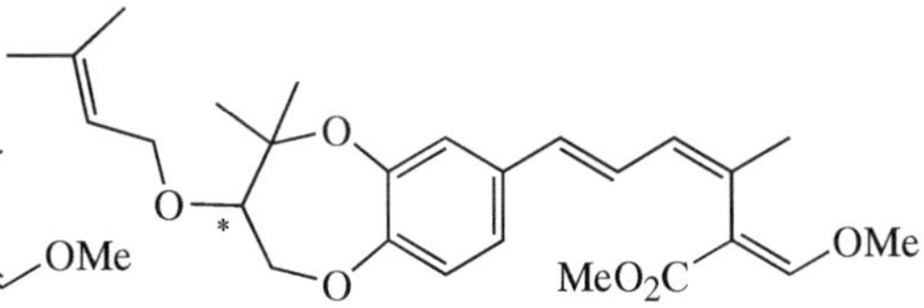

Strobilurin G

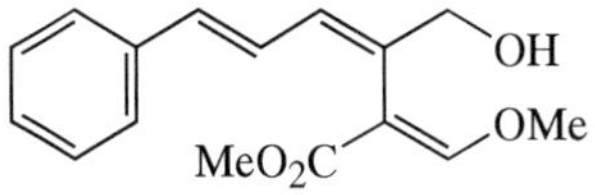

Hydroxystrobilurin A

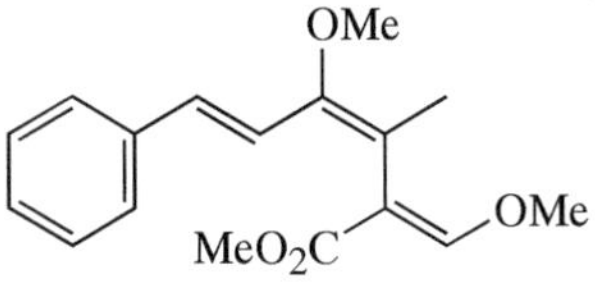

9-Methoxystrobilurin A

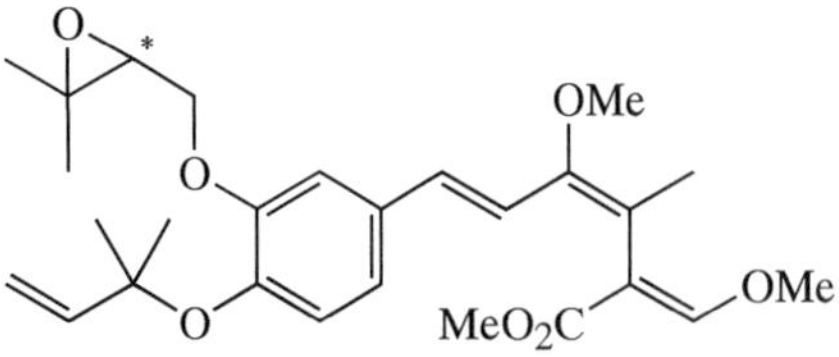

9-Methoxystrobilurin K

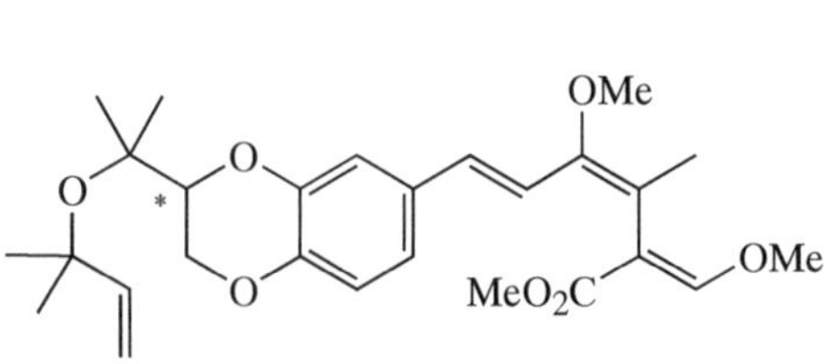

9-Methoxystrobilurin L

Figure 3 The natural strobilurins (*configuration unknown)

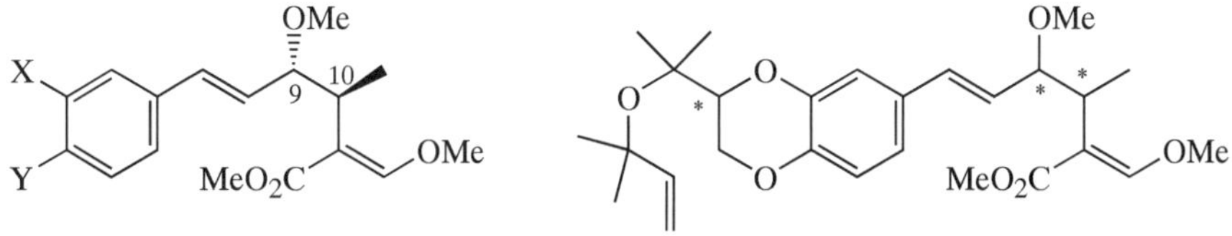

Figure 4 The oudemansins

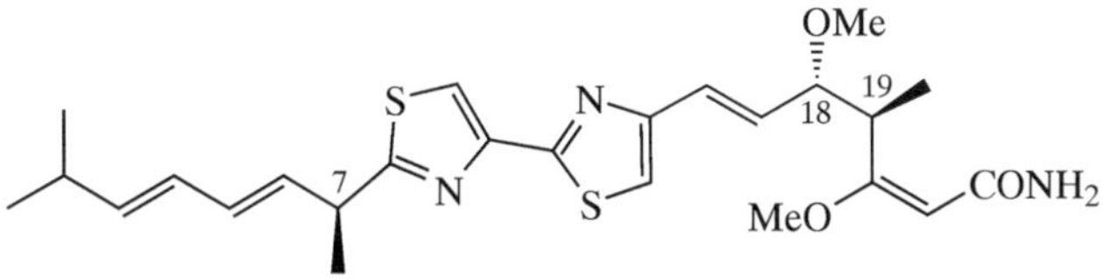

Figure 5 Myxothiazol A

a branched (E, E)-nonadienyl chain. Myxothiazol A, with three chiral centres, has the (7*S*,18*S*,19*R*)-configuration shown. The other myxothiazols show modifications at every part of the structure of myxothiazol A; in some examples the β-methoxyacrylamide unit has been methylated at the α-position or is absent altogether. The most recently reported member of the family, first described in 1994, is the methyl ester corresponding to myxothiazol A.

THE MELITHIAZOLS

Two compounds related to the myxothiazols, named melithiazol A and melithiazol B, have recently been isolated (Figure 6, Table 1). Both compounds are rather unstable.

ECOLOGICAL ROLE

There has been much speculation over many years about why organisms take the trouble to produce secondary metabolites (natural products) which often have considerable structural complexity. Williams and his co-workers propose that 'all such structures serve the producing organisms by improving their survival fitness' (Williams *et al.*, 1989). If this is true, it seems likely that the fungi and bacteria which produce strobilurins, oudemansins, myxothiazols and/or melithiazols give themselves an advantage by doing so, presumably by suppressing other organisms which compete for nutrients in the same environment. In support of this hypothesis, Musílek and his co-workers have reported

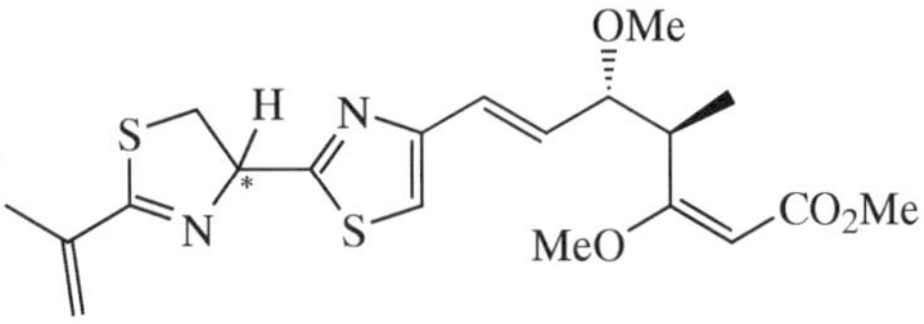

Melithiazol A
(*configuration unknown)

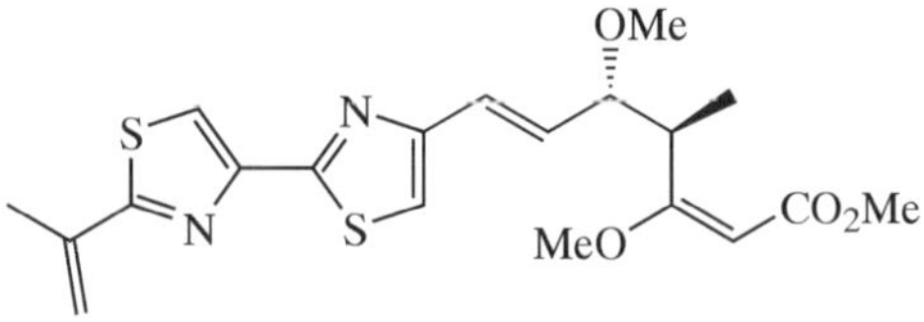

Melithiazol B

Figure 6 The melithiazols

that generally no other types of parasitic fungi appear on beech trees infected with *Oudemansiella mucida* (Musílek *et al.*, 1969).

In the case of the strobilurins and oudemansins there is another interesting twist: these fungicides are produced *by fungi* and the question presents itself as to why the producing organisms do not perish from the effects of their own metabolites. Studies by Brandt and his co-workers (Kraiczy *et al.*, 1996) suggest that at least two different protection mechanisms have evolved. Firstly, the cytochromes b in the strobilurin-producing fungi *Strobilurus tenacellus* and *Mycena galopoda* have altered active sites, rendering them less sensitive to the β-methoxyacrylates. And in addition, *M. galopoda* has an increased rate of respiration in comparison to other Basidiomycetes, which may allow it to grow even if its mitochondria are inhibited to an extent which would kill other fungi.

MODE OF ACTION

The following paragraphs give a summary of what is known about the mode of action of the strobilurins, oudemansins and myxothiazols, as well as the commercial fungicides which they inspired. Further information can be found in the following key papers and reviews: Becker *et al.*, 1981; von Jagow and Link, 1986; von Jagow *et al.*, 1986; Brandt *et al.*, 1988; Mansfield and Wiggins, 1990; Brandt *et al.*, 1991; Brandt and Trumpower, 1994; Trumpower and Gennis, 1994; and Brandt, 1996.

The fungicidal activity of the strobilurins, oudemansins and myxothiazols is a result of their ability to inhibit mitochondrial respiration in fungi. More specifically, it has been established that these compounds interfere with the function of the cytochrome bc_1 complex, located in the inner mitochondrial membrane of fungi and other eukaryotes, by binding at a specific site on cytochrome b. Since the natural products and their synthetic analogues can displace each other from the binding site, it is clear that they are reversibly bound.

Of course, many inhibitors of respiration are known. However, there are many distinct binding sites on the respiration pathway and the importance of the strobilurins, oudemansins and myxothiazols became clear when it was shown that they bind at a site for which no inhibitors had previously been identified. This has meant that the natural products, as well as some of their simple synthetic analogues such as the stilbene **1** (Figure 8), have become important tools for probing the details of the mechanism of mitochondrial respiration. Recent studies describe the crystallization of a cytochrome bc_1 complex, and the solution of its three-dimensional structure to 3.4 Å resolution by X-ray diffraction. The complex was also co-crystallized with myxothiazol, and work is in hand to locate the precise position at which myxothiazol is bound (Yu *et al.*, 1996).

The novel mode of action of the strobilurins, oudemansins and myxothiazols was one of the features which first attracted us to this family of compounds. We realized that if we could discover analogues of the natural products with all of the properties required for an agricultural fungicide, these should not be cross-resistant to established commercial compounds. The way in which we in Zeneca, in parallel with scientists in BASF and Shionogi, were able to discover products of this type is described in the next section of this chapter. Importantly, it has been confirmed that azoxystrobin (Wiggins and Jager, 1994; Baldwin *et al.*, 1996), kresoxim-methyl (Roehl, 1994; Gold *et al.*, 1996) and SSF-126 (Mizutani *et al.*, 1995; Shirane *et al.*, 1995; Mizutani *et al.*, 1996) all bind at the same site on cytochrome b as the strobilurins, oudemansins and myxothiazols.

SYNTHETIC ANALOGUES FOR USE IN AGRICULTURE

INTRODUCTION AND OVERVIEW

New fungicides with novel resistance-breaking modes of action are very important to the agrochemical industry. For this reason, scientists in industry carefully follow the scientific literature for publications describing new fungicidal natural products which might serve as leads for synthesis. In fact, many such natural products are described each year and the main problem for the chemist in the agrochemical industry is to decide which of these constitute leads worth following. In our experience, the only reliable way to find out is to obtain samples for testing on our own screens with commercial fungicides as standards. Where possible, we obtain a sample from the originating scientist, but it is often necessary to resort to re-isolation or even total synthesis of the natural product.

During the early 1980s, most of the fungicide scientists at Zeneca Agrochemicals (then ICI) were working on 1,2,4-triazoles. By the end of 1981, Zeneca's three triazole fungicides, diclobutrazol, flutriafol and hexaconazole, had been prepared and evaluated on a variety of crops. Diclobutrazol was already on sale, and it was clear that the others were promising compounds. We were also aware at that time that some resistance to triazole fungicides was already emerging in the field. Consequently, we made the decision to suspend further triazole chemistry. It was in this context that our colleague Brian Baldwin read a publication (Becker *et al.*, 1981) which drew together for the first time the fungicidal natural products strobilurin A, strobilurin B, oudemansin A and myxothiazol A. Samples were obtained, and when they showed interesting activity in our biological and biochemical assays they became the starting point for the programme of synthesis which was ultimately to lead to azoxystrobin, which we announced in November 1992 (Godwin *et al.*, 1992). An account of this work is given below.

Quite independently, scientists at BASF had also become interested in the natural β-methoxyacrylates, and were working closely with Anke and Steglich along similar lines to our own. Their work, described in more detail below, culminated in the synthesis of kresoxim-methyl, which was also announced in November 1992 (Ammermann *et al.*, 1992). BASF was not aware of our activities in Zeneca until our first patent applications were published in April 1986. Similarly, we at Zeneca were not aware that we had competitors in this area until the first BASF patent applications appeared some seven months later in November 1986.

Shionogi was the third company to announce a development compound of this class, a compound with the code name SSF-126 (Masuko *et al.*, 1993). Shionogi's entry into the area is noteworthy in that it was not prompted by the natural strobilurins or by patents published by other companies. Instead, Shionogi's interest in the strobilurin fungicides arose through speculative chemistry around an isoxazole lead, a programme of chemistry which converged on the strobilurins during optimization of activity. Details of Shionogi's work are described below.

Finally, Richards and his co-workers at AgrEvo have published information about two series of strobilurins, each with a sulfur atom in the side chain, but they have not announced a development compound so far (Cliff *et al.*, 1990; Richards *et al.*, 1994).

Since the early patent publications by Zeneca and BASF, many other agrochemical companies have embarked on strobilurin chemistry, and it is now a very competitive area of research. By May 1997, over 20 companies and research organizations had published a total of more than 400 patent applications claiming a wide variety of strobilurin analogues, mainly as fungicides (Table 2).

Table 2 Companies and research institutes which have published patent applications to strobilurin analogues, listed in order of filing the first patent application[a]

Company or institute	First patent application(s) with priority date		Number of patent applications published to 9 May 1997[b]
Zeneca[c]	EP 178,808 and EP 178,826	19 October 1984	72
BASF	EP 203,606 and EP 203,608	30 May 1985	127
AgrEvo[d]	EP 299,694	11 July 1987	17
Bayer[e]	EP 329,011	18 February 1988	50
Sumitomo	JP 02 121,970	28 October 1988	16
Ciba-Geigy[f,g]	WO 90 07,493	29 December 1988	26
Shionogi[h]	EP 398,692	17 May 1989	19
Roussel Uclaf	EP 402,246	6 June 1989	13
Nihon Nohyaku	EP 414,153	22 August 1989	10
Ube	EP 426,460	2 November 1989	5
Mitsubishi Kasei	EP 433,899	13 December 1989	12
Hokko	JP 04 187,692	22 November 1990	1
Ishihara Sangyo Kaisha (= ISK)	EP 498,396	7 February 1991	3
M. U. R. S. T.[i]	EP 532,126	13 September 1991	5
Isagro	EP 554,957	6 February 1992	4
Sandoz[g]	EP 571,326	13 May 1992	6
Korea R. I. C. T.[j]	WO 94 00,436	25 June 1992	2
Agro Kanesho	JP 06 87,842	9 September 1992	1
Lucky	EP 590,610	28 September 1992	1
Du Pont	WO 95 14,009	19 November 1993	7
Nissan	JP 08 03,002	21 June 1994	1
Nippon Soda	JP 08 12,648	30 June 1994	2
Rohm & Haas	EP 711,759	14 November 1994	2
Rhône-Poulenc	WO 96 33,164	17 April 1995	3

[a] Most of these patent applications claim primarily fungicidal activity. Others claim insecticidal activity or chemical processes.
[b] Numbers of patent applications should be treated as approximate.
[c] Zeneca was formerly ICI (until the biosciences businesses were demerged and renamed Zeneca in 1993).
[d] Several of the early patent applications assigned here to AgrEvo were filed by Schering (the crop-protection businesses of Schering and Hoechst were merged in 1994).
[e] Two of the Bayer patent applications are to Nihon-Bayer.
[f] The first two patent applications assigned here to Ciba-Geigy were filed by Hoffmann-La Roche (Ciba-Geigy acquired Hoffmann-La Roche's agrochemicals interests in 1990).
[g] Ciba-Geigy and Sandoz have recently merged to form Novartis.
[h] Three of the Shionogi patent applications were filed jointly with Sumitomo.
[i] Ministero dell'Universita'e della Ricerca Scientifica e Tecnologica, Rome.
[j] Korea Research Institute of Chemical Technology.

THE DISCOVERY OF AZOXYSTROBIN (ZENECA)

We have described details of the story of the discovery of azoxystrobin in earlier papers (Beautement *et al.*, 1991; Clough *et al.*, 1992, 1994 and 1995; Clough and Godfrey, 1995; Clough *et al.*, 1996) and present the main outline here.

Our interest in the strobilurin fungicides at Zeneca can be traced back to an important paper by Becker, von Jagow, Anke and Steglich which was published in September 1981 (Becker *et al.*, 1981). It reported for the first time that strobilurin A, strobilurin B, oudemansin A and myxothiazol A (the four members of the strobilurin family known at that time) were not only structurally related, but that their fungicidal activity stemmed from the same novel mode of action. We therefore asked ourselves whether this small collection of natural products could be a starting point for the synthesis of new agricultural fungicides. In particular, we were attracted to the strobilurins and oudemansin A because of their structural simplicity. The fact that there were four compounds was also appealing because it indicated that there was at least some scope for structural modification without loss of activity. In addition, the fact that these compounds bind at an enzyme site for which inhibitors had not previously been identified was important because it meant that there should be no cross-resistance between fungicides in current use and the compounds of this class. However, we recognized at the outset of our work that the mode of action, though novel, was not one which we would have chosen in an ideal world since inhibitors of respiration have the potential to be toxic to mammals and, indeed, the acute toxicity of myxothiazol A to mice is high ($LD_{50} = 2\,mg\,kg^{-1}$) (Gerth *et al.*, 1980). Nevertheless, there were strong grounds for believing that selective toxicity towards fungi could be achieved, because strobilurin A had been reported to have a low acute oral toxicity in mice ($LD_{50} = 500\,mg\,kg^{-1}$; Vondráček *et al.*, 1967).

Of course, we were interested to know what levels and spectrum of activity these natural products would show when tested against fungi growing in our own glasshouses. Requests were sent out and in July 1982 we received samples of oudemansin A and myxothiazol A from T. Anke of the University of Kaiserslautern, Germany, and H. Reichenbach of the Gesellschaft für Biotechnologische Forschung, Braunschweig, Germany, respectively. When applied as foliar sprays at concentrations of $33\,mg\,l^{-1}$, these compounds were active against several commercially important fungi growing on plants in the glasshouse. By contrast, strobilurin A, obtained by total synthesis, as described in the following paragraph (Beautement and Clough, 1987), gave quite different results: although active at $25\,mg\,l^{-1}$ against a variety of plant pathogenic fungi growing on agar in subdued light, the compound had no activity when applied as a foliar spray at the same rate to fungi growing on plants in the glasshouse. We were able to show that this lack of activity *in vivo* followed from the photochemical instability and relatively high volatility of strobilurin A, through

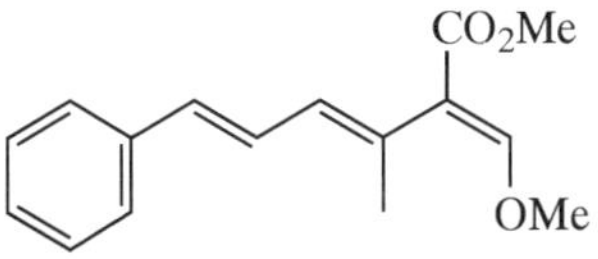

2

Figure 7 The synthetic (all-*E*)-isomer of strobilurin A

which it is lost from the leaf surface. So, to summarize, the results which we had obtained from the natural products were mixed. Oudemansin A and myxothiazol A were both active in the glasshouse as well as in biochemical assays, while strobilurin A did not express its intrinsic activity under glasshouse conditions because of inappropriate physical properties.

It is worth pointing out here that there has been considerable confusion in the past over the stereochemistry of strobilurin A (and, by inference, the other strobilurins). When we began our chemistry in this area in February 1983, there was strong evidence that strobilurin A had the (*all-E*)-configuration, and since we needed a sample for screening, we developed a synthesis which led unambiguously to this (*all-E*)-compound **2** (Figure 7). To our surprise, our synthetic material was different from strobilurin A, although it was plain that the two compounds were closely related. By comparing the spectroscopic data from our synthetic material with those reported for the natural product, we were able to show that strobilurin A and our synthetic material were stereoisomeric, and that strobilurin A, in reality, has the (*E*,*Z*,*E*)-configuration. We went on to confirm this reassignment by synthesizing a sample of the natural (*E*,*Z*,*E*)-isomer, the spectroscopic data from which now matched those from the natural product. We completed this work in February 1984. At about the same time, Steglich and his co-workers independently reassigned the configuration of the natural product, work which they published in September 1984 (Anke *et al.*, 1984) whereupon the information became public knowledge. Nevertheless, our own in-house work had given us a valuable head start over our competitors.

It was clear that the natural products themselves could not be used as agrochemicals. Strobilurin A was inactive in the glasshouse, while oudemansin A, though interesting as a lead, was really quite weak in comparison with commercial fungicides, and it was unlikely that it could be manufactured economically, either by synthesis or fermentation. Nevertheless, we realized that our knowledge of the structures of the natural products, including their conformations, might enable us to design and synthesize more useful analogues. So it was that we initiated a programme of synthesis with the aim of discovering related, more active fungicidal compounds which had suitable physical properties for an agrochemical, and which could be manufactured in a cost-effective way.

We argued that the β-methoxyacrylate unit of strobilurin A was likely to be the toxophore, important for activity; on the other hand, it might be possible to modify the rest of the molecule, which we thought likely to be the part responsible for photoinstability, without loss of activity. In fact, the β-methoxyacrylate and phenylpentadienyl units of strobilurin A are cross-conjugated, and computer modelling showed that these two parts, each roughly planar, would be expected to be strongly twisted with respect to each other, orthogonal in the lowest energy conformation (Beautement *et al.*, 1991). We later realized that this is an unusual structural feature, and a fortunate one in this instance. It means that interactions between the two parts of the molecule are minimized, so that structural changes to the phenylpentadienyl unit have little effect on the β-methoxyacrylate group, and vice versa.

Of our various ideas for stabilising strobilurin A (Beautement *et al.*, 1991), the most fruitful was to lock the (Z)-olefinic bond into a benzene ring to give the stilbene **1** (Figure 8). In contrast to strobilurin A, the stilbene was highly active against fungi growing on plants in the glasshouse, but was nevertheless still too unstable photochemically to express good activity in the field. Further modification produced a more useful compound, the diphenyl ether **3**, which is considerably more photochemically stable than the stilbene. The diphenyl ether controlled a variety of important fungal pathogens of wheat, barley and vines in the field, but it produced unacceptable damage to the crops in some trials. An important property of the diphenyl ether, almost certainly contributing to its activity (as well as its phytotoxicity) in the field, was its systemic movement in plants. Systemicity is a key attribute of modern agricultural fungicides which improves field performance by redistribution of the compound within plant tissue after spraying. So, of course, this was a feature which we wished to retain in further analogues of the strobilurins and oudemansins, provided that phytotoxicity could be avoided.

While studying the scope for the introduction of substituents in the diphenyl ether **3**, we found that highly fungicidal compounds resulted when single atoms or small groups were incorporated at the 3- or 4- (but not at the 2-) position of the phenoxy group. By contrast, the much larger phenoxy substituent could only be accommodated at the 3-position of the phenoxy-group: the tricyclic compound **4** was highly fungicidal, more active in the glasshouse than the diphenyl ether **3**, while the regioisomeric compound **5** was only a very weak fungicide. As one would predict, the introduction of the lipophilic phenoxy substituent to **3** to give **4** resulted in a loss of systemic movement. So while **4** clearly has an excellent shape for binding at cytochrome b, its physical properties prevent it from being redistributed effectively within plants.

As well as preparing derivatives of the diphenyl ether **3**, as described above, we synthesized analogues in which one or both of the benzene rings had been replaced by a heterocycle. For example, phenyl pyridyl ethers of the types **6** and **7** were especially active series. Like the diphenyl ether **3**, many of these heterocyclic compounds had the systemic properties that we sought.

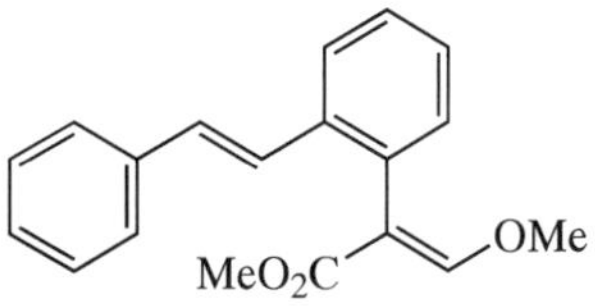

1

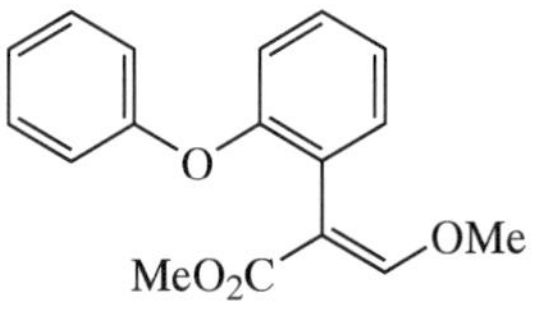

3

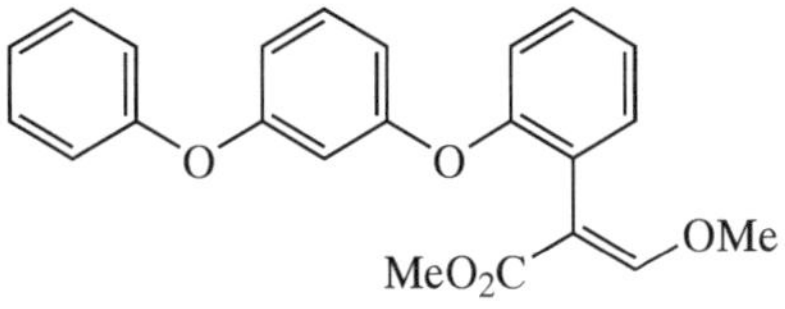

4

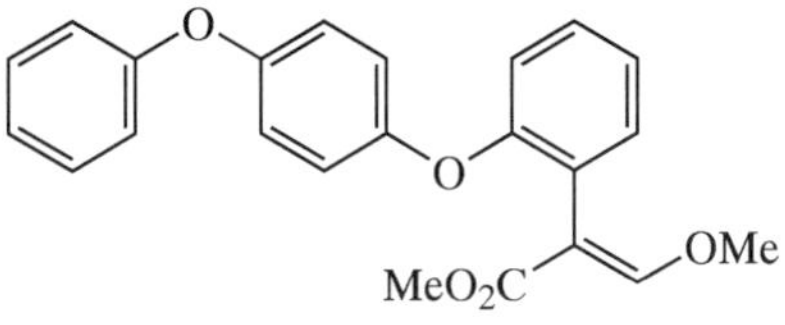

5

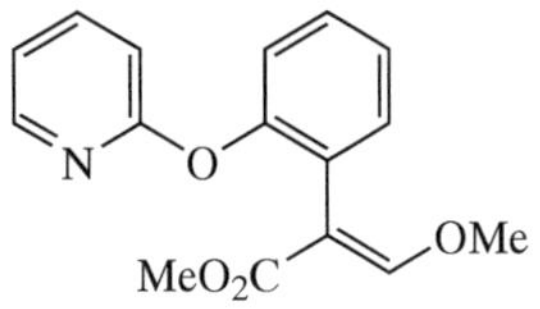

6

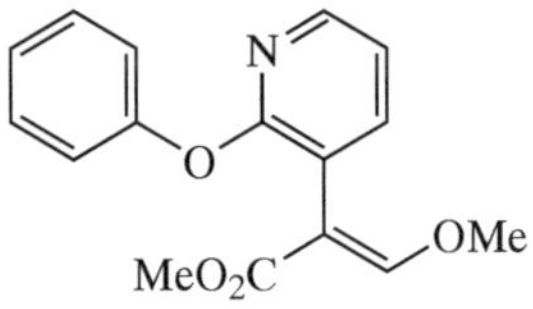

7

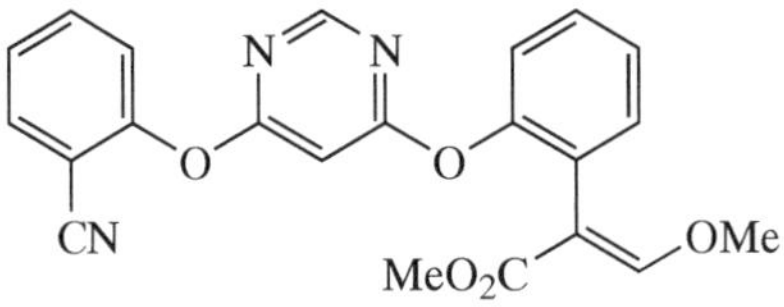

Azoxystrobin

Figure 8 Milestones in the discovery of azoxystrobin

Compounds such as **6** and **7** provided the clue to the best way to lower the lipophilicity of three-ring compounds such as **4**. Only by replacing at least one of the benzene rings in **4** with a heterocycle could the partition coefficient be lowered sufficiently for systemicity to be restored. We worked on the hypothesis that it should be possible to tailor lipophilicity and other important physical properties by the careful selection of suitable rings and substituents on them. Clearly, a large number of compounds, with different combinations of heterocycles and substituents, were possible targets for synthesis, and we did not know at the outset which modifications to the tricyclic compound **4** would be accommodated at the active site, or would be beneficial to binding. Consequently, we selected for synthesis a series of compounds with an appropriate range of physical properties, especially partition coefficient and pK_a. Many compounds were prepared and tested in the glasshouse, and the most

Table 3 Physical properties of azoxystrobin

Melting point	114–116 °C
Density	1.33 g cm^{-3}
Water-solubility	6.0 mg l^{-1} at 20 °C
Log (*n*-octanol/water partition coeff.)	2.5 at 20 °C
Vapour pressure	1.1×10^{-10} Pa at 25 °C

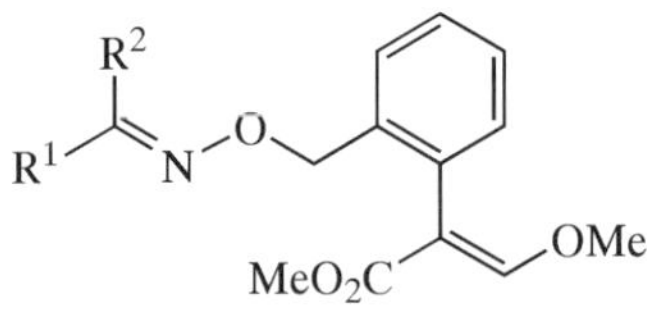

8 (R^1 and R^2 = various groups)

Figure 9 Strobilurins with an oxime ether side chain

promising were taken forward for field trials, where the best showed high fungicidal activity with little or no phytotoxicity. The discovery of azoxystrobin was the culmination of this work. It is a colourless crystalline solid with the physical properties shown in Table 3. Its remarkable biological activity is described later in this chapter.

During the work which led to the discovery of azoxystrobin, many avenues of chemistry were explored and most of these, of course, have not led to commercial fungicides. In the final paragraphs of this subsection, we briefly describe some of the most interesting of these compounds. One series, which in fact has attracted interest from several companies, are compounds **8** with an oxime ether side chain (Figure 9). Zeneca was the first to file a patent application describing these compounds (de Fraine and Martin, 1988; see also de Fraine and Clough, 1995; Bartlett and Stalker, 1996; Bartlett *et al.*, 1996), but Hoffmann-La Roche filed a similar application just five weeks later (Isenring *et al.*, 1988), and related patent applications from Nihon Nohyaku (Tsubata *et al.*, 1989) and Ube (Watanabe *et al.*, 1989; see also Watanabe *et al.*, 1994) followed within a year. The filing dates of these four patent applications show that this family of compounds **8** was independently discovered by the four companies. Other companies have subsequently published further related patent applications.

Certain compounds in which the (*E*)-methyl β-methoxyacrylate group is linked to a heterocyclic ring are good fungicides (one example referred to above is the phenyl pyridyl ether **7**, in which the toxophore is linked to a pyridine ring). Of the wide variety of these compounds which we prepared, pyrroles such as **9** and **10** (Figure 10), in which the β-methoxyacrylate group is linked to the nitrogen atom of the ring, were of particular interest (Beautement *et al.*, 1995).

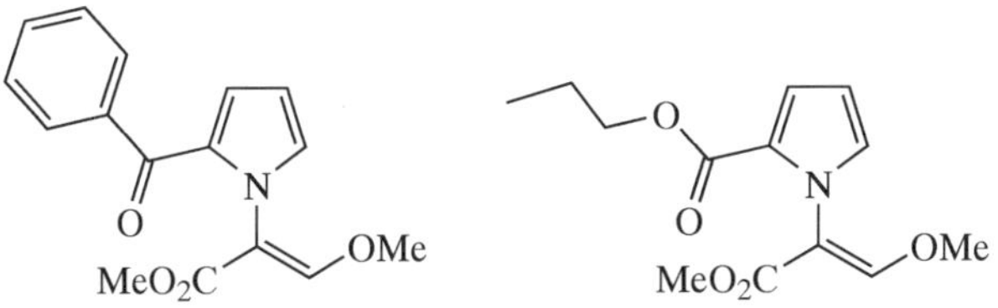

9 **10**

Figure 10 Strobilurins with a pyrrole ring

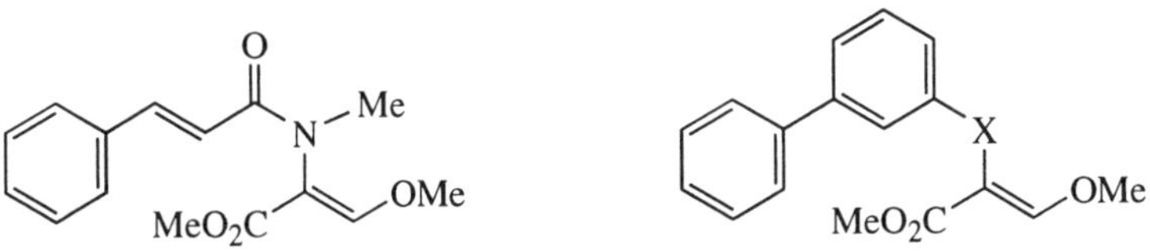

11 **12** (X = CR′R″, NR′, O, S)

Figure 11 Strobilurins in which the toxophore is not linked directly to a ring

These compounds have good activity and are easy to prepare, but in the end proved inferior to the members of our best series.

We also discovered novel active series in which the toxophore is not linked directly to a ring. For example, amides such as **11**, in which the (*Z*)-olefinic bond of strobilurin A has been replaced with an amide group, and biphenyls such as **12**, which map onto strobilurin in a different way, are also fungicidal (Figure 11). Finally, like our counterparts in BASF (see next subsection), we found that there is some scope for variation of the (*E*)-methyl β-methoxyacrylate toxophore without loss of activity (Beautement *et al.*, 1991).

THE DISCOVERY OF KRESOXIM-METHYL (BASF)

Sauter and his co-workers have published accounts of the strobilurin story from the BASF perspective (Ammermann *et al.*, 1992; Sauter *et al.*, 1995 and 1996; Gold *et al.*, 1996) and the following paragraphs are based on these publications. Anke and Steglich, who have worked closely with BASF, have also published an account of their studies (Anke and Steglich, 1989).

Scientists at BASF obtained a sample of strobilurin A from Anke (University of Kaiserslautern, Germany) in July 1983. When applied at the relatively high rate of $250\,mg\,l^{-1}$ to fungi growing on plants in the glasshouse, the natural product showed fungicidal effects against late blight (*Phytophthora infestans*) on tomatoes and powdery mildew (*Erysiphe graminis*) on wheat, but it was weaker than had been expected on the basis of its good activity *in vitro*. Discussions with Steglich (then at the University of Bonn, now at the University

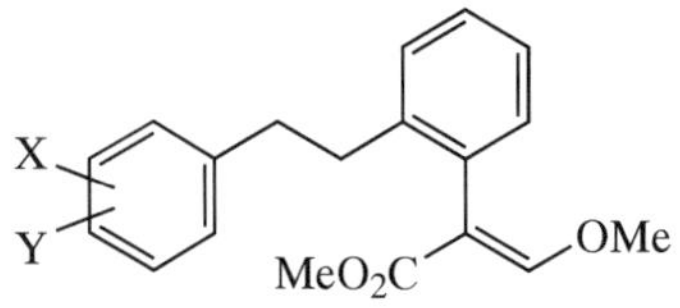

13

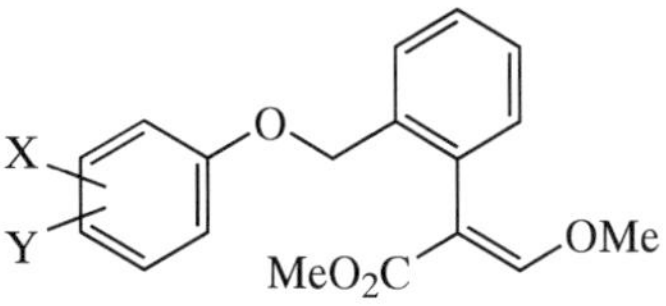

14

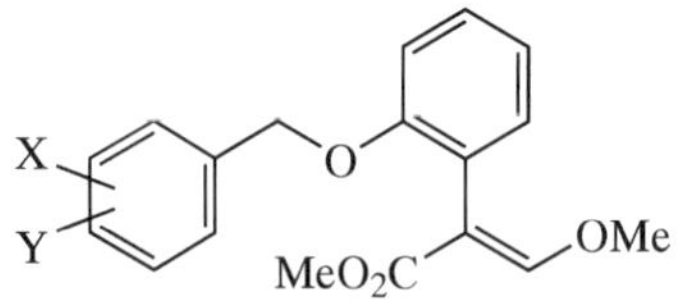

15

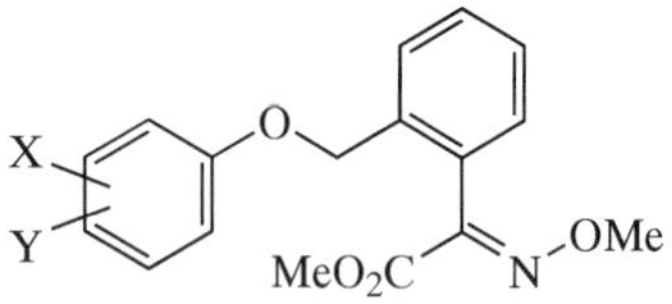

16

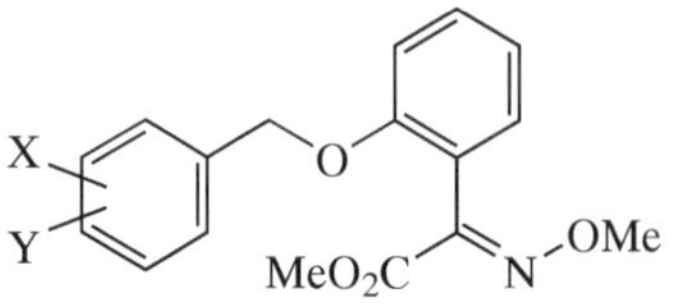

17

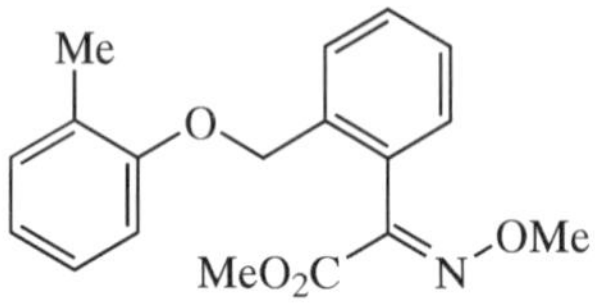

Kresoxim-Methyl

Figure 12 Milestones in the discovery of kresoxim-methyl

of Munich, Germany) led to the suggestion that instability could account for these differences. As at Zeneca, the question was asked as to whether more stable fungicidal analogues could be prepared, and the stilbene **1** was one of a collection of compounds made. Again, there was excitement at the good activity which the compound displayed in the glasshouse, but disappointment at its activity in the field. Laboratory tests confirmed what we had found too, that photostability accounted for much of this discrepancy. Clearly, further work was needed to address this problem.

With this in mind, the chemists at BASF prepared compounds in which the side chain in the stilbene **1** (Figure 8) had been modified. One active series of analogues was obtained by reduction of the styryl side chain to phenethyl, resulting in the compounds **13** (Figure 12). Phenoxymethyl- and benzyloxy-groups were also identified as promising replacements for styryl, leading to the compounds **14** and **15**. All these strobilurin analogues were easy to prepare and showed good activity in the glasshouse. However, the publication of Zeneca's patent application EP 178, 826 in April 1986 (Bushell *et al.*, 1984) put an end to BASF's work in this area since it embraced the three classes of compound **13**, **14** and **15**, as well as other types in which BASF were interested. Consequently, BASF switched the focus of their work to variation of the methoxyacrylate group, the toxophore. An early success was the discovery that compounds with an (*E*)-methyl methoxyiminoacetate group in place of the natural (*E*)-methyl β-methoxyacrylate group, such as **16** and **17**, retain good activity, and a patent application was filed. Again, parallel work was also carried out at Zeneca but, in this case, patenting roles were reversed, and BASF were ahead of us by the narrowest of margins, filing their patent application just two days before we filed ours (BASF: Wenderoth *et al.*, 1986; Zeneca: Anthony *et al.*, 1986).

With their advantageous patent position, scientists at BASF were able to work with confidence on the optimization of activity of compounds with the (*E*)-methyl methoxyiminoacetate toxophore. In addition, they sought further new toxophores which would confer fungicidal activity when linked to an appropriate backbone. During this phase of their programme, *in vitro* assays for inhibition of mitochondrial respiration were used extensively, in parallel with the usual glasshouse screens. It was these *in vitro* assays which highlighted the fact that the compounds **16** with a (substituted) phenoxymethyl side chain were generally intrinsically about 10 times more active than the regioisomers **17** with a benzyloxy side chain. This enabled BASF's chemists to focus on this better series of compounds **16**, and a larger number of compounds of this type were taken forward for field trials as well. Kresoxim-methyl was finally selected on the basis of its performance in the field, together with considerations of its cost of production on a large scale.

Kresoxim-methyl is a colourless crystalline solid with the physical properties shown in Table 4. It can be seen that kresoxim-methyl is more volatile and has a higher *n*-octanol/water partition coefficient than azoxystrobin (Table 3), properties which relate directly to the observed differences in biological activity between the two compounds.

So it was, as Sauter and his colleagues have pointed out (Sauter *et al.*, 1995, 1996), that Zeneca and BASF independently carried out strikingly similar programmes of research. The similarities were especially marked in the early phases of the two programmes, with parallel considerations leading in each case to the stilbene **1**, and the realization that this compound still required further modification before having suitable properties for an agricultural fungicide. The optimization phases of the two programmes were not, in fact, so similar, with the result that azoxystrobin and kresoxim-methyl are substantially different fungicides in terms of both their physical and biological properties. For example, azoxystrobin is a systemic compound with a broad spectrum of fungicidal activity, and will generally be used without mixture with other active

Table 4 Physical properties of kresoxim-methyl (from Ammermann *et al.*, 1992)

Melting point	97.2–101.7 °C
Water-solubility	$2\,\mathrm{mg\,l^{-1}}$ at 20 °C
Log (*n*-octanol/water partition coeff.)	3.4 at 25 °C
Vapour pressure	2.3×10^{-6} Pa at 20 °C

ingredients, while kresoxim-methyl is not systemic, but is relatively volatile and is therefore able to redistribute itself in plants via the vapour phase.

THE DISCOVERY OF SSF-126 (SHIONOGI)

Shionogi was the third company to announce its own strobilurin analogue as a development compound, a compound with the code name SSF-126 (Masuko *et al.*, 1993). SSF-126 has an (*E*)-*N*-methyl methoxyiminoacetamide toxophore, isosteric with the natural (*E*)-methyl β-methoxyacrylate toxophore of the strobilurins and oudemansins. Interestingly, however, SSF-126 originated from a chemistry-led programme of research rather than the natural products which gave rise to azoxystrobin and kresoxim-methyl, and only after Shionogi had performed a substantial amount of chemistry did they realize that their work had to some extent converged with that of Zeneca and BASF.

Hayase and his co-workers at Shionogi report that SSF-126 was discovered in the following way (Hayase *et al.*, 1995). Certain carbamoyl isoxazoles **18** (Figure 13) were found to have fungicidal activity against rice blast (*Pyricularia oryzae*). In the hope of finding compounds with improved activity, various ring-opened analogues of **18** were prepared, including compounds **19** which are formally the result of cleavage of the carbon–carbon double bond of the isoxazole ring. One subset of such compounds was **20**, and these were found to have activity against both rice blast and cucumber grey mould (*Botrytis cinerea*). During extensive optimization work around the structures **20**, scientists at Shionogi became aware of the similarity of their

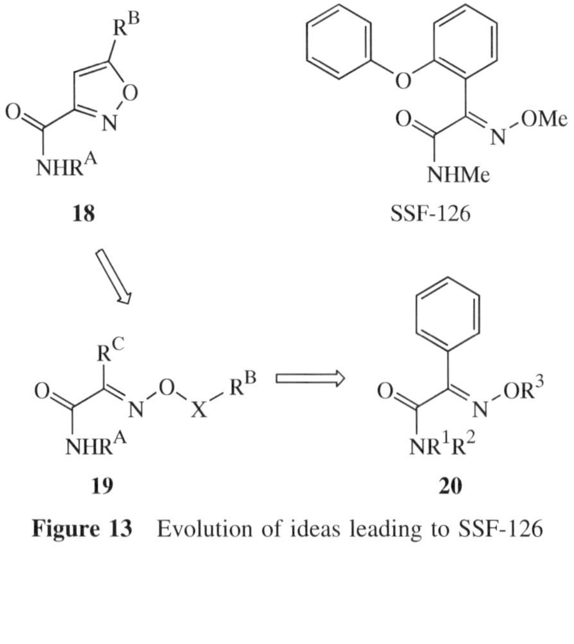

Figure 13 Evolution of ideas leading to SSF-126

Table 5 Physical properties of SSF-126 (from Kume *et al.*, 1996a and Sauter *et al.*, 1996)

Melting point	86.5–87.0 °C
Water-solubility	$128\,mg\,l^{-1}$ at 20 °C
Log (*n*-octanol/water partition coeff.)	2.3 at 20 °C

compounds with those which Zeneca and BASF had described. What they had discovered, however, was a series of compounds with a different but closely related toxophore.

Structure-activity patterns within Shionogi's series were broadly parallel with those for the compounds from Zeneca and BASF, supporting the relationship between compounds with the methoxyiminoacetamide, β-methoxyacrylate and methoxyiminoacetate toxophores. For example, in the toxophore of the compounds **20**, the (*E*)-isomer is preferred over the (*Z*)-isomer, and optimal values for R^1, R^2 and R^3 are methyl, hydrogen and methyl, respectively; the resulting structure has the best overlay with the natural (*E*)-methyl β-methoxyacrylate toxophore found in the strobilurins and oudemansins. Furthermore, the benzene ring bearing the toxophore in **20** is optimally substituted at the *ortho*-position, and good *ortho*-substituents are groups such as phenoxy and phenoxymethyl, the same as those in the series discovered by Zeneca and BASF. The compound which Shionogi finally selected for development was SSF-126 (an isostere of the diphenyl ether **3**, an important milestone in Zeneca's research programme). It is a colourless crystalline solid with the physical properties shown in Table 5. In due course, Shionogi established that SSF-126 binds at the same site on cytochrome b as azoxystrobin and kresoxim-methyl (Mizutani *et al.*, 1995 and 1996). This finally confirmed that there are real parallels at the biochemical level between the structurally related compounds from Shionogi, Zeneca and BASF.

BIOLOGICAL ACTIVITY AND ENVIRONMENTAL PROPERTIES

INTRODUCTION

The best strobilurins are characterized by their exceptional breadth of fungicidal activity, arising from their unique mode of action. This section summarizes the published information about the biological aspects of azoxystrobin, kresoxim-methyl and SSF-126 (relatively little has been reported for the last of these). In addition, we outline what is known about why the strobilurins generally show selective activity against fungi, and the use of strobilurins as herbicides, insecticides and medicinal fungicides.

AZOXYSTROBIN

For leading references, see Godwin *et al.*, 1992; Cohadon *et al.*, 1994; Godwin *et al.*, 1994; Frank and Sanders, 1994; Heaney and Knight, 1994; and Pilling *et al.*, 1996.

Azoxystrobin is a broad spectrum fungicide with activity against all four major groups of plant pathogenic fungi, namely Ascomycetes (e.g. powdery mildews), Basidiomycetes (e.g. rusts), Deuteromycetes (e.g. rice blast) and Oomycetes (e.g. downy mildews). It is intrinsically highly active. For example, in a glasshouse experiment on wheat seedlings, it gave complete control of brown rust (*Puccinia recondita* f.sp. *tritici*) at a concentration of $0.08\,mg\,l^{-1}$. No other commercial agricultural fungicide possesses this combination of breadth of spectrum and high intrinsic activity. In addition, azoxystrobin is systemic. This has been demonstrated on cereals in the glasshouse by marking out four equal-sized zones on the leaves of wheat seedlings. Treatments were then applied to the zone one up from the leaf base and 24 hours later all plants were inoculated with urediospores of wheat brown rust. The compound was taken up and moved within the xylem to the leaf tip thereby giving complete disease control along the leaf length. Azoxystrobin also demonstrates translaminar movement, as well as systemic control of foliar disease when applied as a root drench, nursery box granule, paddy granule or seed treatment.

Azoxystrobin has been rigorously field-tested on more than 60 crops around the world, by both Zeneca and independent cooperators, using a wide range of application methods. In numerous crops, it has demonstrated control of the key pathogens equal or superior to current commercial standards, combined with good crop safety. Table 6 lists some of the diseases controlled by azoxystrobin in field trials to date. In many cases it has provided a spectrum and level of control not achievable with any existing fungicide. For example, excellent control has been demonstrated of both brown patch (*Rhizoctonia solani*) and *Pythium* blight (*P. aphanidermatum*) on turf grass, downy mildew (*Plasmopara viticola*) and powdery mildew (*Uncinula necator*) on grape vines, and blast (*Pyricularia oryzae*) and sheath blight (*R. solani*) on rice.

Azoxystrobin has demonstrated preventative, curative, eradicant and antisporulant activity. However, like all strobilurins, because it is a highly potent inhibitor of spore germination, it is most effectively used prior to infection or in the early stages of disease development. It also demonstrates a good persistence of effect. For example, on European cereals, applications to the upper foliage can give protection for six to eight weeks, thereby maintaining green leaf area until late in the season. Consequently, excellent yields of high quality grain can be achieved. Marked improvements in yield and quality as a result of disease control have also been seen in a number of other crops.

Table 6 Diseases controlled by azoxystrobin on representative crops

Crop or outlet	Disease
Turf grass	Brown patch (*Rhizoctonia solani*) Pythium blight (*Pythium aphanidermatum*) Fusarium patch (*Microdochium nivale*) Melting out (*Drechslera poae*) Grey snow mould (*Typhula incarnata*) Red thread (*Laetisaria fuciformis*) Necrotic ringspot (*Leptosphaeria korrae*) Anthracnose (*Colletotrichum graminicola*) Summer patch (*Magnaporthe poae*) Take-all patch (*Gaeumannomyces graminis*)
Wheat	*Septoria tritici* *Stagonospora (Septoria) nodorum* Brown rust (*Puccinia recondita* f.sp. *tritici*) Yellow rust (*Puccinia striiformis*) Powdery mildew (*Erysiphe graminis* f.sp. *tritici*) Tan spot (*Helminthosporium tritici-repentis*) *Helminthosporium sativum* Sharp eyespot (*Rhizoctonia solani*) Sooty mould (*Cladosporium* spp.) Black point (*Alternaria* spp.)
Rice	Blast (*Pyricularia oryzae*) Sheath blight (*Rhizoctonia solani*) Stem rot (*Sclerotium oryzae*) Panicle blight (*Cochliobolus miyabeanus*)
Grape vines	Downy mildew (*Plasmopara viticola*) Powdery mildew (*Uncinula necator*) Black rot (*Guignardia bidwellii*) Excoriose (*Phomopsis viticola*) Brenner (*Pseudopeziza tracheiphila*)
Cucurbits	Downy mildew (*Pseudoperonospora cubensis*) Powdery mildew (*Sphaerotheca fuliginea*) Anthracnose (*Colletotrichum lagenarium*) Gummy stem blight (*Didymella bryoniae*) Grey mould (*Botrytis cinerea*)
Peach	Brown rot (*Monilinia fructicola/M. fructigena*) Scab (*Cladosporium carpophilum*)
Banana	Black sigatoka (*Mycosphaerella fijiensis*) Yellow sigatoka (*Mycosphaerella musicola*)

As expected on the basis of its novel biochemical mode of action, azoxystrobin is effective against pathogens which have developed reduced sensitivity to other fungicides, such as phenylamides, dicarboximides, benzimidazoles and inhibitors of ergosterol biosynthesis.

Azoxystrobin shows good safety to mammals, with an acute oral LD_{50} (rat) $> 5000\,mg\,kg^{-1}$, and an acute dermal LD_{50} (rat) $> 2000\,mg\,kg^{-1}$. It is Ames-negative. It is degraded in soil by processes that involve both sunlight and microbes. Under field conditions the soil dissipation half life is typically in the range of 7 to 28 days. Importantly, it has been established that the products of both photolytic and microbial processes are themselves readily degraded in soil. Laboratory studies have indicated that azoxystrobin would be of low mobility in soil and this coupled with the rapid degradation indicates that contamination of groundwater is unlikely. This is supported by the results of field dissipation trials where no evidence of mobility has been seen. Azoxystrobin is non-volatile and should not therefore be found in air. Its physical properties indicate that it has a low bioaccumulation potential, which along with its rapid metabolism in animals results in no significant accumulation in the food chain.

The environmental fate data outlined above show that azoxystrobin is readily degraded in the environment and this will help minimize exposure to non-target organisms. Risk to non-target organisms is determined by a combination of the potential exposure and toxicity. Studies have shown that it has low toxicity to both mammals and birds and thus presents no risk to wildlife. In addition, the low toxicity to honey bees, earthworms and a wide range of beneficial arthropods including carabid beetles, parasitic wasps, predatory mites, lycosid spiders, hover-flies and lacewings makes azoxystrobin ideal for use in Integrated Pest Management programmes.

KRESOXIM-METHYL

For leading references, see Ammermann *et al.*, 1992; Gold and Leinhos, 1994 and 1995; Sauter *et al.*, 1995 and 1996; and Gold *et al.*, 1996.

Kresoxim-methyl displays protectant activity against a broad spectrum of fungal pathogens on a wide range of crops including powdery mildew (*Erysiphe graminis*), rusts (*Puccinia recondita* and *P. striiformis*), *Rhynchosporium secalis*, *Septoria nodorum* and *S. tritici*, in cereals; powdery mildew (*Uncinula necator*) and downy mildew (*Plasmopara viticola*) in grape vines; scab (*Venturia inaequalis*) and powdery mildew (*Podosphaera leucotricha*) in apples; blast (*Pyricularia oryzae*) and sheath blight (*Rhizoctonia solani*) in rice; powdery mildew (*Sphaerotheca fuliginea*) in cucurbits; and powdery mildew (*Erysiphe betae*) in sugar beet. Kresoxim-methyl is reported to have eradicative potential against some of these pathogens, particularly powdery mildew and scab on apples. Furthermore, following application, it is sufficiently volatile to be redistributed via the vapour phase, and it has also shown translaminar activity.

Published toxicological data on kresoxim-methyl are highly favourable: it has an acute oral LD_{50} (rat) $>5000\,mg\,kg^{-1}$, an acute dermal LD_{50} (rat) $>2000\,mg\,kg^{-1}$ and it is Ames-negative.

SSF-126

For leading references, see Masuko *et al.*, 1993; Hayase *et al.*, 1995; and Kume *et al.*, 1996a and 1996b.

It appears that Shionogi is developing SSF-126 primarily for the control of rice blast (*Pyricularia oryzae*): the compound is reported to give excellent preventative and curative activities against both panicle blast and leaf blast, and controlled-release formulations, whereby the compound can be applied to paddy water in the form of granules, are especially efficacious. However, SSF-126 also has activity against a variety of other diseases, such as rice sheath blight (*Rhizoctonia solani*), apple scab (*Venturia inaequalis*), and powdery and downy mildews on vegetables.

Published toxicity results for SSF-126 are favourable: it has an acute oral LD_{50} (rat and mouse) $>300\,mg\,kg^{-1}$ and it is Ames-negative.

ACTIVITY AS HERBICIDES, INSECTICIDES AND MEDICINAL FUNGICIDES

Mitochondrial respiration takes place in all eukaryotes: plants, insects and mammals as well as fungi. A variety of strobilurins have been tested against mitochondria isolated from *Botrytis*, yeast, corn, flies and rats, and there was good correlation between activity on the fungal mitochondria and activity on mitochondria from the other sources. The high *in vivo* selectivity for fungi shown by most strobilurins therefore presumably arises from the effects of differential transport and metabolism (Röhl and Sauter, 1994).

Although phytotoxicity has been a problem of some early experimental strobilurins, we are not aware of any strobilurins being taken forward into development as herbicides. However, several patent applications suggest that some strobilurin analogues display interesting levels of insecticidal activity (see, for example, Schuetz *et al.*, 1989).

Interest in the strobilurins as fungicides has been almost exclusively limited to the agricultural scene, with no significant developments in the medicinal area. Strobilurin A itself has been used as a topical antifungal in human and veterinary medicine under the trade name 'Mucidermin Spofa'. However, the important modern human antifungals are mainly systemically active compounds, dosed either orally or intravenously. With this in mind, we have tested representative examples of our large collection of strobilurins for systemic activity in mouse models of fungal infection, but little or no activity was observed (Clough, 1993; Clough *et al.*, 1997).

SYNTHESIS

The characteristic feature of many synthetic strobilurins is an (*E*)-methyl β-methoxyacrylate toxophore, or its bio-isosteric equivalent, attached to a ring with a side chain in an adjacent (*ortho*) position. Since the scope for variation of the side chain (while retaining fungicidal activity) has proved to be very large, the discussion in this section will be limited to published chemistry which has been employed to construct the toxophore itself. Similarly, the use of convergent intermediates containing an intact toxophore will not be described, despite their obvious importance in programmes of analogue synthesis.

The most widely used route to synthesize compounds containing the β-methoxyacrylate toxophore involves a Claisen condensation between a phenylacctate **21** and methyl formate in the presence of a base, such as sodium hydride (Scheme 1). The salt of the condensation product which is formed initially is usually acidified and then subjected to methylation in a separate step, commonly using dimethyl sulphate and potassium carbonate. The advantage of this stepwise approach over a one-pot procedure is that it tends to lead to cleaner products. The overall reaction sequence is highly stereoselective, giving the desired (*E*)-isomer exclusively (Beautement *et al.*, 1991).

It is also possible to prepare the β-methoxyacrylate group by elimination of methanol from acetals in the presence of certain acids or bases. Scheme 2 shows an early application of this approach in the synthesis of the (*all-E*)-isomer of strobilurin A **2** (Figure 7) (Beautement and Clough, 1987). In this case, the required acetal **22** was formed by a Mukaiyama reaction on the *O*-silyl ketene acetal derived from ester **23** by treatment with lithium di-isopropylamide (LDA) and trimethylsilyl chloride. Elimination of methanol was achieved by treatment with a second equivalent of LDA. Later work showed that, in some

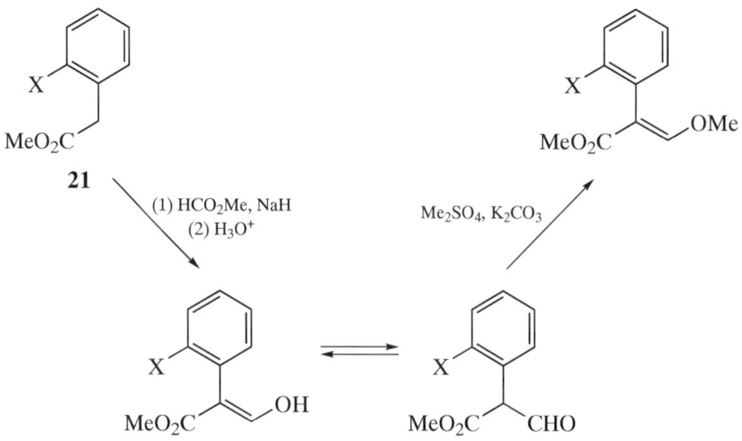

Scheme 1 Synthesis of the β-methoxyacrylate toxophore by Claisen condensation

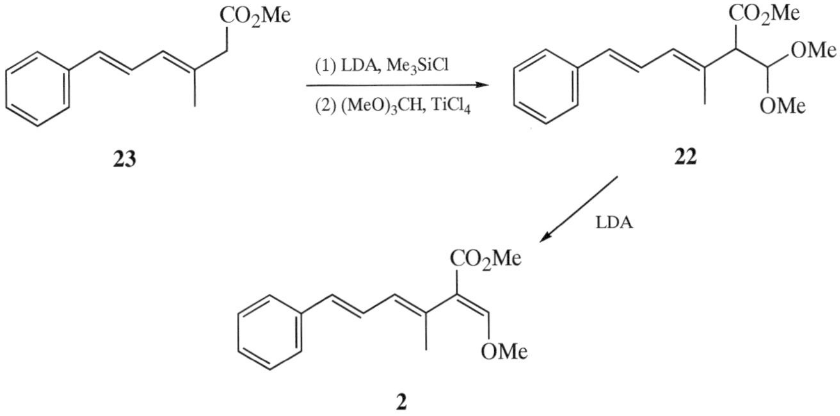

Scheme 2 Synthesis of the β-methoxyacrylate toxophore *via* an intermediate acetal

instances, if the *O*-silyl ketene acetal is prepared using trimethylsilyl triflate and triethylamine, subsequent reaction with trimethyl orthoformate-$TiCl_4$ in the same flask leads directly to the required β-methoxyacrylate, without the need for a separate elimination step. Elimination of methanol from acetals can also be achieved by heating with potassium hydrogen sulphate, provided that the rest of the molecule is sufficiently robust to withstand these conditions.

The Wittig reaction has also been used to prepare methyl β-methoxyacrylates (as well as methyl β-methylthio-acrylates) and Scheme 3 shows the synthesis of the stilbene **1** using this method (Beautement *et al.*, 1991). This approach can be an attractive option during a programme of synthesis if the benzoylformate

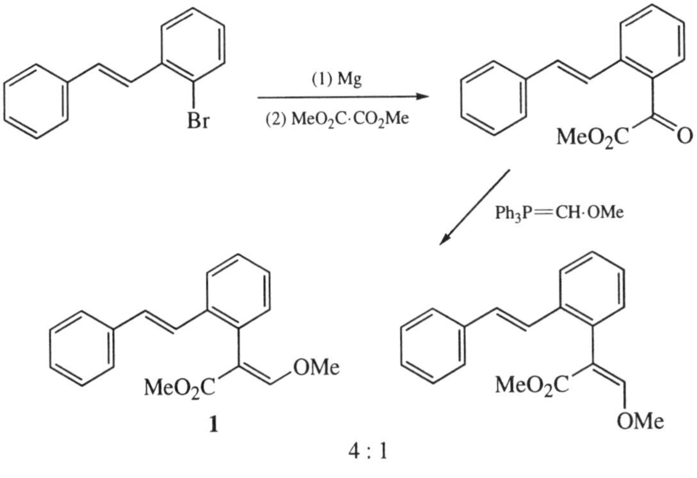

Scheme 3 Synthesis of the β-methoxyacrylate toxophore by a Wittig reaction

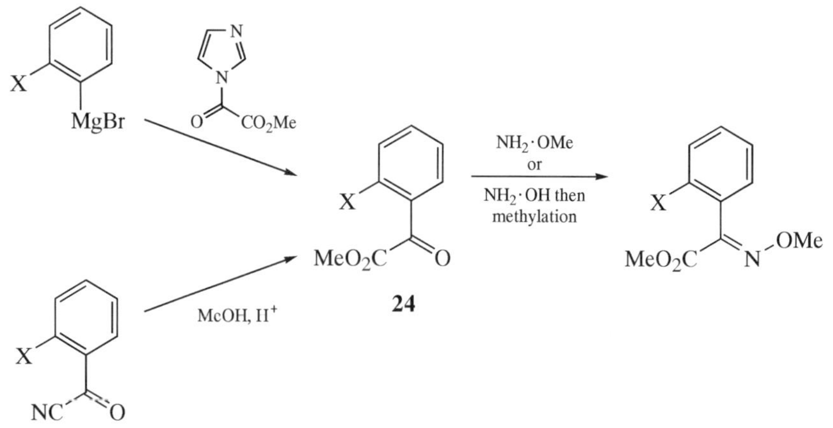

Scheme 4 Synthesis of the methoxyiminoacetate toxophore

precursors are more accessible than the corresponding phenylacetates, or where other substituents are sensitive to the conditions of the Claisen condensation. However, mixtures of stereoisomers are usually formed. Fortunately, these are usually easy to separate by chromatography, and the characteristic chemical shift of the olefinic hydrogen atom of the toxophore in the ^{1}H-NMR spectrum allows unambiguous assignment of the stereochemistry of the products [(*E*)- and (*Z*)-isomers typically display a singlet at around δ 7.5 and δ 6.5 ppm, respectively]. Alternatively, (*Z*)-isomers can be converted into the required (*E*)-isomers by treatment with various electrophilic (Clough and Eshelby, 1990), radical (Clough, 1990) or nucleophilic (Clough *et al.*, 1990) reagents.

Compounds containing the methoxyiminoacetate toxophore, a key feature of kresoxim-methyl, can readily be constructed by reaction of appropriately substituted benzoylformates **24** with methoxylamine (Scheme 4) (Sauter *et al.*, 1996). These, in turn, are obtained by reaction of aryl Grignard or arylzinc reagents with a suitable oxalate derivative, such as methyl oxalyl chloride or imidazolyl methyl oxalate. Alternatively, the required benzoylformates can be prepared by methanolysis of the corresponding benzoyl cyanides. In general, the oxime-forming step produces mixtures of stereoisomers, the ratio of which is dependent on the precise reaction conditions. For example, the desired (*E*)-isomer is favoured in the presence of excess acid. As with the β-methoxyacrylate group, methods for the equilibration of stereoisomers are known, or they may be separated by chromatography (Sauter *et al.*, 1996).

The key intermediates for the preparation of compounds containing the *N*-methyl methoxyiminoacetamide toxophore, a feature of SSF-126, are the appropriate α-keto esters, often again the benzoylformates **24**. These react with methylamine to give the α-ketoamides **25**, which, on treatment with methoxylamine hydrochloride, or successive treatment with hydroxylamine hydro-

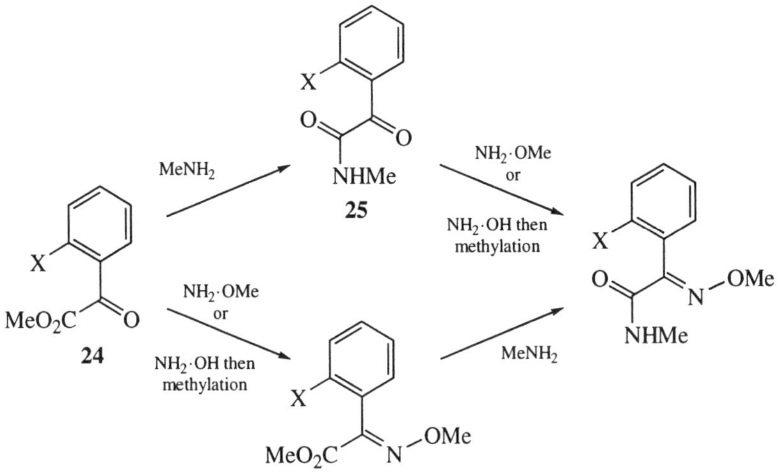

Scheme 5 Synthesis of the methoxyiminoacetamide toxophore

chloride and a methylating agent, afford the target compounds, usually as a mixture of isomers in which the required (*E*)-isomer predominates. Alternatively, the order of the amide-forming and oximation steps can be reversed (Scheme 5) (Hayase *et al.*, 1989 and 1995). Final purification of the *N*-methyl methoxyiminoacetamide is achieved either by crystallization or chromatography. Stereochemical assignment of the products can be made on the basis of ^{1}H-NMR spectroscopy: the chemical shift of the aromatic hydrogen atom adjacent (*ortho*) to the toxophore of a (*Z*)-isomer appears downfield with respect to that of the corresponding (*E*)-isomer. In the case of SSF-126 itself, this signal occurs in the aromatic envelope (δ 6.70–7.47 ppm) while that of the corresponding (*Z*)-isomer is at δ 7.67 ppm, assignments which were confirmed by single crystal X-ray analysis of the two isomers (Hayase *et al.*, 1995).

Finally, an approach which has more recently attracted attention is to incorporate the entire toxophore of a strobilurin in one step by a cross-coupling reaction. Hodgson, working in collaboration with chemists from Roussel Uclaf and AgrEvo, has performed palladium/copper co-catalysed cross-coupling reactions of methyl (*Z*)-2-tributylstannyl-3-methoxypropenoate **26** with aryl and heteroaryl iodides and triflates, as illustrated in Scheme 6 (Brayer *et al.*, 1993; Hodgson *et al.*, 1995). A variety of monosubstituted aromatic partners (including several *ortho*-substituted compounds) successfully underwent this reaction under mild conditions and yields were good (55–92%). More recently, Rossi and co-workers have developed an alternative palladium-catalysed reaction which takes place between substituted arylzincs or arylboronic acids and methyl (*Z*)-2-iodo-3-methoxypropenoate (**27**, X = I) or methyl (*Z*)-2-bromo-3-methoxypropenoate (**27**, X = Br), respectively (Scheme 7) (Rossi

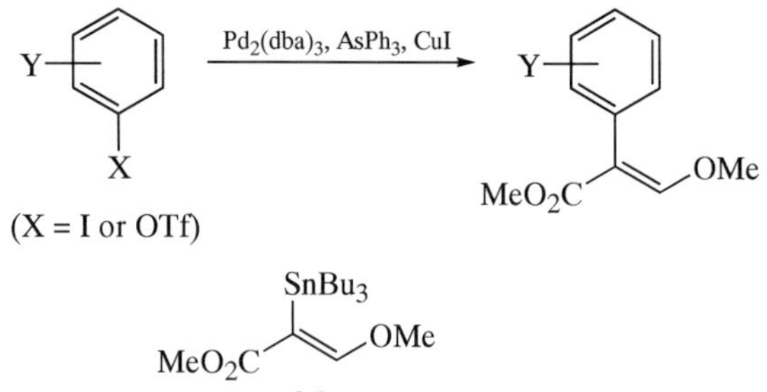

26

Scheme 6 Introduction of the β-methoxyacrylate toxophore by palladium-catalysed cross-coupling—first approach; Y = a variety of substituents

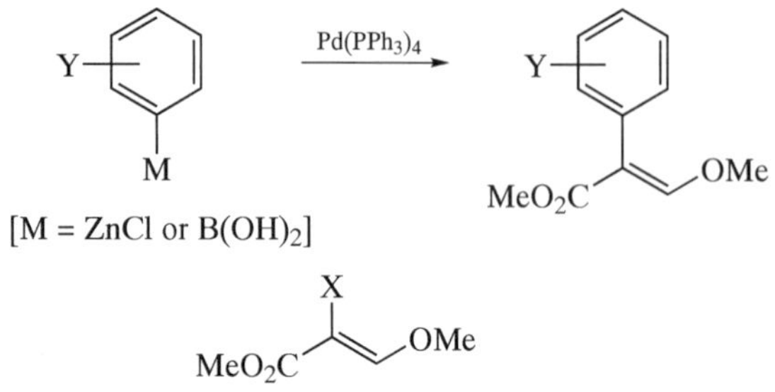

27 (X = I or Br)

Scheme 7 Introduction of the β-methoxyacrylate toxophore by palladium-catalysed cross-coupling—second approach; Y = a variety of substituents

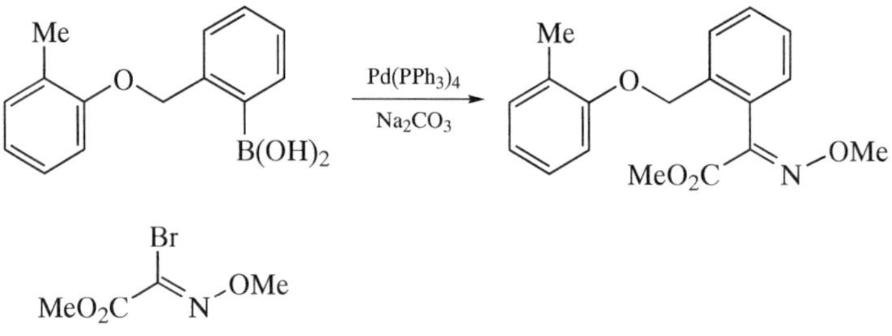

28

Scheme 8 Introduction of the methoxyiminoacetate toxophore by palladium-catalysed cross-coupling

et al., 1996; Rossi and Bellina, 1996). Yields of 35–76% were recorded. Finally, chemists at Ciba-Geigy have published a patent application claiming a process in which the methyl methoxyiminoacetate toxophore can be linked to an aromatic ring in a similar way, by allowing an arylboronic acid to react with the brominated methyl methoxyiminoacetate **28** in the presence of a palladium catalyst. One example describes a preparation of kresoxim-methyl in a yield of about 40% (Scheme 8) (Ziegler *et al.*, 1994).

CONCLUSIONS

The strobilurins, oudemansins and myxothiazols are members of a large family of fungicidal natural products which have been isolated and characterized during the last 35 years. The synthesis of analogues of these compounds in which biological, physical and environmental properties have been optimized has led to a new class of agricultural fungicides which has become known as the strobilurins. These compounds are inhibitors of mitochondrial respiration in fungi. The best examples have high levels and a broad spectrum of fungicidal activity.

The three strobilurins of current commercial interest are azoxystrobin (ICIA5504) from Zeneca, kresoxim-methyl (BAS 490 F) from BASF and SSF-126 from Shionogi, the culmination of many years of work within the three companies. Azoxystrobin and kresoxim-methyl were sold for the first time in 1996. It is almost certain that further development compounds from the strobilurin family will be announced in the future. The strobilurin fungicides represent the most exciting development on the agricultural fungicide scene since the discovery of the 1,2,4-triazoles in the 1970s. We confidently predict that they will form an important part of the global fungicide market for years to come.

Note added after submission of manuscript:

A recent publication suggests that the structures of strobilurin D, hydroxystrobilurin D, 9-methoxystrobilurin K and 9-methoxystrobilurin L should be revised. See Nicholas, G. M., Blunt, J. W., Cole, A. L. J., and Munro, M. H. G. (1997). 'Investigation of the New Zealand Basidiomycete *Favolaschia calocera*: revision of the structures of 9-methoxystrobilurins K and L, strobilurin D, and hydroxystrobilurin D', *Tet. Letts.*, **38**, 7465–7468.

REFERENCES

Agrow (3 January 1997). 'BASF sites fungicide plant in Spain', PJB Publications Ltd, No. 271, 3.

Ammermann, E., Lorenz, G., Schelberger, K., Wenderoth, B., Sauter, H., and Rentzea, C. (1992). 'BAS 490 F—a broad-spectrum fungicide with a new mode of action', in *Brighton Crop Prot. Conf.: Pests and Diseases—Vol. 1*, British Crop Protection Council, Farnham, UK, 403–410.

Anke, T. (1995). 'The antifungal strobilurins and their possible ecological role', *Can J. Bot.*, **73** (Suppl. 1), S940–S945.

Anke, T. and Steglich, W. (1989). 'β-Methoxyacrylate antibiotics: from biological activity to synthetic analogues', in *Biologically Active Molecules: Identification, Characterization and Synthesis*, ed. Schlunegger, U. P., Springer-Verlag, Berlin and Heidelberg, 9–25.

Anke, T., Oberwinkler, F., Steglich, W., and Schramm, G. (1977). 'The strobilurins—new antifungal antibiotics from the basidiomycete *Strobilurus tenacellus* (Pers. ex Fr.) Sing', *J. Antibiot.*, **30**, 806–810.

Anke, T., Hecht, H. J., Schramm, G., and Steglich, W. (1979). 'Antibiotics from basidiomycetes. IX. Oudemansin, an antifungal antibiotic from *Oudemansiella mucida* (Schrader ex Fr.) Hoehnel (Agaricales)', *J. Antibiot.*, **32**, 1112–1117.

Anke, T., Bäuerle, J., Backens, S., and Steglich, W. (1983a). 'Oudemansin B and hydroxystrobilurin D, new antifungal antibiotics from basidiomycetes', Abstract No. O 32, page 244, *Abstr. Ann. Meet. Am. Soc. Microbiol. (1983)* (*Biotech. Abstr.*, 1983, **83**, 3872).

Anke, T., Besl, H., Mocek, U., and Steglich, W. (1983b). 'Antibiotics from basidiomycetes. XVIII. Strobilurin C and oudemansin B, two new antifungal metabolites from *Xerula* species (Agaricales)', *J. Antibiot.*, **36**, 661–666.

Anke, T., Schramm, G., Schwalge, B., Steffan, B., and Steglich, W. (1984). 'Antibiotics from basidiomycetes, XX. Synthesis of strobilurin A and revision of the stereochemistry of natural strobilurins', *Liebigs Ann. Chem.*, 1616–1625.

Anke, T., Werle, A., Bross, M., and Steglich, W. (1990). 'Antibiotics from basidiomycetes. XXXIII. Oudemansin X, a new antifungal *E*-β-methoxyacrylate from *Oudemansiella radicata* (Relhan ex Fr.) Sing', *J. Antibiot.*, **43**, 1010–1011.

Anthony, V. M., Clough, J. M., Godfrey, C. R. A., and Wiggins, T. E. (1986). 'Fungicides', Eur. Pat. Appl. EP 254,426, Imperial Chemical Industries, priority date 18 July 1986.

Augustiniak, H., Gerth, K., Grotjahn, L., Irschik, H., Kemmer, T., Kunze, B., Reichenbach, H., Reifenstahl, G., Trowitzsch, W., and Wray, V. (1978). 'Compounds having the empiric summary formula $C_{25}H_{33}N_3O_3S_2$', WO 80 00,573, Gesellschaft für Biotechnologische Forschung, Germany, priority date 4 September 1978.

Backens, S., Steglich, W., Bäuerle, J., and Anke, T. (1988). 'Antibiotics from basidiomycetes, 28. Hydroxystrobilurin D, an antifungal antibiotic from cultures of *Mycena sanguinolenta* (Agaricales)', *Liebigs Ann. Chem.*, 405–409.

Baldwin, B. C., Clough, J. M., Godfrey, C. R. A., Godwin, J. R., and Wiggins, T. E. (1996). 'The discovery and mode of action of ICIA5504', in *Modern Fungicides and Antifungal Compounds*, eds. Lyr, H., Russell, P. E., and Sisler, H. D., Intercept, Andover, UK, Chapter 8, 69–77.

Bartlett, D. W. and Stalker, A. (1996). 'Metabolism rate of strobilurin analogues in wheat: determination using a bioautography assay', in *Modern Fungicides and Antifungal Compounds*, eds, Lyr, H., Russell, P. E., and Sisler, H. D., Intercept, Andover, UK, Chapter 11, 101–103.

Bartlett, D. W., Howell, A., Pierce, A. J., Rose, S. A., and Fraser, T. E. M. (1996). 'Effects of plant metabolism and photodegradation on the persistence of antifungal activity for a strobilurin analogue in the field' in *Modern Fungicides and Antifungal Compounds*, eds. Lyr, H., Russell, P. E., and Sisler, H. D., Intercept, Andover, UK, Chapter 12, 105–109.

Bäuerle, J. and Anke, T. (1980). 'Antibiotics from the genus *Mycena* and from *Hydropus scabripes*', *Planta Med.*, **39**, 195–196.

Beautement, K., and Clough, J. M. (1987). 'Stereocontrolled syntheses of strobilurin A and its (9*E*)-isomer', *Tet. Letts.*, **28**, 475–478.

Beautement, K., Clough, J. M., de Fraine, P. J., and Godfrey, C. R. A. (1991). 'Fungicidal β-methoxyacrylates: from natural products to novel synthetic agricultural fungicides', *Pestic. Sci.*, **31**, 499–519.

Beautement, K., Clough, J. M., de Fraine, P. J., and Godfrey, C. R. A. (1995). 'Fungicidal β-methoxyacrylates. *N*-Linked pyrroles' in *Synthesis and Chemistry of Agrochemicals IV*, eds. Baker, D. R., Fenyes, J. G., and Basarab, G. S., ACS Symposium Series No. 584, American Chemical Society, Washington, DC, USA, Chapter 29, 326–342.

Becker, W. F., von Jagow, G., Anke, T., and Steglich, W. (1981). 'Oudemansin, strobilurin A, strobilurin B and myxothiazol: new inhibitors of the bc_1 segment of the respiratory chain with an *E*-β-methoxyacrylate system as common structural element', *FEBS Letts.*, **132**, 329–333.

Bedorf, N., Kunze, B., Reichenbach, H., and Höfle, G. (1986). 'Myxothiazols', in *Scientific Annual Report*, Gesellschaft für Biotechnologische Forschung, Braunschweig-Stöckheim, Germany, 14–16.

Brandt, U. (1996). 'Bifurcated ubihydroquinone oxidation in the cytochrome bc_1 complex by proton-gated charge transfer', *FEBS Letts.*, **387**, 1–6.

Brandt, U. and Trumpower, B. (1994). 'The protonmotive Q cycle in mitochondria and bacteria, *Critical Reviews in Biochem. and Molec. Biol.*, **29**, 165–197.

Brandt, U., Schägger, H., and von Jagow, G. (1988). 'Characterisation of binding of the methoxyacrylate inhibitors to mitochondrial cytochrome c reductase', *Eur. J. Biochem.*, **173**, 499–506.

Brandt, U., Haase, U., Schägger, H., and von Jagow, G. (1991). 'Significance of the "Rieske" iron-sulfur protein for formation and function of the ubiquinol-oxidation pocket of mitochondrial cytochrome c reductase (bc_1 complex)', *J. Biol. Chem.*, **266**, 19958–19964.

Brayer, J.-L., Hodgson, D. M., Richards, I. C., and Witherington, J. (1993). 'Novel method for preparing β-alkoxy acrylic acid', WO 94 24,085, Roussel Uclaf, priority date 15 April 1993.

Bushell, M. J., Beautement, K., Clough, J. M., de Fraine, P. J., Anthony, V. M., and Godfrey, C. R. A. (1984). 'Fungicidal phenylacrylic acid derivatives', Eur. Pat. Appl. EP 178,826, Imperial Chemical Industries, priority date 19 October 1984.

Cliff, G. R., Coltman, P. M., Godson, D. H., Holah, D. S., and Richards, I. C. (1990). 'Analogues of the strobilurins: methyl 2-(2-heteroarylthiomethyl)phenyl-3-methoxyacrylates as potential fungicides', Abstract No. 01A-66, Vol. 1, page 78, *7th Internat. Congress of Pestic. Chem.*, Hamburg, 5–10 August 1990.

Clough, J. M. (1990). 'Isomerization process', UK Pat. Appl. GB 2,248,614, Imperial Chemical Industries, priority date 9 October 1990.

Clough, J. M. (1993). 'The strobilurins, oudemansins and myxothiazols, fungicidal derivatives of β-methoxyacrylic acid', *Nat. Prod. Reports*, **10**, 565–574.

Clough, J. M., and Eshelby, J. J. (1990). 'Isomerization process', UK Pat. Appl. GB 2,248,613, Imperial Chemical Industries, priority date 9 October 1990.

Clough, J. M., and Godfrey, C. R. A. (1995). 'Growing hopes', *Chem. in Britain*, 466–469.

Clough, J. M., Godfrey, C. R. A., and Eshelby, J. J. (1990). 'Isomerization process', UK Pat. Appl. GB 2,248,615, Imperial Chemical Industries, priority date 9 October 1990.

Clough, J. M., de Fraine, P. J., Fraser, T. E. M., and Godfrey, C. R. A. (1992). 'Fungicidal β-methoxyacrylates. From natural products to novel synthetic agricultural fungicides', in *Synthesis and Chemistry of Agrochemicals III*, eds. Baker, D. R., Fenyes, J. G., and Steffens, J. J., ACS Symposium Series No. 504, American Chemical Society, Washington, DC, USA, Chapter 34, 372–383.

Clough, J. M., Evans, D. A., de Fraine, P. J., Fraser, T. E. M., Godfrey, C. R. A., and Youle, D. (1994). 'Role of natural products in pesticide discovery. The β-methoxyacrylates', in *Natural and Engineered Pest Management Agents*, eds. Hedin, P. A., Menn, J. J., and Hollingworth, R. M., ACS Symposium Series No. 551, American Chemical Society, Washington, DC, USA, Chapter 4, 37–53.

Clough, J. M., Anthony, V. M., de Fraine, P. J., Fraser, T. E. M., Godfrey, C. R. A., Godwin, J. R., and Youle, D. (1995). 'The synthesis of fungicidal β-methoxyacrylates', *ACS Conference Proceedings Series: Eighth International Congress of Pesticide Chemistry, Options 2000*, Washington, DC, USA, 4–9 July 1994, eds. Ragsdale, N. N., Kearney, P. C., and Plimmer, J. R., American Chemical Society, Washington, DC, USA, 59–73.

Clough, J. M., Godfrey, C. R. A., Godwin, J. R., Joseph, R. S. I., and Spinks, C. (1996). 'Azoxystrobin: a novel broad-spectrum systemic fungicide', *Pestic. Outlook, August Issue*, **7**(4), 16–20.

Clough, J. M., de Fraine, P. J., Godfrey, C. R. A., and Rees, S. B. (1997). 'Strobilurin analogues as inhibitors of mitochondrial respiration in fungi', in *Anti-infectives. Recent Advances in Chemistry and Structure-Activity Relationships*, eds. Bentley, P. H., and O'Hanlon, P. J., Royal Society of Chemistry, Chapter 13, 176–179.

Cohadon, P., Roques, J. F., Godwin, J. R., and Heaney, S. P. (1994), 'ICIA5504, a broad spectrum fungicide with a new mode of action for the control of cereal and vine diseases', *Proceedings, Quatrième Conf. Internationale sur les Maladies des Plantes*, Bordeaux, 6–8 December 1994, 931–938.

Copping, L. G. (1996). 'Introduction', in *Critical Rep. Appl. Chem.*, Copping, L. G., ed., **35** (Crop Protection Agents from Nature: Natural Products and Analogues, Royal Society of Chemistry), xvi–xxv.

Daum, L., Keilhauer, G., Lorenz, G., Ammermann, E., Anke, T., Weber, W., Steglich, W., Steffan, B., and Scherer, A. (1988). 'Strobilurinderivate, ihre Herstellung und Verwendung', Eur. Pat. Appl. EP 342,427, BASF, priority date 6 May 1988.

De Fraine, P. J., and Martin, A. (1988). 'Preparation of substituted methyl 2-[2-[[(methylimino)oxy]methyl]phenyl]-3-methoxy-2-propenoates as pesticides', Eur. Pat. Appl. EP 370, 629, Imperial Chemical Industries, priority date 21 November 1988.

De Fraine, P. J., and Clough, J. M. (1995). 'A new series of broad-spectrum β-methoxyacrylate fungicides with an oxime ether side-chain', *Pestic. Sci.*, **44**, 77–79.

Engler, M., Anke, T., Klostermeyer, D., and Steglich, W. (1995). 'Hydroxystrobilurin A, a new antifungal *E*-β-methoxyacrylate from a *Pterula* species', *J. Antibiot.*, **48**, 884–885.

Frank, J. A. and Sanders, P. L. (1994). 'ICIA5504: a novel broad-spectrum, systemic turfgrass fungicide', in *Brighton Crop Prot. Conf.: Pests and Diseases—Vol. 2*, British Crop Protection Council, Farnham, UK, 871–876.

Fredenhagen, A., Kuhn, A., Peter, H. H., Cuomo, V., and Giuliano, U. (1990a). 'Strobilurins F, G and H, three new antifungal metabolites from *Bolinea lutea*. **I**. Fermentation, isolation and biological activity', *J. Antibiot.*, **43**, 655–660.

Fredenhagen, A., Hug, P., and Peter, H. H. (1990b). 'Strobilurins F, G and H, three new antifungal metabolites from *Bolinea lutea*. **II**. Structure determination', *J. Antibiot.*, **43**, 661–667.

Gerth, K., Irschik, H., Reichenbach, H., and Trowitzsch, W. (1980). 'Myxothiazol, an antibiotic from *Myxococcus fulvus* (Myxobacterales). **I**. Cultivation, isolation, physico-chemical and biological properties', *J. Antibiot.*, **33**, 1474–1479.

Godfrey, C. R. A. (1995). 'Fungicides and bactericides', in *Agrochemicals from Natural Products*, Godfrey, C. R. A., ed., Marcel Dekker, New York, Chapter 7, 311–339.

Godwin, J. R., Anthony, V. M., Clough, J. M., and Godfrey, C. R. A. (1992). 'ICIA5504: a novel, broad spectrum, systemic β-methoxyacrylate fungicide', in *Brighton Crop Prot. Conf.: Pests and Diseases—Vol. 1*, British Crop Protection Council, Farnham, UK, 435–442.

Godwin, J. R., Young, J. E., and Hart, C. A. (1994). 'ICIA5504: effects on development of cereal pathogens', in *Brighton Crop Prot. Conf.: Pests and Diseases—Vol. 1*, British Crop Protection Council, Farnham, UK, 259–264.

Gold, R. E. and Leinhos, G. M. E. (1994). 'Histological studies on the fungicidal activity of the strobilurin BAS 490 F', in *Brighton Crop Prot. Conf.: Pests and Diseases—Vol. 1*, British Crop Protection Council, Farnham, UK, 253–358.

Gold, R. E. and Leinhos, G. M. (1995). 'Fungicidal effects of BAS 490 F on the development and fine structure of plant pathogenic fungi', *Pestic. Sci.*, **43**, 250–253.

Gold, R. E., Ammermann, E., Köhle, H., Leinhos, G. M. E., Lorenz, G., Speakman, J. B., Stark-Urnau, M., and Sauter, H. (1996). 'The synthetic strobilurin BAS 490 F: profile of a modern fungicide', in *Modern Fungicides and Antifungal Compounds*, eds. Lyr, H., Russell, P. E., and Sisler, H. D., Intercept, Andover, UK, Chapter 9, 79–92.

Hayase, Y., Kataoka, T., Takenaka, H., Ichinari, M., Masuko, M., Takahashi, T., and Tanimoto, N. (1989). 'Preparation of substituted phenyl(alkoxyimino)acetamides and their use as fungicides', Eur. Pat. Appl. EP 398,692, Shionogi, priority date 17 May 1989.

Hayase, Y., Kataoka, T., Masuko M., Niikawa, M., Ichinari, M., Takenaka, H., Takahashi, T., Hayashi, Y., and Takeda, R. (1995). 'Phenoxyphenyl alkoxyiminoacetamides. New broad-spectrum fungicides' in *Synthesis and Chemistry of Agrochemicals IV*, eds. Baker, D. R., Fenyes, J. G., and Basarab, G. S., ACS Symposium Series No. 584, American Chemical Society, Washington, DC, USA, Chapter 30, 343–353.

Heaney, S. P. and Knight, S. C. (1994). 'ICIA5504: a novel broad spectrum systemic fungicide for use on fruit, nut and horticultural crops', in *Brighton Crop Prot. Conf.: Pests and Diseases—Vol. 2*, British Crop Protection Council, Farnham, UK, 509–516.

Hodgson, D. M., Witherington, J., Moloney, B. A., Richards, I. C., and Brayer, J.-L. (1995). 'Pd/Cu co-catalysed cross-coupling reactions of methyl (*Z*)-2-tributylstannyl-3-methoxypropenoate: a method for direct introduction of the agrochemically important β-methoxyacrylate toxophore', *Synlett*, 32–34.

Höfle, G., Augustiniak, H., Behrbohm, H., Böhlendorf, B., Herrmann, M., Hölscher, A., Jahn, T., Jansen, R., Kiffe, M., Lautenbach, H., Schlummer, D., Söker, U., Stammermann, T., Steinmetz, H., Washausen, P., and Wray, V. (1994a). 'Isolation, structure elucidation and chemistry', in *Scientific Annual Report*, Gesellschaft für Biotechnologische Forschung, Braunschweig-Stöckheim, Germany, 125–129.

Höfle, G., Reichenbach, H., Böhlendorf, B., and Sasse, F. (1994b). 'Melithiazoles, process for preparing the same, medium with a melithiazole content and *Melittangium lichenicola* DSM 9004 capable of forming melithiazoles', WO 95 26,414, Gesellschaft für Biotechnologische Forschung, Germany, priority date 25 March 1994.

Isenring, H. P., Trah, S., and Weiss, B. (1988). 'Methyl esters of aldimino- or ketiminooxy-*ortho*-tolylacrylic acid, manufacturing process and fungicides containing them', WO 90 07,493, Hoffmann-La Roche, priority date 29 December 1988.

Kohl, W., Witte, B., Kunze, B., Wray, V., Schomburg, D., Reichenbach, H., and Höfle, G. (1985). 'Antibiotics from gliding bacteria, XXVII. Angiolam A—a novel antibiotic from *Angiococcus disciformis* (Myxobacterales)', *Liebigs Ann. Chem.*, 2088–2097.

Kraiczy, P., Haase, U., Gencic, S., Flindt, S., Anke, T., Brandt, U., and von Jagow, G. (1996). 'Thc molecular basis for the natural resistance of the cytochrome bc_1 complex from strobilurin-producing basidiomycetes to center Q_p inhibitors', *Eur. J. Biochem.*, **235**, 54–63.

Kume, R., Tashima, S., Matsumoto, K., Ando, I., and Shiraishi, T. (1996a). 'Effects of adsorbents on release of fungicide SSF-126 from granules', *J. Pestic. Sci.*, **21**, 404–411.

Kume, R., Tashima, S., Matsumoto, K., Ando, I., and Shiraishi, T. (1996b). 'Release of active ingredient from controlled release granules of SSF-126 and its rice blast control activity in pot trial', *J. Pestic. Sci.*, **21**, 438–440.

Mansfield, R. W. and Wiggins, T. E. (1990). 'Photoaffinity labelling of the β-methoxyacrylate binding site in bovine heart mitochondrial cytochrome bc_1 complex', *Biochim. et Biophys. Acta*, **1015**, 109–115.

Masuko, M., Niikawa, M., Kataoka, T., Ichinari, M., Takenaka, H., Hayase, Y., Hayashi, Y., and Takeda, R. (1993). 'Novel antifungal alkoxyiminoacetamide derivatives', Abstract No. 3.7.16, page 91, *Abstracts of the 6th Internat. Congress of Plant Pathology*, 28 July–6 August 1993, Montreal, Canada.

Mizutani, A., Yukioda, H., Tamura, H., Miki, N., Masuko, M., and Takeda, R. (1995). 'Respiratory characteristics in *Pyricularia oryzae* exposed to a novel alkoxyiminoacetamide fungicide', *Phytopathol.*, **85**, 306–311.

Mizutani, A., Miki, N., Yukioka, H., and Masuko, M. (1996). 'Mechanism of action of a novel alkoxyiminoacetamide fungicide SSF-126', in *Modern Fungicides and Antifungal Compounds*, eds. Lyr, H., Russell, P. E., and Sisler, H. D., Intercept, Andover, UK, Chapter 10, 93–99.

Musílek, V. (1965). 'A method of preparing a new antifungal antibiotic material', Czech, Pat. Appl. CS 136,492, priority date 17 December 1965 (= UK Pat. Appl. GB 1,163,910) (*Chem. Abstr.*, 1969, **70**, 18,900y and 1971, **74**, 123,689s).

Musílek, V., Černá, J., Šašek, V., Semerdžieva, M., and Vondráček, M. (1969). 'Antifungal antibiotic of the basidiomycete *Oudemansiella mucida*. I. Isolation and cultivation of a producing strain', *Folia Microbiol. (Prague)*, **14**, 377–387.

Pilling, E. D., Earl, M., and Joseph, R. S. I. (1996). 'Azoxystrobin: fate and effects in the terrestrial environment', in *Brighton Crop Prot. Conf.: Pests and Diseases—Vol. 1*, British Crop Protection Council, Farnham, UK, 315–322.

Reichenbach, H., Gerth, K., Irschik, H., Kunze, B., and Höfle, G. (1988). 'Myxobacteria: a source of new antibiotics', *Trends Biotechnol.*, **6**, 115–121.

Richards, I. C., Milling, R. J., and Pittis, J. E. (1994). 'Achieving biological control of phytopathogenic fungi using β-methoxyacrylates', *Biochem. Soc. Trans.*, **22**, 66S.

Roehl, F. (1994). 'Binding of BAS 490 F to bc_1-complex from yeast', *Biochem. Soc. Trans.*, **22**, 64S.

Röhl, F. and Sauter, H. (1994). 'Species dependence of mitochondrial respiration inhibition by strobilurin analogues', *Biochem. Soc. Trans.*, **22**, 63S.

Rossi, R. and Bellina, F. (1996). 'Selective palladium-mediated carbon-oxygen bond and carbon-sulfur bond forming reactions and their application to the synthesis of structural analogues of strobilurins and some naturally-occurring sulfur-containing carboxyamides', *Korean J. Med. Chem.*, **6**, 317–324.

Rossi, R., Bellina, F., and Carpita, A. (1996). 'New efficient procedures for direct introduction of the agrochemically important β-methoxypropenoate unit into substituted aromatic derivatives', *Synlett*, 356–358.

Sauter, H., Ammermann, E., Benoit, R., Brand, S., Gold, R. E., Grammenos, W., Köhle, H., Lorenz, G., Müller, B., Röhl, F., Schirmer, U., Speakman, J. B., Wenderoth, B., and Wingert, H. (1995). 'Mitochondrial respiration as a target for antifungals: lessons from research on strobilurins', in *Antifungal Agents. Discovery and Mode of Action*, eds. Dixon, G. K., Copping, L. G., and Hollomon, D. W., BIOS Scientific Publishers, Oxford, UK, 173–191.

Sauter, H., Ammermann, E., and Roehl, F. (1996). 'Strobilurins—from natural products to a new class of fungicides', in *Critical Rep. Appl. Chem.*, ed. Copping, L. G., **35** (Crop Protection Agents from Nature: Natural Products and Analogues) 50–81.

Schramm, G., Steglich, W., Anke, T., and Oberwinkler, F. (1978). 'Antibiotics from basidiomycetes, III. Strobilurin A and B, antifungal metabolites from *Strobilurus tenacellus*', *Chem. Ber.*, **111**, 2779–2784.

Schwalge, B. (1986). 'Strobilurin A als Modellverbindung für synthetische Analoga', Dissertation, University of Bonn, Germany.

Schuetz, F., Brand, S., Wild, J., Kuekenhoehner, T., Hofmeister, P., and Kuenast, C. (1989). 'Verfahren zur Schädlingsbekämpfung mit Hilfe von Pyrimidinen', Eur. Pat. Appl. EP 407 873, BASF, priority date 13 July 1989.

Sedmera, P., Musílek, V., Nerud, F., and Vondráček, M. (1981). 'Mucidin: its identity with strobilurin A', *J. Antibiot.*, **34**, 1069.

Shirane, N., Masuko, M., and Takeda, R. (1995). 'Effects of SSF-126, a novel alkoxyiminoacetamide blasticide, on mycelial growth and oxygen consumption of *Pyricularia oryzae*', *Plant Pathol.*, **44**, 636–640.

Tomlin, C. D. S. (ed.) (1994). *The Pesticide Manual*, 10th Edition, British Crop Protection Council, Farnham, UK, and the Royal Society of Chemistry, Cambridge, UK.

Trowitzsch, W., Reifenstahl, G., Wray, V., and Gerth, K. (1980). 'Myxothiazol, an antibiotic from *Myxococcus fulvus* (Myxobacterales). II. Structure elucidation', *J. Antibiot.*, **33**, 1480–1490.

Trowitzsch, W., Höfle, G., and Sheldrick, W. S. (1981). 'The stereochemistry of myxothiazol', *Tet. Letts.*, **22**, 3829–3832.

Trumpower, B. L., and Gennis, R. B. (1994). 'Energy transduction by cytochrome complexes in mitochondrial and bacterial respiration: the enzymology of coupling electron transfer reactions to transmembrane proton translocation', *Annu. Rev. Biochem.*, **63**, 675–716.

Tsubata, K., Niino, N., Endo, K., Yamamoto, Y., and Kanno, H. (1989). 'Preparation of *N*-substituted benzyloxyimine derivatives as agrochemical fungicides', Eur. Pat. Appl. EP 414,153, Nihon Nohyaku, priority date 22 August 1989.

Vondráček, M., Čapková, J., Šlechta, J., Benda, A., Musílek, V., and Cudlín, J. (1967). 'Isolation of a new antifungal antibiotic', Czech. Pat. Appl. CS 136,495, priority date 17 January 1967 (*Chem. Abstr.*, 1971, **75**, 4,029n).

Vondráček, M., Čapková, J., and Čulík, K. (1974). 'Antifungal antibiotic methyl 6-phenyl-3-methyl-2-methoxymethylene-3,5-hexadienoate', Czech. Pat. Appl. CS 180, 775, priority date 29 August 1974 (*Chem. Abstr.*, 1980, **93**, 204,286c).

Vondráček, M., Vondráčkova, J., Sedmera, P., and Musílek, V. (1983). 'Another antibiotic from the basidiomycete *Oudemansiella mucida*', *Collect. Czech. Chem. Commun.*, **48**, 1508–1512.

von Jagow, G. and Link, Th. A. (1986). 'Use of specific inhibitors on the mitochondrial bc_1 complex', *Methods Enzymol.*, **126**, 253–271.

von Jagow, G., Gribble, G. W., and Trumpower, B. L. (1986). 'Mucidin and strobilurin A are identical and inhibit electron transfer in the cytochrome bc_1 complex of the mitochondrial respiratory chain at the same site as myxothiazol', *Biochem.*, **25**, 775–780.

Watanabe, M., Tanaka, T., Kobayashi, H., and Yokoyama, S. (1989). 'Oxime ether derivative, preparation thereof and germicide containing the same', Eur. Pat. Appl. EP 426,460, Ube, priority date 2 November 1989.

Watanabe, M., Tanaka, T., Yokoyama, S., and Kobayashi, H. (1994). 'Methoxyacrylates having oximino side chains', *Biochem. Soc. Trans.*, **22**, 67S.

Weber, W., Anke, T., Bross, M., and Steglich, W. (1990a). 'Strobilurin D and strobilurin F: two new cytostatic and antifungal (*E*)-β-methoxyacrylate antibiotics from *Cyphellopsis anomala*', *Planta Med.*, **56**, 446–450.

Weber, W., Anke, T., Steffan, B., and Steglich, W. (1990b). 'Antibiotics from basidiomycetes. XXXII. Strobilurin E: a new cytostatic and antifungal (*E*)-β-methoxyacrylate antibiotic from *Crepidotus fulvotomentosus* Peck', *J. Antibiot.*, **43**, 207–212.

Wenderoth, B., Rentzea, C., Ammermann, E., Pommer, E.-H., Steglich, W., and Anke, T. (1986). 'Oximether und enthaltende Fungizide', Eur. Pat. Appl. EP 253,213, BASF, priority date 16 July 1986.

Wiggins, T. E., and Jager, B. J. (1994). 'Mode of action of the new methoxyacrylate antifungal agent ICIA5504', *Biochem. Soc. Trans.*, **22**, 68S.

Williams, D. H., Stone, M. J., Hauck, P. R., and Rahman, S. K. (1989). 'Why are secondary metabolites (natural products) biosynthesised?', *J. Nat. Prod.*, **52**, 1189–1208.

Wood, K. A., Kau, D. A., Wrigley, S. K., Beneyto, R., Renno, D. V., Ainsworth, A. M., Penn, J., Hill, D., Killacky, J., and Depledge, P. (1996). 'Novel β-methoxyacrylates of the 9-methoxystrobilurin and oudemansin classes produced by the basidiomycete *Favolaschia pustulosa*', *J. Nat. Prod.*, **59**, 646–649.

Wood Mackenzie Consultants Limited, Edinburgh and London (1997a). 'Agrochemical products. Part 1: the key agrochemical product groups', in *Agrochemical Service, Update of the Products Section*, May 1997, 1–74.

Wood Mackenzie Consultants Limited, Edinburgh and London (1997b). 'UK approval for Zeneca's Amistar', in *Agrochemical Monitor, News Update*, 13 May 1997, No. 139, 10–14.

Yu, C.-A., Xia, J.-Z., Kachurin, A. M., Yu, L., Xia, D., Kim, H., and Deisenhofer, J. (1996). 'Crystallization and preliminary structure of beef heart mitochondrial cytochrome-bc$_1$ complex', *Biochim. et Biophys. Acta*, **1275**, 47–53.

Zapf, S., Werle, A., Anke, T., Klostermeyer, D., Steffan, B., and Steglich, W. (1995). '9-Methoxystrobilurins—a link between strobilurins and oudemansins', *Angew. Chem. Int. Edn. Engl.*, **34**, 196–198.

Ziegler, H., Neff, D., and Stutz, W. (1994). 'Process for the preparation of arylacetic ester derivatives *via* palladium-catalyzed cross coupling reaction', WO 95 20,569, Ciba-Geigy, priority date 27 January 1994.

6 Biological Control of Fungal Diseases

R. P. LARKIN, D. P. ROBERTS and J. A. GRACIA-GARZA
USDA, Beltsville, Maryland, USA

INTRODUCTION

Biological control of plant diseases can be generally defined as the reduction of inoculum or disease-causing activity of a pathogen accomplished by or through one or more organisms other than man (Baker, 1987; Cook and Baker, 1983).

Fungicidal Activity. Edited by D. H. Hutson and J. Miyamoto

The incorporation of appropriate biological controls for the management of plant pathogens may allow substantial reductions in the use of chemical pesticides, maintain greater biological balance and diversity, lead to more sustainable long-term production practices, and in some cases, achieve better disease control than with current conventional control methods. This potential of biological control has generated much interest and research activity over the past three decades (Cook, 1991). However, the transition from control observed in the laboratory and greenhouse to the practical and effective implementation of biocontrol in commercial production agriculture has been difficult.

Although there are several success stories, many attempts at biological control have resulted in inconsistent or unsatisfactory disease control under varying environmental conditions and locations. Because of these difficulties, a substantial role for biocontrol in commercial disease management practices has often been questioned (Campbell, 1994; Cook, 1991; Deacon and Berry, 1993). However, much of the inconsistency in biocontrol work may be traced to a general lack of understanding of how these biological control systems work, and under what conditions they can or cannot be expected to provide disease control. This lack of knowledge has resulted in the introduction of biological control agents into environments for which they are often ecologically unsuited (Deacon, 1991). Due to the complexity of biological control mechanisms and their microbial interactions, considerable time and research efforts are still needed to understand basic biocontrol processes. Recent applications of molecular, biochemical and ecological approaches are making substantial progress towards understanding key factors in several biocontrol systems and may lead to the development of better, more consistent biological control. Although biocontrol has not yet made much impact on standard agronomic practices for most crops, as agriculture continues to move toward greater sustainability, reduced pesticide use, and long-term productivity, biocontrol must be an integral part of the management of our agricultural ecosystems and soil resources.

A reduction in the use of chemical pesticides is one of the primary motivations for developing biological controls. Concerns about the health, safety and environmental effects of agricultural chemicals and pesticides in our water, soil and food require that these inputs be minimized. In addition, biological control may be especially important for use in pathosystems in which chemical controls are not economical or effective, and for which there are currently no effective controls. Biological control may also lessen other problems associated with certain chemical controls, such as the development of pathogen-resistance to chemicals, reductions in beneficial organism populations and the creation of biological vacuums (Cook and Baker, 1983). Biological control generally has more specific effects, with only the target pathogen organism(s) being adversely affected, leaving other beneficial organisms and a diverse soil microbial community intact to provide for healthier plants and roots. Thus, biological control can be safer for humans, the crop and the environment. Biological control has the potential to be more stable and longer lasting than some other controls and is compatible with the concepts and goals of integrated pest management and sustainable agriculture.

Sustainable agriculture strives to maintain the long-term economic viability and productivity of agriculture, while enhancing environmental quality and the natural resource base. Emphasis is placed on minimizing chemical and energy inputs as much as possible through more effective and efficient use of non-renewable and onsite resources, and the integration of biological cycles, processes and controls. Understanding and implementing biological cycling and control systems is one of the cornerstones for developing more sustainable agricultural systems. Although biological controls are not intended to completely replace chemical controls, and chemical controls will always play an important role in pest and disease control, biological control can be incorporated with other control strategies as part of an integrated pest management system. Overall, biological control has much potential, but more research is needed so that it can be implemented to a wider degree for effective control of more pathogens and diseases.

The foundation for the development of contemporary biological control was laid by the two landmark books by Cook and Baker (1983); Baker and Cook (1974). In addition, there have been numerous books and reviews on more recent developments in biological control concepts and implementation (Andrews, 1992; Baker, 1987; Baker and Dunn, 1990; Campbell, 1989, 1994; Chet, 1987, 1994; Cook, 1993; Fravel, 1988; Fravel and Engelkes, 1994; Handelsman and Stabb, 1996; Hornby, 1990; Lumsden *et al.*, 1993; Tjamos *et al.*, 1992; Weller, 1988; Wilson and Wisniewski, 1994). The objectives of this chapter are to highlight the current status of biological control work, including its problems and limitations, as well as to outline potential solutions and future directions of research to optimize biocontrol capabilities for the control of fungal pathogens.

BIOLOGICAL CONTROL ORGANISMS AND APPROACHES TO BIOLOGICAL CONTROL

A wide variety of soil microorganisms, including bacteria, actinomycetes, fungi, viruses and protozoa, have been shown to have biocontrol activity against various fungal pathogens or the diseases they cause, and have been studied as biological control agents in several pathosystems. Some of the best-known examples are included in Table 1. Some of these biocontrol agents are available commercially, while others have only been tested experimentally. Within the last few years, substantial numbers of biocontrol agents have been released commercially and are becoming available for use. As of 1996, more than 40

Table 1 Organisms with demonstrated success as potential biological control agents of fungal diseases

Organism	Target pathogen/disease	References
Bacteria		
Bacillus subtilis	*Rhizoctonia, Fusarium, Alternaria*	Krebs *et al.* (1993)
B. cereus	*Phytophthora*	Osburn *et al.* (1995)
Burkholdeira cepacia	*Rhizoctonia, Pythium, Fusarium,*	Hebbar *et al.* (1992), McLoughlin *et al.* (1992)
	Aphanomyces / damping-off, root rots	King and Parke (1993)
Enterobacter cloacae	*Pythium* / damping-off, seedling rot	Nelson (1988)
Erwinia herbicola	*Sclerotinia, Pythium* / white mold, rot	Yuen *et al.* (1994), Nelson (1988)
Pseudomonas fluorescens	*Pythium, Rhizoctonia, Gaeumannomyces*	Weller and Cook (1983)
P. putida	/damping-off, take-all	Bakker *et al.* (1986)
P. syringae	*Penicillium, Botrytis* / post-harvest decay	Bull *et al.* (1997), Janiesiwicz and Marchi (1992)
Serratia marcescens	*Magnaporthe*/summer patch disease	Kobayashi *et al.* (1995)
Streptomyces, spp.	*Streptomyces, Fusarium, Alternaria, Botrytis, Pythium*/potato scab, seed rot, wilt	Liu *et al.* (1995) Yuan and Crawford (1995)
Xanthomonas maltophilia	*Botrytis, Magnaporthe*	Kobayashi *et al.* (1995)
Fungi		
Ampelomyces quisqualis	Powdery mildew	Falk *et al.* (1995)
Candida oleophila	*Botrytis, Penicillium/post-harvest decay*	McGuire (1994)
Chaetomium globosum	*Fusarium, Botrytis*	Hubbard *et al.* (1982)
Cladorrhinum foecundissimum	*Rhizoctonia*/damping-off	Lewis *et al.* (1995)
Coniothyrium minitans	*Botrytis, Sclerotinia*	McLaren *et al.* (1994), Whipps and Gerlagh (1992)
Fusarium oxysporum	Fusarium/Fusarium wilt	Alabouvette *et al.* (1993)
F. solani		Larkin *et al.*(1996)
Gliocladium roseum	*Rhizoctonia, Pythium*/ damping-off, rots	Sutton *et al.* (1997)
G. virens		Lumsden and Locke (1989) Lewis *et al.* (1993)
Glomus spp.	*Fusarium*	Datnoff *et al.* (1995)
Laetisaria arvalis	*Rhizoctonia*, Pythium	Lewis and Papavizas (1992)
Myrothecium verrucaria	*Rhizoctonia, Botrytis*	Peng and Sutton (1991)
Paecilomyces lilacinus	*Rhizoctonia*	Cartwright and Benson (1995)
Penicillium spp.	*Rhizoctonia*	Kaiser and Hannon (1984)
Phlebia giganteum	*Heterobasidion* / wood decay	Rishbeth (1975)

(*continued*)

Table 1 (*continued*)

Organism	Target pathogen/disease	References
Pythium oligandrum	*Pythium*	Martin and Hancock (1987)
P. nunn	*Pythium*	Paulitz and Baker (1987), Fang and Tsao (1995)
Rhizoctonia (binucleate)	*Rhizoctonia*	Villajuan-Abgona *et al.* (1996), Herr(1995), Harris *et al.* (1993, 1994)
Stilbella aciculosa	*Rhizoctonia*	Lewis and Papavizas (1993)
Sporodesmium sclerotivorum	*Sclerotinia*/lettuce drop	Adams and Ayers (1982), Adams and Fravel (1990)
Talaromyces flavus	*Verticillium, Sclerotinia*/wilt	Marois *et al.* (1982), McLaren *et al.* (1994)
Trichoderma harzianum *T. hamatum* *T. viride*	*Botrytis, Fusarium, Rhizoctonia, Pythium, Sclerotium*/seedling diseases, wilt, rots	Harman (1991), Taylor *et al.* (1994) Lewis and Papavizas (1991)
Verticillium biguttatum	*Verticillium*, Rhizoctonia	Jeger and Velvis (1986), Boogert and Velvis (1992)
Others		
Viruses (dsRNA)	*Cryphonectria, Monosporascus, Rhizoctonia*	Heininger and Rigling (1994), Park *et al.* (1996)
Protozoa: Amoeba	*Pythium*/damping-off	Homma *et al.* (1979)
Fungus gnats, Collembola	*Sclerotinia, Rhizoctonia*	Gracia-Garza *et al.* (1997), Lartey *et al.* (1994)
Earthworms	*Rhizoctonia*	Stephens *et al.* (1994)

commercial biocontrol products were available world wide (Fravel and Larkin, 1996; Fravel *et al.*, 1997; Lumsden *et al.*, 1995). Of these, 12 are currently registered by the EPA for use in the USA to control fungal diseases. Although there are still relatively few biocontrol products currently available or used on a widespread basis, more and better products and formulations are on the way. There are undoubtedly many more types of organisms that may have potential for biological control, but have not yet been isolated, screened or tested. There still remains an enormous, essentially untapped pool of potential biocontrol organisms in natural ecosystems (Handelsman and Stabb, 1996). To date, primarily only organisms that are easily isolated, readily identifiable, and abundantly produced in culture have been tested extensively.

Finding promising antagonistic organisms with potential to control pathogens or disease is the first step toward the development of effective biological control. Efficacy of the resultant biocontrol is dependent on the efficacy of the antagonist organisms involved. However, identification of a good antagonist strain does not guarantee successful biological control. There are many other factors that determine whether an organism will be successful. Also, the search for new antagonists is a continual process. We should not be satisfied when one or a few potentially good antagonists are isolated and tested. As advocated by Cook (1993), there should be constant scrutiny to find and develop better antagonists that will be effective under a variety of different conditions, or that are locally adapted to the specific requirements of a particular environment, site, soil type, or crop plant. In general, however, it has not been the lack of effective antagonists that has limited biocontrol efforts in the past, rather, to a large degree, it has been our inability to properly implement biocontrol to achieve adequate disease control in the field.

One of the problems of many attempts at biological control in the past has been the overall approach often used in the isolation, development and testing of biocontrol strains. For the most part, biocontrol has been approached according to a ‘chemical paradigm’ rather than the appropriate ‘biological paradigm’ (Cook, 1993). Perhaps this is due to the dominance of chemical control as a disease control strategy, or to the fact that chemical control is considered the standard that biocontrol must be measured against. Whatever the reason, biocontrol has typically been expected to function in the same way as an applied chemical pesticide. Most often, a researcher would isolate some soil organism which inhibited the pathogen in culture, then test it by adding large quantities of the antagonist to soil or soilless mixtures in the greenhouse. If successful, the potential biocontrol organism was then tested in numerous fields at different locations. The biocontrol was considered inconsistent or unreliable if this inundative application of one organism did not control the disease in every field. In many cases, the cause of failure was not even determined.

This type of approach, in which the biocontrol strain is broadly applied and expected to be effective against one or several diseases on multiple crops in several different environments, is generally inappropriate for biological control. This type of approach ignores the unique nature of biocontrol and does not give it a chance to work, particularly in the soil ecosystem, which is extremely complex. Biological control involves the use of living organisms, and is not remotely similar to the action of chemical pesticides. It represents an entirely different class of control and requires an entirely different approach and implementation than chemical pesticides. Biocontrol agents must be ecologically fit to survive, become established and function within the particular conditions of the ecosystem. In general, organisms introduced into an established ecosystem do not persist (Cook, 1993). Any individual biocontrol organism can only be expected to perform within a limited set of physical, biological and environmental conditions. Yet, in many cases, these conditions are not adequately defined. Consider, for example, the development and use of resistant crop cultivars, which is treated entirely within a biological paradigm.

One cultivar is not expected to function equally well in all environments; numerous cultivars that are adapted to function best in a variety of different environments are generally used. Likewise, it may be necessary to develop locally adapted organisms as biocontrol agents for a variety of different environmental conditions, regions, or crop plants.

In order to understand the processes occurring in the soil and rhizosphere, as well as how to implement and manage biological control interactions, adopting a more ecological approach to biological control is essential. An ecological approach focuses on studying the interactions among the pathogen, host, antagonists, and the surrounding microbial communities in the soil, rhizosphere, roots, and other plant parts. Through an ecological approach, the specific interactions and mechanisms responsible for biological control, as well as the conditions and requirements necessary for biological control processes to function can be determined. Through an understanding of these interactions, we can establish the limitations as well as the full potential for biocontrol in a given system, and develop strategies for optimizing biocontrol capabilities to provide the most effective control.

This need for specialized knowledge regarding the ecological interactions of the biocontrol organism before effective implementation of biocontrol can be realized is the primary disadvantage of biological control compared to other control methods. Because of this, effective biocontrol agents are currently available for only a few pathogens and disease systems. In addition, biological control generally needs to be handled, managed and maintained differently than conventional systems, and may require some changes in farming or production practices. The specificity of some biological control can also be a problem. It is not as broad-spectrum in its applications as other approaches, resulting in more limited and potentially specialized uses. In addition, there are still questions as to how effective and reliable biological control will be. In general, biological control does not completely eliminate the pathogen, may not work as fast as chemical methods, and may provide only a partial level of control. For these reasons, biological control will be best realized when incorporated with other control strategies as part of an integrated pest management system.

Fortunately, biocontrol research has been moving towards a more ecological approach during the past several years, with much more emphasis being placed on microbial communities and interactions among organisms. This approach combined with the availability of innovative molecular and biochemical techniques for studying organisms, populations and community interactions, has enabled great strides to be made in the understanding of important components of rhizosphere ecology, microbial interactions, and specific mechanisms of action. However, although much has been accomplished in recent years, there are still crucial gaps in our understanding of biocontrol processes in the soil and rhizosphere which limit our abilities to develop and implement biological control to its full potential.

CRITICAL RESEARCH AREAS FOR EFFECTIVE BIOLOGICAL CONTROL

The following sections outline five major areas of research that are critical for the further development and implementation of effective biological control. These areas are:

1. Identification of traits, conditions and requirements necessary for optimal performance of specific biocontrol mechanisms and interactions;
2. Use of multiple antagonists and multiple mechanisms of action;
3. Influence of the host plant on microbial communities;
4. Integration of biological control with other disease control strategies; and
5. Improved formulations and delivery systems.

IDENTIFICATION OF TRAITS, CONDITIONS AND REQUIREMENTS NECESSARY FOR OPTIMAL PERFORMANCE OF SPECIFIC BIOCONTROL MECHANISMS AND INTERACTIONS

Determination of the specific traits, conditions and requirements of biocontrol reveal how, where, when and why biocontrol works. With this information, conditions that will enable and enhance biological control can be established, as well as indications of any possible limitations to its effectiveness. Understanding the mechanism(s) of action involved in biocontrol processes is of primary importance in establishing these characteristics, and will provide much insight into where and when the interaction occurs, and how the pathogen will be affected. The mechanism of action will also, to a large degree, determine how the antagonist will need to be implemented and managed, as well as suggest the strategies that will improve control.

Almost all biocontrol interactions can be placed into one or more of four general types of mechanisms. These categories are competition, antibiosis, predation/parasitism, and induced resistance. These mechanisms can be used in various ways to implement the three basic control strategies used in disease management: the reduction of pathogen inoculum; protection of the infection court (or prevention of infection); and, limiting disease development after pathogen infection.

Mechanisms of Action

Competition

Competition has been defined as the active demand in excess of the immediate supply of material on the part of two or more organisms (Clarke, 1965). The result is a restriction on population size or microbial activity of one or more of the competitors (Paulitz and Baker, 1987). Competition between microorganisms generally refers to competition for nutrients, such as available carbon, nitrogen, iron, or trace elements, or competition for space, such as for colonization or infection sites on the root or seed surface. Nutrients from roots and seeds support microbial growth and other activities in the spermosphere and rhizosphere (Curl and Truelove, 1986; Paulitz, 1990). Nutrients from roots and seeds are derived from several sources including exudates, secretions, lysates and mucilages (Bowen and Rovira, 1976). These nutrients are chemically diverse and include carbohydrates, amino acids, peptides, organic acids, and other plant metabolites (Curl and Truelove, 1986).

Competition can be an effective biocontrol mechanism when the antagonist organism is present in sufficient quantities at the correct time and location, and can utilize limited nutrients or other resources more efficiently than the pathogen. An example of this interaction was observed in Fusarium wilt-suppressive soils in France (Alabouvette and Couteaudier, 1992; Alabouvette *et al.*, 1993; Couteaudier, 1992; Lemanceau, 1989). High populations of saprophytic strains of *Fusarium oxysporum* effectively competed with the pathogenic strains of *F. oxysporum* for reduced carbon sources in these soils, resulting in an inhibition of pathogen propagule germination, reduced saprophytic growth of the pathogen, and low levels of disease. Another example of competition as a biocontrol mechanism is competition among bacterial biocontrol agents and bacterial plant pathogens on leaf surfaces. Competition for nutrients was demonstrated between non-ice-nucleating and ice-nucleating strains of *Pseudomonas syringae*, resulting in antagonism indicated by reduced colonization by ice-nucleating strains of *P. syringae* (Lindow, 1987; Wilson and Lindow, 1994a). Wilson and Lindow (1994b) determined that coexistence of bacteria of leaf surfaces was inversely correlated with the similarity in reduced carbon utilization between the interacting strains. Strains with high niche overlap were antagonistic to each other through competition for limited nutritional resources. Conversely, coexistence of bacterial species on leaf surfaces was mediated through nutritional niche differentiation, which is the utilization of different nutrients by coexisting strains (Wilson and Lindow, 1994b).

Competition for iron between biocontrol bacteria and plant pathogens has been well documented (Leong, 1986). Iron is required for growth by microbes, but is typically limited in availability in soil. Strains of certain biocontrol bacteria, such as *Pseudomonas fluorescens*, produce siderophores, which have a high affinity for soluble ferric iron (Fe^{+3}). Siderophores bind iron, allowing these biocontrol bacteria to effectively compete with pathogens for iron. Production of siderophores and the limited availability of iron in soil have been associated with disease suppression by several bacterial antagonists (Bakker *et al.*, 1986; Buysens *et al.*, 1996; Kloepper *et al.*, 1980; Loper, 1988; Scher and Baker, 1982).

Antibiosis

Antibiosis refers to the inhibition or destruction of the pathogen by a metabolic product of the antagonist, such as the production of specific toxins, antibiotics, or enzymes. This interaction can result in suppression of activity of the pathogen or destruction of pathogen propagules. To be effective, antibiotics must be produced *in situ* in sufficient quantities at the precise time and place where they will interact with the pathogen. The production of antibiotics by various strains of *P. fluorescens*, which include phenazine compounds and 2,4-diacetylphloroglucinol (Figure 1), has been shown to be important in biocontrol through mutational analysis. Mutant strains of *P. fluorescens* that do not produce these antibiotics have reduced efficacy as biocontrol agents (Keel *et al.*, 1992; Thomashow and Weller, 1988).

Antibiosis has also been shown to be important in biocontrol interactions between biocontrol fungi and fungal plant pathogens. *Gliocladium virens*, an important biocontrol fungus, has activity against several soil-borne plant pathogens including *Pythium ultimum* and *Rhizoctonia solani* (Howell, 1982; Lumsden and Locke, 1989; Papavizas and Lewis, 1989). Different strains of *G. virens* produce a variety of metabolites, including gliotoxin and gliovirin, which are toxic or inhibitory to several fungal pathogens. Production of gliotoxin has been shown to be associated with biocontrol efficacy of strain Gl-21, the active component of the commercial product Soil Gard^{TM}. Mutants of Gl-21 that do not produce gliotoxin are reduced in their ability to control damping-off caused by *P. ultimum* (Wilhite *et al.*, 1994). In addition, gliotoxin has been detected in a number of soils colonized by *G. virens* and quantities of gliotoxin in these soils have been correlated with disease suppression (Lumsden *et al.*, 1992).

Another example of antibiosis is demonstrated by *Talaromyces flavus*, a biocontrol fungus used to control Verticillium wilt of eggplant and potatoes, caused by *Verticillium dahliae*. Antibiosis of *V. dahliae* is due to the evolution of hydrogen peroxide in the soil (Fravel and Roberts, 1991; Kim *et al.*, 1988, 1990a, 1990b). Microsclerotia of *V. dahliae* are 100 times more sensitive to hydrogen peroxide than propagules of many other fungi (Kim *et al.*, 1990b). Hydrogen peroxide is one of the products of a reaction catalyzed by glucose oxidase, an extracellular enzyme produced by *T. flavus*, in the presence of glucose. Glucose oxidase is the only protein in the culture filtrates of *T. flavus* responsible for inhibition of germination of microsclerotia of *V. dahliae* (Stosz *et al.*, 1996). Rhizosphere studies demonstrated that hydrogen peroxide produced in a reaction catalyzed by glucose oxidase was sufficient to kill microsclerotia of *V. dahliae in situ*. In addition, a variant strain of *T. flavus* with reduced glucose oxidase production also showed reduced suppression of Verticillium wilt of eggplant in natural soil (Fravel and Roberts, 1991).

It is important to note that even in biocontrol interactions in which antibiosis is the mechanism of action, biocontrol is substantially different from the broadcast (or even seed) application of chemical fungicides. Use of a microorganism to deliver an antibiotic toxic to the pathogen involves a precise and limited distribution of the antibiotic in the environment. For example, microorganisms colonizing the root produce antibiotics in extremely small quantities and only in the rhizosphere. This minimizes exposure of other soil organisms to the antibiotic. In addition, antibiotic-producing biocontrol agents deliver the compound to the precise location that needs to be protected from the pathogen, in this case, the root.

Minimizing exposure of soil organisms to antibiotics is particularly important in reducing the potential for the development of resistance by the pathogen. The risk of development of resistance is substantially reduced in the case of biological control compared to chemical controls because total exposure of pathogen populations (and other microorganisms) to the antibiotic is low, and selection pressure is minimized. In addition, since quantities of the antibiotic are very low, the antibiotic generally will not cause problems with runoff or contamination of soil and ground water. Thus, the risk associated with a biocontrol antibiotic is much diminished compared with a chemical fungicide (Lumsden and Walter, 1995). Thomashow *et al.*, (1990), working with *P. fluorescens* strain 2–79, which is suppressive to *Gaeumannomyces graminis* var. *tritici* (the pathogen responsible for take-all of wheat) primarily due to the production of the antibiotic phenazine-1-carboxylate, estimated that this strain produces about 55–80 mg/ha of the antibiotic when established as an introduced microorganism in the rhizosphere of wheat. This is a very small amount of antibiotic and represents several orders of magnitude less than what would be applied with a typical chemical seed treatment (such as triadimefon) used to control take-all. In addition, biocontrol agents often produce more than one antibiotic that has activity against fungal pathogens, which further reduces the chances of resistance developing in pathogens. None the less, more research is needed to determine the potential for the development of resistance to biocontrol antibiotics and how best to avoid this resistance from developing (Handelsman and Stabb, 1996).

Parasitism/Predation

Parasitism and/or predation occurs when the antagonist feeds on or within the pathogen, resulting in the direct destruction or lysis of propagules and structures. When one fungus parasitizes another fungus, it is called mycoparasitism (Lumsden, 1992). This mechanism employs the control strategy of reducing pathogen inoculum. An example of mycoparasitism is demonstrated by the fungal antagonist *Sporodesmium sclerotivorum*, which is an effective mycoparasite of *Sclerotinia minor*, cause of lettuce drop and other diseases (Adams and Ayers, 1982; Adams and Fravel, 1990). This fungus is an obligate parasite, reproduces on the host sclerotia, and systematically destroys the colonized sclerotia over time (Adams and Ayers, 1983).

Trichoderma spp. are well-known mycoparasites of a wide variety of fungi (Papavizas, 1985). Trichoderma coil around and parasitize fungal mycelia. This complex process involves the activities of extracellular enzymes produced by *Trichoderma* spp. Biocontrol isolates of *T. harzianum* produce a number of different chitinases and glucanases in culture that degrade major components of cell walls of plant pathogenic fungi (Haran *et al.*, 1995; Lorito *et al.*, 1994). Glucanase and chitinase activities have been detected in sterile soil containing mycelium of *Sclerotium rolfsii*, *Rhizoctonia solani*, or *Pythium aphanidermatum* and inoculated with *T. harzianum* (Elad *et al.*, 1982). These enzymes have also been shown to inhibit germination of conidia and germ tube elongation of several plant pathogenic fungi *in vitro* (Lorito *et al.*, 1993, 1994). However, the role mycoparasitism plays in effective biological control by this fungus has been questioned for specific pathogen interactions (Howell, 1987), and has not been definitely determined.

Induced resistance

Induced resistance occurs when an antagonist induces defense responses within the host plant resulting in resistance to disease through reducing, restricting, or blocking the ability of the pathogen to produce disease. This can happen by prevention of infection, restricting pathogen growth within the plant, or some other mechanism of defense activation within the plant. Induction of systemic resistance may involve activation of multiple potential defense mechanisms, including increased activity of chitinases, β-1,3-glucanases, peroxidases, and other pathogenesis-related (PR) proteins, accumulation of antimicrobial compounds, such as phytoalexins, and formation of protective biopolymers, such as lignin, callose and hydroxyproline-rich glycoproteins (Kloepper *et al.*, 1996). Examples of this interaction are presented by certain rhizobacteria (such as strains of *P. fluorescens*) that can induce systemic resistance to a number of different pathogens (Liu *et al.*, 1995; Van Peer *et al.*, 1991; Wei *et al.*, 1991; Zhou and Paulitz, 1994). On cucumber, induction of resistance by a single rhizobacterial strain provided protection against several different pathogens, including fungi, bacteria and viruses, as well as reductions in insect feeding (Kloepper *et al.*, 1996; Wei *et al.*, 1991). Protection of crops by induced resistance has also been demonstrated in the field for some crops (Tuzun *et al.*, 1992; Wei *et al.*, 1996). Another example of induced systemic resistance (ISR) is that incited by certain nonpathogenic or avirulent strains of *F. oxysporum*, which are able to parasitically colonize roots without causing disease, and induce resistance to Fusarium wilt in several crops (Kroon *et al.*, 1991; Larkin *et al.*, 1996; Leeman *et al.*, 1995; Mandeel and Baker, 1991). Because there is generally an induction period of one to several days required after exposure of the plant to the antagonist before ISR occurs, this mechanism is most effective when the antagonist is applied prior to exposure to the pathogen, such as with a seed treatment, or in transplant operations.

The effective use of microbial antagonists to induce host resistance requires a better understanding of how resistance in plants is induced. Although the mechanism of induction is unknown, a role for salicylic acid has been confirmed in some ISR reactions. Treatment of plants with exogenous salicylic acid induced PR protein synthesis and enhanced resistance to a number of pathogens. In addition, endogenous salicylic acid levels rose specifically during resistance responses in plants (Chen *et al.*, 1995). Studies have also been directed towards understanding the inducing molecules produced by bacterial antagonists. Leeman *et al.* (1995) provided evidence that lipopolysaccharides from the outer membrane of *P. fluorescens* may function as an eliciting molecule for ISR. However, it is unclear if lipopolysaccharides play a similar role in other interactions. Although there is still much to learn about induced resistance, the use of this defense response as a mechanism of biocontrol has much potential (Kloepper *et al.*, 1996; Tuzun and Kloepper, 1995).

Combinations of mechanisms

Many effective antagonist organisms combine two or more different mechanisms of action, providing multiple means of antagonism. *Trichoderma* and *Gliocladium* spp. are known to use various combinations of competition, antibiosis and mycoparasitism in their interactions with pathogens, which may contribute to their effectiveness in controlling a variety of different pathogenic fungi (Papavizas, 1985). Mandeel and Baker (1991) determined that a single isolate of nonpathogenic *F. oxysporum* combined three different mechanisms of action in the control of Fusarium wilt of cucumber: saprophytic competition for nutrients, parasitic competition for infection sites on the root, and ISR within the host plant. Certain rhizobacterial strains produce antibiotics which inhibit soil-borne pathogens and also are capable of inciting ISR in the host plant (Maurhofer *et al.*, 1994).

Differences among mechanisms of action are also important to consider in the search for effective antagonists. The attributes of the pathogen determine the particular mechanisms of action most desirable for the control of that pathogen. The methods used to screen for antagonists can be tailored to select for antagonists utilizing the desired mechanism of action. For example, to isolate organisms that parasitize sclerotia of a given pathogen, sclerotia could be collected from naturally infested fields (and preferably fields that have had a long history of the disease) and examined for colonization and decay by soil microorganisms. Thus, determination of the mechanism of action can be important from the initial screening stages. Many known biocontrol agents utilize antibiosis as a mechanism of action. This may be primarily because antibiosis has been routinely screened for by using *in vitro* inhibition assays and other measures of direct antagonism. The use of multiple screening approaches may aid in finding antagonists utilizing varied and multiple mechanisms of action.

Identification of Specific Traits

Role in biocontrol

Identification of specific traits which contribute to a particular mechanism of action is important and can be very helpful in selecting, understanding, or improving successful biological control organisms. Experimental approaches for identifying traits involved in biocontrol must be able to accommodate the inherently complex environment that biocontrol organisms inhabit (Handelsman and Stabb, 1996). Traits can be identified using techniques from molecular biology, analytical chemistry and biochemistry. The techniques are used to demonstrate that a correlation exists between the expression of the trait and biocontrol and that the trait is expressed *in situ*. A highly effective approach has used molecular techniques to construct genetically modified strains. These strains can be studied in simplified laboratory systems or in the field (Handelsman and Stabb, 1996). The construction of mutant strains, deficient in the expression of a specific trait, using transposon mutagenesis, permits the evaluation of the importance of that trait to biocontrol *in situ* in comparison tests with wild-type strains. With an alternate approach, the relevance of traits to biocontrol can be evaluated in a second strain after the gene responsible for expression of the trait has been isolated by molecular cloning. Other techniques have been used to demonstrate that a trait, such as production of antibiotics or enzymes, is expressed *in situ* through the detection of these antibiotics or enzymes in the spermosphere or rhizosphere.

As an example, Thomashow and Weller (1988) determined that the production of the antibiotic phenazine-1-carboxylate by *P. fluorescens* 2-79 was important for control of *G. graminis* var. *tritici*. Mutant strains of *P. fluorescens* were constructed by random transposon mutagenesis with transposon Tn5. Each of these mutants was deficient in the production of phenazine-1-carboxylate and had reduced biocontrol efficacy when compared with the wild-type strain. Phenazine-1-carboxylic acid was detected on roots and in rhizosphere soil of wheat seedlings grown in steamed or natural soil in growth chambers and field studies when *P. fluorescens* strain 2-79R10, a rifampicin-resistant derivative of strain 2-79, was applied. This antibiotic was not detected in treatments containing *P. fluorescens* strain 2-79-B46, a phenazine-deficient transposon mutant. The antibiotic was also detected in the rhizosphere of wheat colonized by a derivative strain 2-79-B46 which had phenazine-1-carboxylate production restored by genetic complementation (Thomashow *et al.*, 1990). Mutant strains of other biocontrol bacteria have been constructed by insertional inactivation to determine the importance of the antibiotics oomycin A, 2,4-diacetylphloroglucinol, hydrogen cyanide, pyrrolnitrin, and phenazine in other biocontrol interactions (Hill *et al.*, 1994; Howie and Suslow, 1991; Keel *et al.*, 1992; Pierson and Thomashow, 1992; Voisard *et al.*, 1989). Chemical structures of some of the antibiotics produced by biocontrol

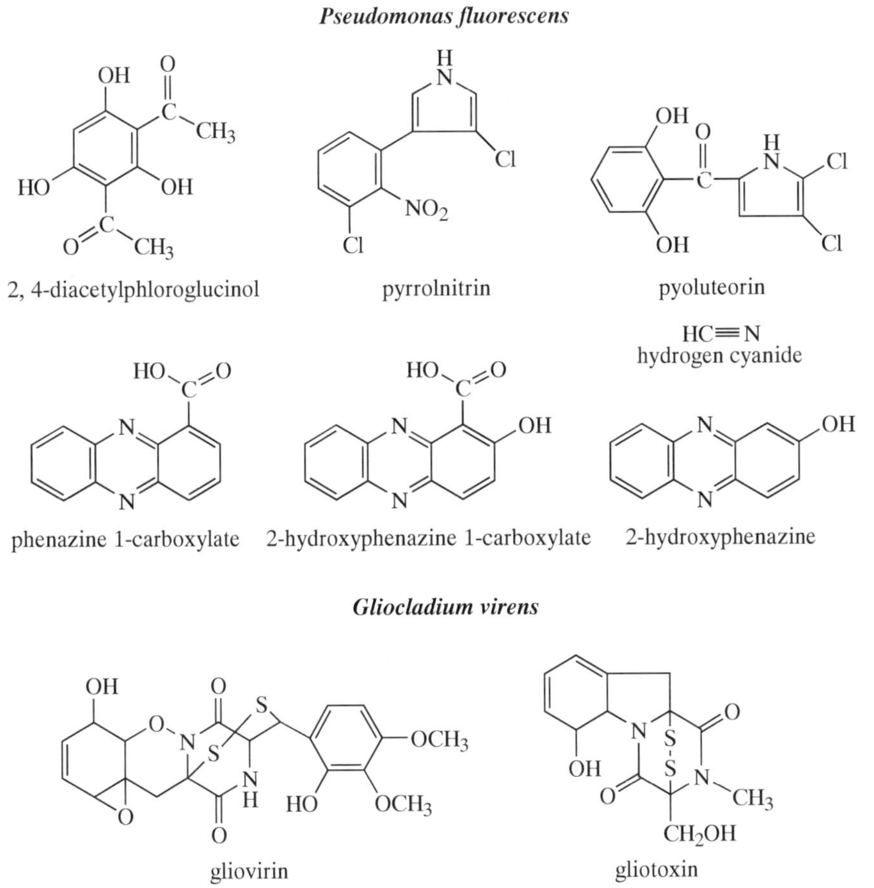

Figure 1 Chemical structures of antibiotics produced by selected bacterial and fungal organisms that are known to be involved in biological control. *Pseudomonas fluorescens* strain PF-5 produces pyoluteorin, pyrrolnitrin, 2,4-diacetylphloroglucinol, and hydrogen cyanide (Kraus and Loper, 1995); strain CHA0 produces pyoluteorin, 2,4-diacetylphloroglucinol, and hydrogen cyanide (Maurhofer *et al.*, 1992; Keel *et al.*, 1992); and strain 2-79 produces the phenazine derivatives (Thomashow and Weller, 1988). *Gliocladium virens* strains G-3 and G-9 produce gliovirin and strain Gl-21 produces gliotoxin (Lumsden *et al.*, 1992).

organisms that are known to be involved in biocontrol interactions are depicted in Figure 1.

Analytical chemistry approaches, such as those used by Thomashow and Weller (1988) are limited by requirements for detectable quantities of these antibiotics or enzymes in samples. Recent approaches have incorporated reporter genes to detect expression of genes involved in antibiotic or enzyme synthesis (Lindow, 1995). Reporter systems have demonstrated transcriptional activation of genes involved in antibiotic or siderophore biosynthesis in bacteria recovered from subterranean portions of plants (Georgakopoulos *et al.*, 1994; Howie and Suslow, 1991; Kraus and Loper, 1995; Loper and Lindow, 1994). Results from these studies demonstrate that genes involved in biocontrol interactions are expressed *in situ*.

In general, however, biocontrol is not due to the expression of a single trait by the biocontrol agent. Current research findings indicate that biocontrol is due to the expression of a variety of traits concurrently or in sequence (Nelson and Maloney, 1992). The expression of multiple traits can support individual mechanisms or a combination of mechanisms ultimately responsible for suppression of disease. For example, *P. fluorescens* strain CHA0 is a potential biocontrol agent for black root rot of tobacco caused by *Thielaviopsis basicola*. This strain produces the antibiotics hydrogen cyanide, 2,4-diacetylphloroglucinol, and pyoluteorin in culture (Keel *et al.*, 1990, 1992; Laville *et al.*, 1992; Voisard *et al.*, 1989). The antibiotics hydrogen cyanide and 2,4-diacetylphloroglucinol but not pyoluteorin are inhibitory to *T. basicola in vitro*. Mutant strains, deficient in the production of hydrogen cyanide or 2,4-diacetylphloroglucinol, had reduced ability to suppress disease caused by *T. basicola* under gnotobiotic conditions. Disease suppression was increased to near wild-type levels when production of these antibiotics was restored through genetic complementation (Keel *et al.*, 1992; Voisard *et al.*, 1989). These findings demonstrate a role for these antibiotics and antagonism based on antibiosis in this biocontrol interaction. However, these mutant strains were reduced, but not deficient in, disease suppression capabilities (Keel *et al.*, 1992; Voisard *et al.*, 1989), indicating that other traits and possibly mechanisms other than antibiosis are involved in this biocontrol interaction. *Pseudomonas fluorescens* CHA0 produces siderophores and the extracellular enzymes protease, phospholipase C, and lipase in culture (Sacherer *et al.*, 1994). These metabolites can potentially function in competition for nutrients with *T. basicola* and the destruction of pathogen hyphae, respectively.

In a number of other genetic studies involving other biocontrol interactions, the deletion of a trait involved in biocontrol resulted in decreased biocontrol but not a complete loss of biocontrol capabilities (Howie and Suslow, 1991; Pierson and Thomashow, 1992). This indicates that multiple traits, and potentially multiple mechanisms of antagonism, are involved in many biocontrol interactions.

Manipulation of traits

Identification of specific traits and genes involved in biocontrol can also be used to improve or enhance existing levels of biocontrol, or to screen for new and better antagonist strains that utilize the same mechanism. Incorporation of multiple copies or overexpression of a biocontrol gene may enhance biocontrol in some cases. Overproduction of the antibiotics phenazine and phloroglucinol by a recombinant strain of *P. fluorescens* resulted in increased biocontrol capabilities against fungal diseases of cucumber, but not of wheat, and phytotoxic effects due to the increased levels of antibiotics where observed on tobacco and sweet corn (Maurhofer *et al.* 1992, 1995). Once identified, genes for specific biocontrol traits can be inserted into other similar organisms that are already well adapted to the environment or niche desired for biocontrol, such as being a good root colonizer. Or, multiple biocontrol genes could be combined in a single organism. For example, Fenton *et al.* (1992) inserted the gene for the production of phloroglucinol into a nonproducing biocontrol strain of *P. fluorescens*, resulting in a recombinant strain that produced the antibiotic and was more suppressive to damping-off diseases than the parent strain. Vincent *et al.* (1991) also observed increased inhibition of a variety of fungal pathogens after the transfer of biosynthetic genes for phlorogucinol from one strain to another. However, genetic manipulation of biocontrol agents has other disadvantages, including problems in testing, registering, releasing, and public acceptance of genetically engineered microorganisms.

Genetic probes to detect the biocontrol gene or genes can also be used to screen other potential antagonist strains to find organisms that utilize the same mechanism of action but may be better suited ecologically for a desired locality or environmental niche. Phloroglucinol-producing strains of fluorescent pseudomonads have been identified from a variety of geographically diverse locations and environments in this way (Keel *et al.*, 1996), as have isolates of *Bacillus cereus* that produce the biocontrol antibiotic zwittermicin-A (Stabb *et al.*, 1994). Recently, Raaijmakers *et al.* (1997) used colony hybridization and PCR analysis techniques to determine that phloroglucinol-producing strains of *P. fluorescens* were abundant in three soils that naturally suppress take-all of wheat, but were undetectable in disease-conducive soils, whereas phenazine-producing strains were not detected in any of the soils. Thus, these techniques can also be used to determine the role of biocontrol traits in natural systems.

Conditions and Requirements of Biocontrol

Once the mechanism(s) of action and/or the specific traits responsible for biocontrol are determined, specific conditions and requirements that make the biocontrol interaction successful can be evaluated. These conditions and requirements include: the environmental and physical conditions, such as temperature, moisture, soil type, etc.; the biological conditions, including the composition of microbial communities and interactions which affect biocontrol; as well as the chemical and nutritional conditions, such as pH, soil fertility and chemistry, carbon and nitrogen status, etc. Determination of the influence of these various factors on the antagonist organism, mechanism of action and biocontrol traits will be crucial to the optimization of biocontrol processes and the development of improved methodologies for the implementation and management of the biological control system.

Each mechanism and corresponding control strategy has different general conditions and requirements. Competition and antibiosis both function by means of protection of the infection court from pathogen infection. Thus, these mechanisms are of primary importance in the rhizosphere and spermosphere, or on leaf surfaces, which act as the primary infection courts where most pathogens penetrate plant tissue. By targeting crucial stages of the pathogen-plant interaction in this way, disease can be controlled without broad-scale application or establishment of the biocontrol agent. Parasitism, on the other hand, is generally not very active in the rhizosphere, but occurs wherever the sclerotia or other survival structures are located, often aggregated with decaying organic matter. Parasitism is effective primarily for pathogens that have large overwintering survival structures, such as sclerotia, as the main source of pathogen inoculum. For this approach to be effective, the mycoparasite must be disseminated so that sclerotia or other survival structures will be adequately colonized and parasitized. One disadvantage of this mechanism is that it is generally slow to operate, and may take several weeks, or even several seasons to effectively reduce the pathogen population (Adams, 1990). Thus, it will only be suitable for certain diseases that have one infection cycle per season (monocyclic) and the overwintering inoculum provides the primary source of inoculum.

Role of colonization

The establishment and redistribution of populations of biocontrol agents from seeds or other application media to the rhizosphere is thought to be essential for successful biocontrol in most applications where protection of the infection court is the strategy (Weller, 1988). Bull *et al.* (1991) demonstrated a positive correlation between population sizes of *P. fluorescens* in wheat rhizosphere and biocontrol of *G. graminis* var. *tritici*. Extensive colonization may be necessary for biocontrol of root-infecting fungi when using biocontrol agents that directly suppress the pathogen through competition for nutrients or production of antibiotics. Many root-infecting pathogens infect root tissues through regions that are relatively distant from the point of introduction of the biocontrol agent into the soil. In addition, the plant can be susceptible to infection by these fungi for long periods of time possibly requiring the long-term persistence of the biocontrol agent. However, the role of colonization in biocontrol has rarely been studied (Handelsman and Stabb, 1996).

Only with certain damping-off diseases has the role of rhizosphere colonization by bacterial biocontrol agents been demonstrated to be unimportant. Plants are susceptible to damping-off pathogens, such as *P. ultimum*, for several hours to a few weeks. Therefore, the susceptibility of the juvenile plant tissue is limited in time and space (Paulitz, 1992). For example, cucumber seedlings are susceptible to infection by *P. ultimum* for 6 to 12 hours after the onset of germination. The cucumber seedling becomes increasingly resistant to infection by *P. ultimum* after this time (Nelson and Maloney, 1992; Nelson *et al.*, 1986). Due to the inundative delivery of high populations of bacterial biocontrol agents at the time of seed treatment to a spatially limited infection court, the spatial redistribution of these bacteria is unimportant (Paulitz, 1992). In addition, biocontrol of *P. ultimum* damping-off of cucumber was demonstrated with a rhizosphere colonization-deficient bacterium (Roberts *et al.*, 1997).

Biocontrol agents can colonize spatially distinct portions of plants and soil during successful biocontrol interactions. Biocontrol bacteria colonize at least a portion of the rhizosphere when applied as treatments of seed or seed pieces (Bahme and Schroth, 1987; Bull *et al.*, 1991; Chao *et al.*, 1986) and can become widely distributed in the rhizosphere (Bahme and Schroth, 1987; Weller, 1988). These bacteria do not form a complete sheath around the root system. Instead, their distribution is highly variable along and among roots (Bahme and Schroth, 1987; Loper *et al.*, 1984). Populations of these biocontrol bacteria typically decrease with increasing distance (horizontally and vertically) from the point of introduction into the soil (Bahme and Schroth, 1987; Chao *et al.*, 1986; Loper *et al.*, 1984) or as the plant develops (Weller, 1983).

Competition with the indigenous microflora for limiting resources such as nutrients is thought to be the major factor limiting colonization by biocontrol bacteria (Chao *et al.*, 1986; Cook, 1992; Weller and Cook, 1983). For example, the biotic status of the soil drastically affected the ability of the biocontrol bacterium *Enterobacter cloacae* to colonize pea rhizosphere (Chao *et al.*, 1986). Populations of *E. cloacae* were not detected more than 3 cm below the seed in pea rhizosphere in natural soil. In contrast, in autoclaved soil, populations of this bacterium were detected 7 to 9 cm below the seed, the point of introduction of this bacterium. Colonization was again limited to the upper region of pea rhizosphere when the native bacterial flora were added back to the autoclaved soil (Chao *et al.*, 1986).

Other factors related to the physical and chemical properties of the soil environment can greatly affect the population and activity of biocontrol organisms. Ownley *et al.* (1992) demonstrated that a wide variety of soil factors, including soil texture, pH, organic matter content, and availability of C, N, S, NH_3, Fe and Mg, affected the biocontrol activity of antibiotic-producing fluorescent pseudomonads. Hoper *et al.* (1995) demonstrated the role of clay type and pH in the disease suppressiveness and biological activity of Fusarium wilt-suppressive soils. Duffy and Défago (1996) observed that the addition of Zn to the growing medium containing the antagonistic bacterium *P. fluorescens* CHA0 further reduced crown and root rot in tomatoes caused by *F. oxysporum* f.sp. *radicis-lycopersici*. Reduction in production of fusaric acid by the pathogen and increased antibiotic production by the bacterium appeared to be responsible for the increased disease control.

Role of the surrounding microbial community

Little is known about the role that the surrounding microbial community plays in biocontrol processes, but these communities may be especially important in establishing and maintaining effective biological control. Biocontrol organisms are no doubt affected by microbial communities, and they in turn may influence these communities. In one instance, disease suppression resulting from the introduction of a biocontrol agent was more closely related to changes in the microbial community structure than to the direct effects of the biocontrol agent. *Bacillus cereus* UW85 increased emergence of soybean without becoming established in high populations in soybean rhizosphere or becoming a dominant member of the heterotrophic rhizosphere community (Halverson *et al.*, 1993). However, the introduction of this bacterium caused dramatic changes in the composition of the rhizosphere community (Gilbert *et al.*, 1993). Other reports also suggest that direct manipulation of microbial communities may be an effective means of biocontrol (Boehm and Hoitink, 1992; English and Mitchell, 1988).

USE OF MULTIPLE ANTAGONISTS AND MULTIPLE MECHANISMS OF ACTION

The use of multiple antagonist organisms with several mechanisms of action is another area of research that requires more attention. Most cases of effective naturally occurring biocontrol (i.e. disease-suppressive soils, organic amendments, composted soils) are the result of the combined activity of a community of organisms. However, the vast majority of biocontrol efforts still focus on one antagonist and one disease. Much inconsistent performance by biocontrol preparations may be due to the use of individual biocontrol agents which are not active in all environments in which they are applied or against all pathogens causing disease of the crop plant (Weller, 1988). A potential solution is the use of combinations of biocontrol organisms in biocontrol treatments (Duffy *et al.*, 1996; Lemanceau *et al.*, 1993; Pierson and Weller, 1994). Biocontrol preparations containing combinations of organisms for the control of one or more plant pathogens is a more ecologically sound approach than using individual organisms. Combinations of biocontrol organisms may overcome inconsistencies in performance due to the biological, physical or chemical heterogeneity of the rhizosphere and soil. A combination of biocontrol agents is more likely to have a greater variety of traits responsible for the suppression of one of several plant pathogens (Lemanceau *et al.*, 1992; Lemanceau and Alabouvette, 1991; Pierson and Weller, 1994). It is also more likely that one or more of these traits will be expressed over a wide range of environmental conditions. A combination of biocontrol agents with a wide range of environmental adaptations should also colonize the dynamic and heterogeneous rhizosphere environment more consistently (Pierson and Weller, 1994). Moreover, the combination of biocontrol agents would allow the strategic targeting of organisms in biocontrol preparations against multiple pathogens or the purposeful use of multiple mechanisms against one target pathogen.

In some cases, combining antagonists can result in a much greater effect than that of either antagonist alone (synergistic interaction). For example, this occurs in the interaction of nonpathogenic *F. oxysporum* and the biocontrol bacterium, *P. fluorescens*, for the control of Fusarium wilt (Lemanceau and Alabouvette, 1991). The nonpathogenic *F. oxysporum* strain Fo47 used alone reduced disease somewhat, whereas the bacterium, *P. fluorescens* strain C7, did not significantly reduce disease when applied alone. However, when both biocontrol organisms were combined, substantial improvements in disease control were observed. The mechanism involves competition for carbon by *F. oxysporum* and competition for iron by the bacterium. Evidence suggests that the presence of the siderophore produced by *P. fluorescens* increases the intensity of antagonism provided by Fo47, making the pathogen more sensitive to glucose competition by the nonpathogenic *F. oxysporum* (Lemanceau *et al.*, 1992, 1993). A similar response with different isolates of nonpathogenic *F. oxysporum* and *P. fluorescens* was observed by Park *et al.* (1988) and Leeman *et al.* (1996). Other examples of antagonists that are more effective in controlling soilborne pathogens when used in combination have been reported by Pierson and Weller (1994) using mixtures of *P. fluorescens*; Duffy and Weller (1995) using *G. graminis* combined with *P. fluorescens*; by Duffy *et al.* (1996) using fluorescent pseudomonads combined with *Trichoderma harzianum*; and by Gracia-Garza *et al.* (1997) using *T. hamatum* combined with insects (fungus gnats) to control *Sclerotinia sclerotiorum*. Other studies reported no detectable benefits to combining organisms (Dandurand and Knudsen, 1993; Larkin and Fravel, 1996; Minuto *et al.*, 1995a). Ultimately, even when there is no obvious enhancement in the level of control observed, there still may be a advantages to combinations related to control of multiple pathogens or control over a wider range of environmental conditions.

Disadvantages of using multiple antagonists include that they are more difficult to study and that antagonists must be compatible with each other. Also, the use of multiple antagonists complicates their production, formulation, implementation and product registration. These are areas that need to be addressed before multiple antagonist formulations, or even ‘antagonist soups’ containing numerous ill-characterized organisms can be used as biocontrol agents. However, combinations of multiple antagonists have much potential for strengthening the efficacy of biocontrol and overcoming some of the limitations of single antagonist biological control interactions.

INFLUENCE OF THE HOST PLANT AND PLANT GENOTYPE ON MICROBIAL COMMUNITIES

Another area that has received very little attention, but could be very important in the improvement of biological control interactions, is the recognition and use of the influence of different plant genotypes on microbial communities in the spermosphere, rhizosphere and phyllosphere. This approach can be utilized in biological control through the use of host plants that tend to support microbial communities that are more antagonistic to certain pathogens. Substantial evidence exists that plants vary in their ability to attract and support certain biocontrol organisms and communities of antagonistic microorganisms (Atkinson *et al.*, 1975; Azad *et al.*, 1985; Howie and Echandi, 1983; Latour *et al.*, 1996; Lemanceau *et al.*, 1995; Neal *et al.*, 1973). These variations have been observed among different plant species as well as among cultivars within species. Certain plants are known to stimulate or increase populations of antagonistic organisms, or to decrease pathogen populations in the soil and rhizosphere (Bourbos and Skoudridakis, 1987). Presumably, these differences are a result of the quality and quantity of root exudates, secretions and lysates. Rotation crops that select for antagonist organisms at the expense of pathogen populations may be feasible in some cases. Velvetbean, which is used as a rotation crop for cotton and peanut, increases populations of organisms antagonistic to pathogenic fungi and nematodes, resulting in lower disease (Rodriguez-Kabana *et al.*, 1992).

However, there are also cases in which a particular cultivar of the host plant tends to promote suppressiveness to the pathogen after a period of time, due to the build up of antagonistic organisms in the plant rhizosphere. Larkin *et al.* (1993) observed a cultivar-specific response to Fusarium wilt of watermelon. One cultivar of watermelon (Crimson Sweet-CS) promoted disease suppression in soils after monoculture for several years, whereas several other cultivars did not. They observed that after a number of successive plantings, the cultivar Florida Giant (FG) tended to increase pathogen populations in the soil and rhizosphere, whereas planting to CS produced no change in pathogen populations. However, populations of nonpathogenic *F. oxysporum*, mainly saprophytes and antagonists, showed the reverse trend; saprophytic populations were unchanged when planted to FG, whereas these antagonistic populations increased substantially when planted to CS. It was observed that planting to CS also resulted in significantly higher populations of bacteria, actinomycetes and fluorescent pseudomonads, all of which are generally antagonistic to Fusarium wilt and many other fungi (Larkin *et al.*, 1993). This research demonstrated the effect different cultivars can have on microbial communities, and that they can play a significant role in biocontrol. Much more research is needed to explore the genetic diversity within plant species and its role on interactions with microbial communities (Handelsman and Stabb, 1996; O'Connell *et al.*, 1996). If this type of plant attribute, in which the host plant selects for effective disease antagonists in the soil, could be selected for and incorporated into a breeding program and implemented along with applications of these antagonists, the successful establishment and maintenance of the biological control agents in the soil and rhizosphere would be much easier and would provide more effective and lasting disease control.

INTEGRATION OF BIOLOGICAL CONTROL WITH OTHER CONTROL STRATEGIES (IPM)

Due to the nature and properties of biological control, biocontrol of plant diseases should be perceived as a complementary tool for the management of plant diseases, and not as a complete replacement for any current management strategies. Integrated pest management (IPM) incorporates numerous control strategies, including cultural, chemical and biological methods, coordinated to provide more efficient and effective control, and designed to combine the best aspects of all these approaches while minimizing the negative qualities of each.

Elaborate IPM programs are already underway in several parts of the world, where there has been a commitment to reduce the input of chemical pesticides by 50% by the year 2002 (Hall, 1995). Many practices used in IPM programs were developed by ancient cultures hundreds of years ago. Incorporation of agricultural residues, manure and composted organic material was practised by the Aztecs to make soil suppressive to common diseases (Rojas-Rabiela, 1983). The chinampa agricultural system, developed extensively by the Aztecs and still maintained in Mexico today, is considered one of the most productive and intensive farming methods ever devised (Lumsden *et al.*, 1987, 1990). In China, Japan, and other parts of the Orient, composted materials have been used to improve agriculture for centuries (Kelman and Cook, 1977). Due to the recent political and economic changes in the former Soviet Union, Cuba has been forced to change its agricultural practices. Currently, all pest control in Cuba is accomplished with biocontrol agents and IPM systems, many of which have been very successful (Vandermeer *et al.*, 1993). In the USA, a network of professionals and university professors maintain updated information available to producers about the potential for outbreak of disease based on inoculum densities and weather data, results of cultural practices on survival of infective propagules of plant pathogens, utilization of new and existing pesticides, and biocontrol agents currently available for the control of pests.

Chemical Controls

The combination of biological controls with reduced rates of established chemical controls has been successful in many cases. Elad and Shtienberg (1996) reported on the compatibility of the biocontrol product Trichodex, a formulation of the mycoparasite *Trichoderma harzianum*, in combination with fungicides for the control of grey mold of strawberry, vegetable crops and grapes, caused by *Botrytis cinerea*. Trichodex effectively controlled *Botrytis* diseases in greenhouse crops and in vineyards when applied alone or in combination with chemical fungicides. Alternation of biocontrol applications with reduced application rates of chemical fungicides resulted in disease suppression as effective as that achieved by applying the fungicide alone at full application rates, and was more consistent than applications with the biocontrol agent alone. *B. cinerea* is known to rapidly develop resistance to repeated applications of potent fungicides, such as the benzimidazoles in the 1970s and then dicarboximides in the 1980s (Katan,1982; Washington *et al.*, 1992). Thus, any methods which reduce fungicide use for control of this pathogen may help prolong fungicide effectiveness. Saksirirat *et al.* (1996) observed improved control of *Sclerotium rolfsii* on tomatoes by combining *T. harzianum* with the fungicide mancozeb. Sivan and Chet (1993) also combined the use of *T. harzianum* with reduced rates of methyl bromide or soil solarization to get improved control of Fusarium crown and root rot of tomatoes. Minuto *et al.* (1995a) improved control of Fusarium wilt of cyclamen by combining applications of antagonistic *F. oxysporum* strains with benzimidazole fungicides.

Sublethal Stressing

Another practice given much attention during the past several years is the combination of biocontrol agents with the application of stress to resting propagules of plant pathogens. These stresses include sublethal heating, sublethal doses of chemical fungicides or fumigants, or the addition of amendments to soil which produce chemicals detrimental to pathogens. However, it is important to consider that the same stress is usually also being applied to the beneficial organisms, unless the biocontrol agent is added after the stress treatment. Thus, the biocontrol agent must be more tolerant of the stressing agent than the pathogen in order to be effective. Therefore, as stated by Katan *et al.* (1996), 'our aim should be to apply the minimal dosage necessary to achieve economically desirable, but not necessarily maximal disease control, an approach which leads to minimizing adverse effects of the killing agent on the environment.' Cells exposed to elevated temperatures may result in death or inactivation of the cell, damage to wall components and membranes, damage to nucleic acids (including ribosomal RNA), or loss of pathogenicity (Barbercheck and Von Broembsen, 1986; Hurst, 1984; Pullman *et al.*, 1981). In fungi, sublethal temperatures delay germination of spores and resting hyphae for varying periods, depending on the temperature and the duration of treatment. This delay appears to be cumulative up to the point when the cells are no longer capable of germination. Multicellular structures seem to be more tolerant to the stress (De Vay and Katan, 1991).

Soil solarization heats the soil to mild levels resulting in sublethal heating, especially in deeper soil layers (Katan *et al.*, 1996). Gamliel and Katan (1991) found that populations of *Pseudomonas* spp. increased in the rhizosphere of plants grown in soils previously solarized, and an enhanced growth response was observed in tomato plants. These results were similar to those obtained by Stapleton and De Vay (1984) in which fluorescent pseudomonads colonized the roots of beets and radishes up to six times more effectively when grown in solarized soils compared to non-solarized soils. These bacteria have been correlated with supressiveness of solarized soils. Populations of antagonistic fungi, such as *Trichoderma* spp. and *T. flavus*, have also been reported to increase in solarized soil and the rhizosphere of plants grown in them (Elad *et al.*, 1980; Munnecke *et al.*, 1976; Tjamos and Paplomatas, 1987). Tjamos and Niklis (1990) found that treating solarized soils with *T. harzianum* effectively reduced Fusarium wilt of beans, and a synergistic effect was observed between the two treatments. Tjamos and Fravel (1995) also observed a synergistic effect between sublethal heating and *T. flavus* treatments in the control of *V. dahliae*. Several other researchers also demonstrated improved control using combinations of biological control agents with soil solarization and a variety of different organisms in several pathosystems (Chet *et al.*, 1982; Minuto *et al.*, 1995b; Ristaino *et al.*, 1991). Fravel (1996), working with the biocontrol agents *T. flavus* and *Gliocladium roseum*, was able to further reduce Verticillium wilt of eggplant when these mycoparasites were combined with sublethal doses of metham sodium.

Combination of soil solarization with other methods of control such as chemicals (methyl bromide, metham sodium and benomyl) and cultural practices (incorporation of cruciferous amendments) also resulted in successful strategies to reduce damage caused by several plant pathogens (Ben-Yephet *et al.*, 1988; Katan *et al.*, 1996; Ramirez-Villapudua and Munnecke, 1987; Tjamos and Paplomatas, 1988).

Cultural Controls

The application of different composted materials has also suppressed diseases caused by several plant pathogens (Chung *et al.*, 1988; Lumsden *et al.*, 1983). Suppression of diseases is affected, among other factors, by the compost type, maturity level and the composting method. Activity of antagonists involved in biological control is affected by the nutrients present in the compost. Suppressiveness is also affected by specific physical, chemical and biological properties of the compost (Grebus *et al.*, 1994; Hoitink and Fahy, 1986). Suppression of plant pathogens in the composted material is generally due to exposure to high temperatures produced during the composting process, the release of toxic products, or microbial antagonism (Hoitink and Fahy, 1986). As with solarization, sublethal heat stress and weakening of the pathogen to subsequent microbial antagonism is also common. Other factors in composted materials which affect the incidence of diseases caused by soil-borne plant pathogens are particle size, nitrogen content, cellulose and lignin content, electrical conductivity (soluble salt content), pH, and inhibitors released by compost (Hoitink and Fahy, 1986). Populations of organisms antagonistic to plant pathogens have been reported to increase in some composted materials (Kuter *et al.*, 1983; Nelson *et al.*, 1983). In addition, certain beneficial antagonistic organisms can also be added to composted material to improve or enhance the disease suppressive characteristics (Hoitink and Fahy, 1986).

Cruciferous plant residues, which have a high content of volatile sulfur-containing compounds toxic to many fungal pathogens (Gamliel and Stapleton, 1993), have been used as amendments to reduce a variety of fungal diseases (Subbarao and Hubbard, 1996). When cruciferous plant residues were used in combination with plastic mulching, toxic volatiles were retained in the soil longer, resulting in improved control of soil-borne pathogens (Ramirez-Villapudua and Munnecke, 1987, 1988).

The type of tillage used can also have a major effect on soil pathogens, and tillage practices can be coordinated with biological control to establish conditions which will be more favorable to the antagonists than to the pathogen(s). Reduced and no-till systems may suppress some pathogens, but enhance the development of others (Bailey and Duczek, 1996). Several recent reviews have dealt with the effects of tillage on plant diseases (Blevins and Frye, 1993; Conway, 1996; Rothorock, 1992; Watkins and Boosalis, 1994). More research is needed on the changes that occur when conventional tillage is replaced with no- or minimum-till practices, as well as how biological control can be incorporated in these systems.

IMPROVED FORMULATIONS AND DELIVERY SYSTEMS

Formulations

Formulations are an important, but often overlooked, component in the development of an effective biological control system (Lewis, 1991). Their development and application can have a major impact on the success of a biocontrol agent. Formulations must deliver the antagonist to the appropriate location, in sufficient quantities, and in the proper form and physiological state to be effective. The activity and growth of the biocontrol agent depend greatly on the endogenous nutritional status of the formulated propagule, the energy source provided in the formulation, and the appropriate placement relative to a susceptible host. An effective formulation provides nutrients for germination and sporulation of the biocontrol organism, withstands harsh environmental conditions, interacts with other organisms to the advantage of the biocontrol agent, can be integrated with cultural practices, and can be applied with existing machinery or methods (i.e. similar procedures as application of chemicals, green amendments, etc.). The formulation must also be inexpensive, easy to produce, and have an appropriate shelf life (Lumsden *et al.*, 1995).

Currently, there are over 40 different formulations of biocontrol agents commercially available around the world for the control of various plant diseases (Fravel *et al.*, 1997). Biocontrol agents formulated include fungi, bacteria and actinomycetes. The composition of the formulation varies greatly with the intended use (Fravel *et al.*, 1997). Fungi and actinomycetes are currently on the market as pellets or 'crumbles' for soil incorporation, alginate prills for use with soilless potting mix, dowels for insertion into wood, wettable powders, granules, microgranules, water dispersible granules, and dusts. Fungi are also impregnated in sticks, and produced on grain. Actinomycetes have been specifically produced as powders for drenches or sprays, or added through irrigation systems. Bacteria have been formulated as aqueous suspensions of fermentor biomass to spray, liquid suspensions for drench or for drip irrigation, dry powders, and wettable powders. Bacteria or fungi can also be applied to seed with a sticker and a peat or clay carrier (Fravel *et al.*,1997).

The nutrient base used in formulations is known to affect the performance of the biocontrol organism. Fravel *et al.* (1995) found that *T. flavus* significantly delayed symptoms of Verticillium wilt of eggplant when *T. flavus* was formulated using pyrophyllite clay (Pyrax) and corn cobs as organic carriers, but not when a variety of other agricultural byproducts, such as peanut hulls, wheat bran, neem cake, chitin, or fish meal, were used. Both pyrophyllite clay and corn cobs had significantly greater C/N ratios than the other organic carriers tested. Food sources low in readily available nutrients for other organisms in the soil may account for the success of these formulations. Other reports have established the importance of the C or N source in the adequate performance of the formulated biocontrol agent (Engelkes *et al.*, 1997; Mungnier and Jung, 1985, Stack *et al.*, 1987). Specialized food sources that can be utilized by the biocontrol organism, but cannot be readily used by the pathogen or other soil organisms may also be used to enhance biocontrol performance.

Some nutritional sources in formulations may enhance the population growth or general activity of other organisms in soil. Indigenous organisms may be capable of colonizing and utilizing the nutrients, altering the performance of the biocontrol organism, either by depriving the biocontrol organism of needed nutrients, or by increasing the populations of organisms which may be antagonistic to the biocontrol organism. In a preliminary test of several formulations containing various food bases and *F. oxysporum* as a potential biocontrol agent, some formulations favored specific organisms in the soil (Gracia-Garza, unpublished). Granular extrusion formulations made of wheat and rice flour were more heavily colonized by bacteria and actinomycetes than alginate pellets of the same food base. The attraction of certain organisms may be detrimental to the biocontrol agent in the formulation. However, the same attraction to certain other organisms (i.e. insects) may indeed be beneficial. In additional tests with a mycoherbicidal strain of *F. oxysporum*, it was observed that some formulations were very attractive to seed harvester ants (Gracia-

Garza *et al.*, 1996). Some formulations based on a wheat flour food source, were preferred by the ants, especially if they contained oils. Granules were removed by ants within 24 hours of placement in the field and they were taken inside their nests, increasing the number of propagules in the lower levels of the soil, where roots of susceptible plants are present. Other insects such as honey bees have been reported to successfully deliver biocontrol agents such as *G. roseum*, *Epicoccum* sp., *P. fluorescens*, and *Erwinia herbicola* (Johnson *et al.*, 1993; Sutton, 1995; Thomson *et al.*, 1992). The use of other organisms to enhance the delivery of biocontrol agents deserves more research.

Many other practical aspects related to fermentation, the logistics of 'scaling-up' formulations, the interest of industry to develop biocontrol products, and the regulatory procedures to register a biocontrol agent, require still more efforts by all sectors involved to fully develop biocontrol agents for use in commercial agriculture. In the past, much of the work related to developing formulations, scale-up, large-scale performance testing, and product registration has been done by private industry interested in producing and selling the biocontrol product (Mintz and Walter, 1993). However, because many individual biocontrol agents may not have broad-scale applications and will be used in relatively small, specialized markets, researchers can not always rely on the support and participation of private industry to develop a biological control agent for commercial use. Increasingly, the original researchers and developers will need to play a greater role to continue the development of biological control agents through the formulation testing, scale-up, and large-scale performance phases of research to bring these products to the marketplace (Cook, 1993).

Approaches to Enhance Delivery Systems

The populations and activities of biocontrol agents typically decrease over time in the rhizosphere (Schroth and Becker, 1990). In addition, populations of biocontrol agents and the spatial distribution of these populations are not always adequate for protection of roots from root-infecting fungi. Genetic approaches, alone or in combination with specific formulations, offer potential for enhancing colonization and disease suppression activity of biocontrol agents.

It is possible to increase populations and activities of bacterial biocontrol agents using nutritional supplements targeted for use by these bacteria. For example, populations of a colonization-deficient bacterium on cucumber seedlings were stimulated through the addition of casamino acids at the time of seed treatment (Roberts *et al.*, 1996). Exotic carbon sources offer the potential to selectively increase populations and activity by biocontrol agents genetically engineered with catabolic capabilities to utilize these compounds as sources of reduced carbon (Colbert *et al.*, 1993). Plasmid pNAH7, containing genes for salicylate catabolism, was mobilized into the bacterial biocontrol agent *P. putida* by conjugation. The resultant *P. putida* strain was capable of growth on salicylate in culture, on sugarbeet seed, and in field soil. Salicylate amendments to soil resulted in increased respiration and populations of this genetically engineered *P. putida* strain (Colbert *et al.*, 1993). Finally, once genes important for biocontrol have been identified it may be possible to supplement formulations with chemicals required for expression of these genes. This approach, where specific compounds are added to seed treatments or other formulations, can be expected to be useful only over relatively short time periods with pathogens that infect the plant close to the point of introduction of the biocontrol agent.

In a second approach, plants were genetically engineered to exude compounds from their roots that only a particular biocontrol agent or select group of biocontrol agents could utilize as sources of reduced carbon. Tobacco plants transformed with *Agrobacterium tumefaciens* genes that conferred the biosynthesis of mannityl opines exuded these compounds from roots, stems, and leaves (Savka and Farrand, 1992). Since only a small group of soil microbes can use these mannityl opines for nutrition these genetically engineered tobacco plants can provide selective nutrition to biocontrol agents genetically altered with genes conferring catabolism of these opines. This approach has the potential to overcome problems with the spatial distribution of biocontrol organisms in the environment and with their declining biocontrol activity during the growing season.

SUMMARY AND CONCLUSIONS

Biological control of fungal diseases constitutes a very broad category of control, consisting of a wide variety of different organisms, mechanisms, interactions and processes. Biological control has enormous potential to supplement and complement existing disease control strategies. However, biological control also has very different properties, requirements and constraints than previous conventional controls and needs to be properly implemented and integrated with current production strategies. Biological control depends on the effective functioning of the appropriate antagonist strains within each particular plant-pathogen ecosystem. Identifying the appropriate antagonist strains is generally the first stage in this process. Understanding how, where, when and why the biocontrol works may also be crucial to successful development of the biological control system. Because of the complexity of soil microbial communities and the role of the biocontrol organism within these communities, an ecological approach to the development of the biological control system is recommended. Evaluating the ecological interactions of the biocontrol organism with the pathogen, host plant, surrounding microbial community, and the environment will be useful in developing the best strategies for the implementation and management of the biological control system. For maximum effectiveness, biocontrol organisms that are locally adapted to the particular environments and pathosystems where they are needed may need to be developed.

Biocontrol may function through competition, antibiosis, parasitism, induced resistance, or a combination of mechanisms. These various mechanisms may control disease by the reduction or inhibition of the pathogen inoculum, protection of the infection court (prevention of pathogen infection), or by limiting disease development after pathogen infection. Each organism, mechanism, and activity has different traits, conditions, and requirements associated with it that are essential to biocontrol activity. Overall, the more we know about the specific mechanisms, traits, interactions, and requirements of biological control agents, the better we can establish strategies to optimize biocontrol for that pathosystem. Other areas of research that may be critical to the further development and improvement of biological control systems are the development and use of multiple antagonists incorporating several different mechanisms of action, incorporating the influence of the host plant on microbial communities to enhance biocontrol organisms and activity, better integration of biocontrol with other disease control strategies, and improved formulations and delivery systems to provide the optimal starting point from which biocontrol can develop.

There is no question that biological control can and does work—in natural systems it has been working for thousands of years. The challenge facing us is to be able to adapt biocontrol mechanisms and principles to allow them to function fully and effectively within commercial production systems, as well as for those production systems to be flexible enough to make the changes necessary to adapt current agronomic practices to enable biological control to function properly. It is clear that there are many roles and applications of biological control for the management of fungal diseases. Through further implementation of the directions of research outlined here, along with the further application of innovative research and development techniques, biological control finally may be able to fulfill its potential for the effective management of plant pathogens within IPM and sustainable agriculture systems.

REFERENCES

Adams, P. B. (1990). 'The potential of mycoparasites for biological control of plant diseases', *Annu. Rev. Phytopathol.*, **28**, 59–72.

Adams, P. B., and Ayers, W. A. (1982). 'Biological control of Sclerotinia lettuce drop in the field by *Sporodesmium sclerotivorum*', *Phytopathology*, **72**, 485–488.

Adams, P. B. and Ayers, W. A. (1983). 'Histological and physiological aspects of infection of sclerotia of two *Sclerotinia* species by two mycoparasites', *Phytopathology*, **73**, 1072–1076.

Adams, P. B. and Fravel, D. R. (1990). 'Economical control of Sclerotinia lettuce drop by *Sporodesmium sclerotivorum*', *Phytopathology*, **80**, 1120–1124.
Alabouvette, C. and Couteaudier, Y. (1992). 'Biological control of Fusarium wilts with nonpathogenic fusaria', in *Biological Control of Plant Diseases: Promises and Challenges for the Future* (eds E. C. Tjamos, G. C. Papavizas and R. J. Cook), pp. 415–426, Plenum Press, New York.
Alabouvette, C., Lemanceau, P., and Steinberg, C. (1993). 'Recent advances in the biological control of *Fusarium* wilts', *Pest. Sci.*, **37**, 365–373.
Andrews, J. H. (1992). 'Biological control in the phyllosphere', *Annu. Rev. Phytopathol.*, **30**, 603–635.
Atkinson, T. G., Neal, J. L., and Larson, R. L. (1975). 'Genetic control of the rhizosphere microflora of wheat', in *Biology and Control of Soil-Borne Plant Pathogens* (ed. G. W. Bruehl), pp. 116–122, APS Press, St Paul, MN.
Azad, H. R., Davis, J. R., and Kado, C. I. (1985). 'Relationships between rhizoplane and rhizosphere bacteria and Verticillium wilt resistance in potato', *Arch. Microbiol.*, **140**, 347–351.
Bahme, J. B. and Schroth, M. N. (1987). 'Spatial-temporal colonization patterns of a rhizobacterium on underground organs of potato', *Phytopathology*, **77**, 1093–1100.
Bailey, K. L., and Duczek, L. J. (1996). 'Managing cereal diseases under reduced tillage', *Can. J. Plant Pathol.*, **18**, 159–167.
Baker, K. F. (1987). 'Evolving concepts of biological control of plant pathogens', *Annu. Rev. Phytopathol.*, **26**, 67–85.
Baker, K. F. and Cook, R. J. (1974). *Biological Control of Plant Pathogens*, Freeman, San Francisco (Reprinted 1982, American Phytopathology Society, St Paul, MN).
Baker, R. and Dunn, P. E., eds (1990). *New Directions in Biological Control: Alternatives for Suppressing Agricultural Pests and Diseases*, Alan R. Liss, New York.
Bakker, P. A. H. M., Lamers, J. G., Bakker, A. W., Marugg, J. D., Weisbeek, P. J., and Schippers, B. (1986). 'The role of siderophores in potato tuber yield increase by *Pseudomonas putida* in a short rotation of potato', *Neth. J. Plant Pathol.*, **92**, 249–256.
Barbercheck, M. E. and Von Broembsen, S. L. (1986). 'Effects of soil solarization on plant-parasitic nematodes and *Phytophthora cinnamomi* in South Africa', *Plant Dis.*, **70**, 945–950.
Ben-Yephet, J., Melero-Vera, J. M., and De Vay, J. E. (1988). 'Interaction of soil solarization and metham sodium in the destruction of *Verticillium dahliae* and *Fusarium oxysporum* f. sp. *vasinfectum*', *Crop Prot.*, **7**, 327–331.
Blevins, R. L. and Frye, W. W. (1993). 'Conservation tillage: an ecological approach to soil management', *Adv. Agron.*, **51**, 33–78.
Boehm, M. J. and Hoitink, H. A. J. (1992). 'Sustenance of microbial activity in potting mixes and its impact on severity of Pythium root rot of poinsettia', *Phytopathology*, **82**, 259–264.
Boogert, P. H. J. F. and Velvis, H. (1992). 'Population dynamics of the mycoparasite *Verticillium biguttatum* and its host, *Rhizoctonia solani*', *Soil Biol. Biochem.*, **24**, 157–164.
Bourbos, V. A. and Skoudridakis, M. T. (1987). 'Das verhalten einiger pilzlicher antagonisten in der rhizosphäre resistenter and anfälliger gewächshaustomaten', *J. Phytopathol.*, **120**, 193–198.
Bowen, G. D. and Rovira, A. D. (1976). 'Microbial colonization of plant roots', *Annu. Rev. Phytopathol.*, **14**, 121–144.
Bull, C. T., Stack, J. P., and Smilanick, J. L. (1997). '*Pseudomonas syringae* strains ESC-10 and ESC-11 survive in wounds on citrus and control green and blue molds of citrus', *Biol. Control*, **8**, 81–88.

Bull, C. T., Weller, D. M., and Thomashow, L. S. (1991). 'Relationship between root colonization and suppression of *Gaeumannomyces graminis* var. *tritici* by *Pseudomonas fluorescens* strain 2-79', *Phytopathology*, **81**, 954–959.

Buysens, S., Heungens, K., Poppe, J., and Höfte, M. (1996). 'Involvement of pyochelin and pyoverdin in suppression of *Pythium*-induced damping-off of tomato by *Pseudomonas aerguginosa* 7NSK2', *Appl. Environ. Microbiol.*, **62**, 865–871.

Campbell, R. (1989). *Biological Control of Microbial Plant Pathogens*, Cambridge University Press, Cambridge.

Campbell, R. (1994). 'Biological control of soil-borne diseases: some present problems and different approaches', *Crop Prot.*, **13**, 4–13.

Cartwright, D. K. and Benson, D. M. (1995). 'Biological control of Rhizoctonia stem rot of poinsettia in polyfoam rooting cubes with *Pseudomonas cepacia* and *Paecilomyces lilacinus*', *Biol. Control*, **5**, 237–244.

Chao, W. L., Nelson, E. B., Harman, G. E., and Hoch, H. C. (1986). 'Colonization of the rhizosphere by biological control agents applied to seeds', *Phytopathology*, **76**, 60–65.

Chen, Z. X., Malamy, J., Henning, J., Conrath, U., Sanchezcasas, P., Silva, H., Ricigliano, J., and Klessig, D. F. (1995). 'Induction modification, and transduction of the salicylic acid signal in plant defense responses', *Proc. Nat'l. Acad. Sci. USA*, **92**, 4134–4137.

Chet, I., ed. (1987). *Innovative Approaches to Plant Disease Control*, John Wiley, New York.

Chet, I. (1994). 'Biological control of fungal pathogens', *Appl. Biochem. Biotechnol.*, 48, 37–43.

Chet, I., Elad, Y., Kalfon, A., Hadar, Y., and Katan, J. (1982). 'Integrated control of soilborne pathogens in iris', *Phytoparasitica*, **10**, 229–236.

Chung, Y. R., Hoitink, H. A. J., Dick, W. A., and Herr, L. J. (1988). 'Effects of organic matter decomposition level and cellulose amendment on the inoculum potential of *Rhizoctonia solani* in hardwood bark media', *Phytopathology*, **78**, 836–840.

Clarke, F. E. (1965). 'The concept of competition in microbial ecology' in *Ecology of Soil-borne Plant Pathogens* (eds K. F. baker and W. C. Snyder), pp. 339–345, University of California Press, Berkeley, CA.

Colbert, S. F., Hendson, M., Ferri, M., and Schroth, M. N. (1993). 'Enhanced growth and activity of a biocontrol bacterium genetically engineered to utilize salicylate', *Appl. Environ. Microbiol.*, **57**, 1504–2076.

Conway, K. E. (1996). 'An overview of the influence of sustainable agricultural systems on plant diseases', *Crop Prot.*, **15**, 223–228.

Cook, R. J. (1991). 'Twenty-five years of progress towards biological control', in *Biological Control of Plant Pathogens* (ed. D. Hornby), pp. 1–14, CAB International, Oxon, UK.

Cook, R. J. (1992). 'A customized approach to biological control of wheat root diseases', in *Biological Control of Plant Diseases: Progress and Challenges for the Future* (eds E. C. Tjamos, G. C. Papavizas and R. J. Cook), pp. 211–222, Plenum Press, New York.

Cook, R. J. (1993). 'Making greater use of introduced microorganisms for biological control of plant pathogens', *Annu. Rev. Phytopathol.*, **31**, 53–80.

Cook, R. J., and Baker, K. F. (1983). *The Nature and Practice of Biological Control of Plant Pathogens*, American Phytopathology Society, St Paul, MN.

Couteaudier, Y. (1992). 'Competition for carbon in soil and rhizosphere; a mechanism involved in biological control of Fusarium wilts', in *Biological Control of Plant Diseases: Progress and Challenges for the Future* (eds E. C. Tjamos, G. C. Papavizas and R. J. Cook), pp. 99–104, Plenum Press, New York.

Curl, E. A. and Truelove, B. (1986). *The Rhizosphere*, Springer-Verlag, New York.

Dandurand, L. M. and Knudsen, G. R. (1993). 'Influence of *Pseudomonas fluorescens* on hyphal growth and biocontrol activity of *Trichoderma harzianum* in the spermosphere and rhizosphere of pea', *Phytopathology*, **83**, 265–270.

Datnoff, L. E., Nemec, S., and Pernezny, K. (1995). 'Biological control of Fusarium crown and root rot of tomato in Florida using *Trichoderma harzianum* and *Glomus interadices*', *Biol. Control*, **5**, 427–431.

Deacon, J. W. (1991). 'Significance of ecology in the development of biocontrol agents against soil-borne pathogens', *Biocontrol Sci. Technol.*, **1**, 5–20.

Deacon, J. W. and Berry, L. A. (1993). 'Biocontrol of soilborne plant pathogens: concepts and their application', *Pesticide Sci.*, **37**, 417–426.

De Vay, J. E. and Katan, J. (1991). 'Mechanisms of pathogen control in solarized soils', in *Soil Solarization* (eds J. Katan and J. E. De Vay), pp. 88–101, CRC Press, Boca Raton.

Duffy, B. and Défago, G. (1996). 'Effect of trace-minerals on biocontrol of *Fusarium oxysporum* f.sp. *radicis-lycopersici* on tomato using *Pseudomonas fluorescens* CHA0', (Abstr.) *Phytopathology*, **86** (36).

Duffy, B. K. and Weller, D. M. (1995). 'Use of *Gaemannomyces graminis* var. *graminis* alone and in combination with fluorescent *Pseudomonas* spp. to suppress take-all of wheat', *Plant Dis.*, **79**, 907–911.

Duffy, B. K., Simon, A., and Weller, D. M. (1996). 'Combination of *Trichoderma koningii* with fluorescent pseudomonads for control of take-all on wheat', *Phytopathology*, **86**, 188–194.

Elad, Y. and Shtienberg, D. (1996). '*Trichoderma harzianum* T36 (TRICHODEX) integrated with fungicides for the control of grey mould of strawberry, vegetable greenhouse-crops and grapes', in *Advances in Biological Control of Plant Diseases* (eds T. Wenhua, R. J. Cook and A. Rovira), pp. 310–319, China Agricultural University Press. Beijing, China.

Elad, Y., Katan, J. and Chet, I. (1980). 'Physical, biological, and chemical control integrated for soilborne diseases in potatoes', *Phytopathology*, **70**, 418–422.

Elad, Y., Chet, I., and Henis, Y. (1982). 'Degradation of plant pathogenic fungi by *Trichoderma harzianum*', *Phytopathology*, **28**, 719–725.

Engelkes, C. A., Nuclo, R. L., and Fravel, D. R. (1997). 'Effect of carbon, nitrogen, and C:N ratio on growth, sporulation, and biocontrol efficacy of *Talaromyces flavus*', *Phtopathology*, **87**, 500–505.

English, J. T. and Mitchell, D. J. (1988). 'Influence of an introduced composite of microorganisms on infection of tobacco by *Phytophthora parasitica* var. *nicotianaè*, *Phytopathology*, **78**, 1484–1490.

Falk, S. P., Gadoury, D. M., Pearson, R. C., and Seem, R. C. (1995). 'Partial control of grape powdery mildew by the mycoparasite *Ampelomyces quisqualis*', *Plant Dis.*, **79**, 483–490.

Fang, J. G. and Tsao, P. H. (1995). 'Evaluation of *Pythium nunn* as a potential biocontrol agent against Phytophthora root rots of azalea and sweet orange', *Phytopathology*, **85**, 29–36.

Fenton, A. M., Stephens, P. M., Crowley, J., O'Callaghan, M., and O'Gara, F. (1992). 'Exploitation of gene(s) involved in 2,4-diacetylphloroglucinol biosynthesis to confer a new biocontrol capability to a *Pseudomonas* strain', *Appl. Environ. Microbiol.*, **58**, 3873–3878.

Fravel, D. R. (1988). 'Role of antibiosis in the biocontrol of plant diseases', *Annu. Rev. Phytopathol.*, **26**, 75–91.

Fravel, D. R. (1996). 'Interaction of biocontrol fungi with sublethal rates of metham sodium for control of *Verticillium dahliae*', *Crop Prot.*, **15**, 115–119.

Fravel, D. R. and Engelkes, C. A. (1994). 'Biological management', in *Epidemiology and Management of Root Diseases* (eds C. L. Campbell and D. M. Benson), pp. 293–308, Springer-Verlag, New York.

Fravel, D. R. and Larkin, R. P. (1996). 'Availability and application of biocontrol products', *Biol. Cult. Tests*, **11**, 1–7.

Fravel, D. R., and Roberts, D. P. (1991). '*In situ* evidence for the role of glucose oxidase in the biocontrol of verticillium wilt by *Talaromyces flavus*', *Biocontrol Sci. Technol.*, **1**, 91–99.

Fravel, D. R., Connick, W. J. Jr., and Lewis, J. A. (1997). 'Formulation of microorganisms to control plant diseases', in *Formulation of Microbial Biopesticides, Beneficial Microorganisms and Nematodes* (ed. H. D. Burges), Chapman and Hall, London (in press).

Fravel, D. R., Lewis, J. A., and Chittams, J. L. (1995). 'Alginate prill formulations of *Talaromyces flavus* with organic carriers for biocontrol of *Verticillium dahliae*', *Phytopathology*, **85**, 165–168.

Gamliel, A. and Katan, J. (1991). 'Involvement of fluorescent pseudomonads and other microorganisms in increased growth response of plants in solarized soils', *Phytopathology*, **81**, 494–502.

Gamliel, A. and Stapleton, J. J. (1993). 'Characterization of antifungal volatile compounds evolved from solarized soil amended with cabbage residues', *Phytopathology*, **83**, 899–905.

Georgakopoulos, D. G., Hendson, M., Panopoulos N. J., and Schroth, M. N. (1994). 'Analysis of expression of a phenazine biosynthesis locus of *Pseudomonas aureofaciens* PGS12 on seeds with a mutant carrying a phenazine biosynthesis locus-ice nucleation reporter gene fusion', *Appl. Environ. Microbiol.*, **60**, 4573–4579.

Gilbert, G. S., Parke, J. L., Clayton, M. K., and Handelsman, J. (1993). 'Effects of an introduced bacterium on bacterial communities on roots', *Ecology*, **74**, 840–854.

Gracia-Garza, J. A., Bailey, B. A., Paulitz, T. C., Lumsden, R. D., Reeleeder, R. D., and Roberts, D. P. (1997). 'Effect of sclerotial damage of *Sclerotinia sclerotiorum* on the mycoparasitic activity of *Trichoderma hamatum*', *Biocontrol Sci. Technol.*, **7**, 401–403.

Gracia-Garza, J. A., Fravel, D. R. Bailey, B. A., and Hebbar, P. K. (1996). 'Effect of formulation on distribution of the mycoherbicide *Fusarium oxysporum* f.sp. *erythroxyli* in the field', (Abstr.) *Phytopathology*, **86**, S23.

Grebus, M. E., Watson, M. E., and Hoitink, H. A. J. (1994). 'Biological, chemical and physical properties of composted yard trimmings as indicators of maturity and plant disease suppression', *Compost Sci. Utilization*, **2**, 57–71.

Hall, R. (1995). 'Challenges and prospects of integrated pest management', in *Novel Approaches to Integrated Pest Management* (ed. R. Reuveni), pp. 1–19, Lewis Publishers, Boca Raton, FL.

Halverson, L. J., Clayton, M. K., and Handelsman, J. (1993). 'Population biology of *Bacillus cereus* UW85 in the rhizosphere of field-grown soybeans', *Soil Biol. Biochem.*, **25**, 485–493.

Handelsman, J. and Stabb, E. V. (1996). 'Biocontrol of soilborne pathogens', *Plant Cell*, **8**, 1855–1869.

Harman, G. E. (1991). 'Seed treatments for biological control of plant disease', *Crop Prot.*, **10**, 166–171.

Haran, S., Schickler, H., Oppenheim, A., and Chet, I. (1995). 'New components of the chitinolytic system of *Trichoderma harzianum*', *Mycol. Res.*, **99**, 441–446.

Harris, A. R., Schisler, D. A., Neate, S. M., and Ryder, M. H. (1994). 'Suppression of damping-off caused by *Rhizoctonia solani*, and growth promotion, in bedding plants by binucleate Rhizoctonia spp'., *Soil Biol. Biochem.*, **26**, 263–268.

Harris, A. R., Schisler, D. A., and Ryder, M. H. (1993). 'Binucleate *Rhizoctonia* isolates control damping-off caused by *Pythium ultimum* var. *sporangiiferum* and promote growth in *Capsicum* and *Celosia* seedlings in pasteurized potting medium', *Soil Biol. Biochem.*, **25**, 909–914.

Hebbar, K. P., Atkinson, D., Tucker, W., and Dart, P. J. (1992). 'Suppression of *Fusarium moniliforme* by maize root-associated *Pseudomonas cepacia*', *Soil Biol. Biochem.*, **24**, 1009–1020.

Heininger, U., and Rigling, D. (1994). 'Biological control of chestnut blight in Europe', *Annu. Rev. Phytopathol.*, **32**, 581–599.
Herr, L. J. (1995). 'Biological control of *Rhizoctonia solani* by binulceate *Rhizoctonia* spp. and hypovirulent *R. solani* agents', *Crop Prot.*, **14**, 179–186.
Hill, D. S., Stein, J. I., Torkewitz, N. R., Morse, A. M., Howell, C. R., Pachlatko, J. P., Becker, J. O., and Ligon, J. M. (1994). 'Cloning of genes involved in the synthesis of pyrrolnitrin from *Pseudomonas fluorescens* and role of pyrrolnitrin synthesis in biological control of plant disease', *Appl. Environ. Microbiol.*, **60**, 78–85.
Hoitink, H. A. J., and Fahy, P. C. (1986). 'Basis for the control of soilborne plant pathogens with compost', *Annu. Rev. Phytopathol.*, **24**, 93–114.
Homma, Y., Sitton, J. W., Cook, R. J., and Old, K. (1979). 'Perforation and destruction of pigmented hyphae of *Gaeumannomyces graminis* by vampyrellid amoebae from Pacific Northwest wheat field soils', *Phytopathology*, **69**, 1118.
Hoper, H. Steinberg, C., and Alabouvette, C. (1995). 'Involvement of clay type and pH in the mechanisms of soil suppressiveness to Fusarium wilt of flax', *Soil Biol. Biochem.*, **27**, 955–967.
Hornby, D., ed. (1990). *Biological Control of Plant Pathogens*, CAB Intl., Oxon, UK.
Howell, C. R. (1982). 'Effect of Gliocladium virens on *Pythium ultimum*, *Rhizoctonia solani*, and damping-off of cotton seedlings', *Phytopathology*, **72**, 496–498.
Howell, C. R. (1987). 'Relevance of mycoparasitism in the biological control of *Rhizoctonia solani* by *Gliocladium virens*', *Phytopathology*, **77**, 992–994.
Howie, W. J. and Echandi, E. (1983). 'Rhizobacteria: influence of cultivar and soil type on plant growth and yield of potato', *Soil Biol. Biochem.*, **15**, 127–132.
Howie, W. J. and Suslow, T. V. (1991). 'Role of antibiotic biosynthesis in the inhibition of *Pythium ultimum* in the cotton spermosphere and rhizosphere by *Pseudomonas fluorescens*', *Molec. Plant-Microbe Interact.*, **4**, 393–399.
Hubbard, J. P., Harman, G. E., and Eckenrode, C. J. (1982). 'Interaction of a biological control agent, *Chaetomium globosum*, with seed coat microflora', *Can. J. Microbiol.*, **28**, 431–437.
Hurst, A. E. (1984). 'Reversible heat damage', in *Repairable Lesions in Microorganisms* (eds A. Hurst and A. Nasim), pp. 303–318, Academic Press, New York.
Janiesiwiscz, W. J. and Marchi, A. (1992). 'Control of storage rots on various pear cultivars with a saprophytic strain of *Pseudomonas syringae*', *Plant Dis.*, **76**, 555–560.
Jeger, G. and Velvis, H. (1986). 'Biological control of *Rhizoctonia solani* on potatoes by antagonists', *Neth. J. Plant Pathol.*, **92**, 231–238.
Johnson, K. B., Stockwell, V. O., Burgett, D. M., Sugar, D., and Loper, J. E. (1993). 'Dispersal of *Erwinia herbicola* and *Pseudomonas fluorescens* by honey bees from hives to apple and pear blossoms', *Phytopathology*, **83**, 478–484.
Kaiser, W. J. and Hannon, R. M. (1984). 'Biological control of seed rot and preemergence damping-off of chickpea with *Penicillium oxalicum*', *Plant Dis.*, **68**, 806–811.
Katan, T. (1982). 'Resistance to 3,5-dichlorophenyl-*N*-cyclicimide (dicarboximide) fungicides in the grey mould pathogen *Botrytis cinerea* in protected crops', *Plant Pathol.*, **31**, 133–141.
Katan, J., Ginzburg, C., and Assaraf, M. (1996). 'Pathogen weakening as a component of integrated control', in *Advances in Biological Control of Plant Diseases* (eds T. Wenhua, R. J. Cook and A. Rovira), pp. 320–326, China Agricultural University Press. Beijing, China.
Keel, C., Schnider, U., Maurhofer, M., Voisard, C., Laville, J., Burger, U., Wirthner, P., Haas, D., and Defago, G. (1992). 'Suppression of root diseases by *Pseudomonas fluorescens* CHAO: Importance of the bacterial secondary metabolite 2,4-diacetylphloroglucinol', *Molec. Plant-Microbe Interact.*, **5**, 4–13.

Keel, C., Weller, D. M., Natsch, A., Defago, G., Cook, R. J., and Thomashow, L. (1996). 'Conservation of the 2,4-diacetylphloroglucinol biosynthesis locus among fluoresecnt Pseudomonas strains from diverse geographoc locations', *Appl. Environ. Microbiol.*, **62**, 552–563.

Keel, C., Wirthner, P. H., Oberhansli, T. H., Voisard, C., Burger, U., Haas, D., and Defago, G. (1990). 'Pseudomonads as antagonists of plant pathogens in the rhizosphere: role of the antibiotic 2,4-diacetylphloroglucinol in the suppression of black root rot of tobacco', *Symbiosis*, **9**, 327–341.

Kelman, A., and Cook, R. J. (1977). 'Plant pathology in the People's Republic of China', *Annu. Rev. Phytopathol.*, **17**, 409–429.

Kim K. K., Fravel, D. R., and Papavizas, G. C. (1988). 'Identification of a metabolite produced by *Talaromyces flavus* as glucose oxidase and its role in the biocontrol of *Verticillium dahliae*', *Phytopathology*, **78**, 488–492.

Kim, K. K., Fravel, D. R., and Papavizas, G. C. (1990a). 'Production, purification, and properties of glucose oxidase from the biocontrol fungus *Talaromyces flavus*', *Can. J. Microbiol.*, **36**, 199–205.

Kim, K. K., Fravel, D. R., and Papavizas, G. C. (1990b). 'Glucose oxidase as the antifungal principle of talaron from *Talaromyces flavus*', *Can. J. Microbiol.*, **36**, 760–764.

King, E. B. and Parke, J. L. (1993). 'Biocontrol of Aphanomyces root rot and Pythium damping-off by *Pseudomonas cepacia* AMMD on four pea cultivars', *Plant Dis.*, **77**, 1185–1188

Kloepper, J. W., Leong, J., Teintze, M., and Schroth, M. N. (1980). 'Pseudomonas siderophores: a mechanism explaining disease-suppressive soils', *Curr. Microbiol.*, **4** 317–320.

Kloepper, J. W., Zehnder, G. W., Tuzun, S., Murphy, J. F., Wei, G., Yao, C., and Raupach, G. (1996). 'Toward agricultural implementation of PGPR-mediated induced systemic resistance against crop pests', in *Advances in Biological Control of Plant Diseases* (eds T. Wenhua, R. J. Cook and A. Rovira), pp. 165–174, China Agricultural University Press. Beijing, China.

Kobayashi, D. Y., Guglielmoni, M., and Clarke, B. B. (1995). 'Isolation of the chitinolytic bacteria *Xanthomonas maltophilia* and *Serratia marcescens* as biological control agents for summer patch disease of turfgrass', *Soil Biol. Biochem.*, **27**, 1479–1487.

Kraus, J. and Loper, J. E. (1995). 'Characterization of a genomic region required for production of the antibiotic pyoluteorin by the biological control agent *Pseudomonas fluorescens* Pf-5', *Appl. Environ. Microbiol.*, **61**, 849–854.

Krebs, B., Junge, H., Ockhardt, A., Hoding, B., Heubner, D., and Erben, U. (1993). '*Bacillus subtilis*—and effective biocontrol agent', *Pestic. Sci.*, **37**, 427–429.

Kroon, B. A. H. M., Scheffer, R. J., and Elgersma, D. M. (1991). 'Induced resistance in tomato plants against fusarium wilt invoked by *Fusarium oxysporum* f.sp. *dianthi*', *Neth. J. Plant Pathol.*, **97**, 401–408.

Kuter, G. A., Nelson, E. B., Hoitink, H. A. J., and Madden, L. V. (1983). 'Fungal populations in container media amended with composted hardwood bark suppressive and conducive to *Rhizoctonia* damping-off, *Phytopathology*, **73**, 1450–1456.

Larkin, R. P. and Fravel, D. R. (1996). 'Efficacy of various biocontrol organisms in the control of Fusarium wilt of tomato (Abstract), *Phytopathology*, **86**, S83.

Larkin, R. P., Hopkins, D. L., and Martin, F. N. (1993). 'Effect of successive watermelon plantings on *Fusarium oxysporum* and other microorganisms in soils suppressive and conducive to Fusarium wilt of watermelon', *Phytopathology*, **83**, 1097–1105.

Larkin, R. P., Hopkins, D. H., and Martin, F. N. (1996). 'Suppression of Fusarium wilt of watermelon by nonpathogenic *Fusarium oxysporum* and other microorganisms recovered from a disease-suppressive soil', *Phytopathology*, **86**, 812–819.

Lartey, R. T., Curl, E. A., and Peterson, C. M. (1994). 'Interactions of mycophagous Collembola and biological control fungi in the suppression of *Rhizoctonia solani*', *Soil Biol. Biochem.*, **26**, 81–88.
Latour, X., Corberand, T. S., Laguerre, G., and Lemanceau, P. (1996). 'The composition of pseudomonad populations associated with roots is influenced by plant and soil type', *Appl. Environ. Microbiol.*, **62**, 2449–2456.
Laville, J., Voisard, C., Keel, C., Maurhofer, M., Defago, G., and Haas, D. (1992). 'Global control in *Pseudomonas fluorescens* mediating antibiotic synthesis and suppression of black root rot of tobacco', *Proc. Nat'l Acad. Sci. USA*, **89**, 1562–1566.
Leeman, M., Den Ouden, F. M., Van Pelt, J. A., Cornelissen, Matamala-Garros, A, Bakker, P. A. H. M., and Schippers, B. (1996). 'Suppression of fusarium wilt of radish by co-inoculation of fluorescent *Pseudomonas* spp. and root-colonizing fungi', *Eur. J. Plant Pathol.*, **102**, 21–31.
Leeman, M., Van Pelt, J. A., Den Ouden, F. M., Heinsbrook, M., Bakker, P. A. H. M., and Schippers, B. (1995). 'Induction of systemic resistance by *Pseudomonas fluorescens* in radish cultivars differing in susceptibility to fusarium wilt using a novel bioassay', *Eur. J. Plant Pathol.*, **101**, 655–664.
Lemanceau, P. (1989). 'Role of competition for carbon and iron in mechanisms of soil suppressiveness to Fusarium wilts', in *Vascular Wilt Diseases of Plants* (eds E. C. Tjamos and C. Beckman), pp. 385–395, Springer-Verlag, New York.
Lemanceau, P. and Alabouvette, C. (1991). 'Biological control of fusarium diseases by fluorescent *Pseudomonas* and non-pathogenic *Fusarium*', *Crop Prot.*, **10**, 279–286.
Lemanceau, P., Bakker, P. A. H. M., de Kogel, W. J., and Alabouvette, C. (1992). 'Effect of pseudobactin 358 production by *Pseudomonas putida* WCS358 on suppression of Fusarium wilt of carnations by nonpathogenic *Fusarium oxysporum* Fo47', *Appl. Environ. Microbiol.*, **58**, 2978–2982.
Lemanceau, P., Bakker, P. A. H. M., de Kogel, W. J., Albouvette, C., and Schippers, B. (1993). 'Antagonistic effect on nonpathogenic *Fusarium oxysporum* strain Fo47 and pseudobaction 358 upon pathogenic *Fusarium oxysporum* f.sp. *dianthi*', *Appl. Environ. Microbiol.*, **59**, 74–82.
Lemanceau, P., Corberand, T., Gardan, L., Latour, X., Laguerre, G., Boeufgras, J. M., and Alabouvette, C. (1995) 'Effect of two plant species, flax (*Linum usitatissimum* L.) and tomato (*Lycopersicon esculentum* Mill.), on the diversity of soilborne populations of fluorescent pseudomonads', *Appl. Environ. Microbiol.*, **61**, 1004–1012.
Leong, J. (1986). 'Siderophores: their chemistry and possible role in the biocontrol of plant pathogens', *Annu. Rev. Phytopathol.*, **24**, 187–209.
Lewis, J. A. (1991). 'Formulation and delivery systems of biocontrol agents with emphasis on fungi', in *The Rhizosphere and Plant Growth* (eds D. L. Keister and P. B. Cregan), pp. 279–287, Kluwer Academic Publishers, Dordrecht.
Lewis, J. A. and Papavizas, G. C. (1991). 'Biocontrol of cotton damping-off caused by *Rhizoctonia solani* in the field with formulation of *Trichoderma* and *Gliocladium virens*', *Crop Prot.*, **10**, 396–402.
Lewis, J. A. and Papavizas, G. C. (1992). 'Potential of *Laetisaria arvalis* for the biocontrol of *Rhizoctonia solani*', *Soil Biol. Biochem.*, **24**, 1075–1079.
Lewis, J. A. and Papavizas, G. C. (1993). '*Stilbella aciculosa*: a potential biocontrol fungus against *Rhizoctonia solani*', *Biocontrol Sci. Technol.*, **3**, 3–11.
Lewis, J. A., Fravel, D. R., and Papavizas, G. C. (1995). '*Cladorrhinum foecundissimum*: a potential biological control agent for the reduction of *Rhizoctonia solani*', *Soil Biol. Biochem.*, **27**, 863–869.
Lewis, J. A., Papavizas, G. C., and Hollenbeck, M. D. (1993). 'Biological control of damping-off of snapbeans caused by *Sclerotium rolfsii* in the greenhouse and field with formulations of *Gliocladium virens*, *Biol. Control*, **3**, 109–115.

Lindow, S. E. (1987). 'Competitive exclusion of epiphytic bacteria by Ice$^-$ *Pseudomonas syringae* mutants', *Appl. Environ. Microbiol.*, **53**, 2520–2527.

Lindow, S. E. (1995). 'The use of reporter genes in the study of microbial ecology', *Molec. Ecol.*, **4**, 555–566.

Liu, L., Kloepper, J. W., and Tuzun, S. (1995). 'Induction of systemic resistance in cucumber against Fusarium wilt by plant growth-promoting rhizobacteria', *Phytopathology*, **85**, 695–698.

Loper, J. E. (1988). 'Role of fluorescent siderophore production in biological control of *Pythium ultimum* by a *Pseudomonas fluorescens* strain', *Phytopathology*, **78**, 166–172.

Loper, J. E. and Lindow, S. E. (1994). 'A biological sensor for iron available to bacteria in their habitats on plant surfaces', *Appl. Environ. Microbiol.*, **60**, 1934–1941.

Loper, J. E., Suslow, T. V., and Schroth, M. N. (1984). 'Longnormal distribution of bacterial populations in the rhizosphere', *Phytopathology*, **74** 1454–1460.

Lorito, M., Harman, G. E., Hayes, C. K., Broadway, R. M., Tronsmo, A., Woo, S. L., and Di Pietro, A. (1993). 'Chitinolytic enzymes produced by *Trichoderma harzianum*: antifungal activity of purified endochitinase and chitobiosidase', *Phytopathology*, **83**, 302–307.

Lorito, M., Hayes, C. K., Di Pietro, A., Woo, S. L., and Harman, G. E. (1994). 'Purification, characterization, and synergisitic activity of a glucan 1,3-β-glucosidase and an *N*-acetyl-β-glucosaminidase from *Trichoderma harzianum*', *Phytopathology*, **84**, 398–405.

Lumsden, R. D. (1992). 'Mycoparasitism of soilborne plant pathogens', in *The Fungal Community: Its Organization and Role in the Ecosystem* (eds G. C. Carroll and D. T. Wicklow), pp. 275–293, Marcel Dekker, New York.

Lumsden, R. L. and Locke, J. C. (1989). 'Biological control of damping-off caused by *Pythium ultimum* and *Rhizoctonia solani* with *Gliocladium virens* in soilless mix', *Phytopathology*, **79**, 361–366.

Lumsden, R. D. and Walter, J. F. (1995). 'Development of the biocontrol fungus *Gliocladium virens*: risk assessment and approval for horticultural use', in *Biological Control: Benefits and Risks* (eds H. M. T. Hokkanen and J. M. Lynch), pp. 263–269, Cambridge University Press, Cambridge.

Lumsden, R. D., Garcia-E., R., Lewis, J. A., and Frias-T., G. A. (1987). 'Suppression of damping-off caused by *Pythium* spp. in soil from the indigenous Mexican chinampa agricultural system', *Soil Biol. Biochem.*, **19**, 501–508.

Lumsden, R. D., Garcia-E., R., Lewis, J. A., and Frias-T., G. A. (1990). 'Reduction of damping-off disease in soils from indigenous Mexican agroecosystems', in *Agroecology* (ed. S. R. Gleissman), pp. 83–103, Springer-Verlag, New York.

Lumsden, R. D., Lewis, J. A., and Fravel, D. R. (1995). 'Formulation and delivery of biocontrol agents for use against soilborne plant pathogens', in *Biorational Pest Control Agents: Formulation and Delivery* (eds F. R. Hall and J. W. Barry), pp. 166–182, Amer. Chem. Soc., Washington D. C.

Lumsden, R. D., Lewis, J. A., and Locke, J. C. (1993). 'Managing soilborne plant pathogens with fungal antagonists', in *Pest Management: Biologically Based Technologies* (eds R. D. Lumsden and J. L. Vaughn), pp. 196–203, Amer, Chem. Soc., Washington D. C.

Lumsden, R. D., Lewis, J. A., and Millner, P. D. (1983). 'Effect of composted sewage sludge on several soilborne pathogens and diseases', *Phytopathology*, **73**, 1543–1548.

Lumsden, R. D., Locke, J. C., Adkins, S. T., and Ridout, C. J. (1992). 'Isolation and localization of the antibiotic gliotoxin produced by *Gliocladium virens* from alginate prill in soil and soilless media,' *Phytopathology*, **89**, 230–235.

Mandeel, Q. and Baker, R. (1991). 'Mechanisms involved in biological control of Fusarium wilt of cucumber with strains of nonpathogenic *Fusarium oxysporum*', *Phytopathology*, **81**, 462–469.

Marois, J. J., Johnston, S. A., Dunn, M. D., and Papavizas, G. C. (1982). 'Biological control of Verticillium wilt of eggplant in the field', *Plant Dis.*, **66**, 1166–1168.

Martin, F. N. and Hancock, J. C. (1987). 'The use of *Pythium oligandrum* for biological control of pre-emergence damping-off caused by *Pythium ultimum*', *Phytopathology*, **77**, 1013–1020.

Maurhofer, M., Keel, C., Schneider, U., Voisard, C., Hass, D., and Defago, G. (1992). 'Influence of enhanced antibiotic production in *Pseudomonas fluorescens* strain CHA0 on its disease suppressive capacity', *Phytopathology*, **82**, 190–195.

Maurhofer, M., Hase, C., Meuwly, P., Metraux, J.-P., and Defago, G. (1994). 'Induction of systemic resistance of tobacco to tobacco necrosis virus by the root-colonizing *Pseudomonas fluorescens* strain CHA0: influence of the *gacA* gene and of pyoverdine production', *Phytopathology*, **84**, 139–146.

Maurhofer, M., Keel, C., Haas, D., and Defago, G. (1995). 'Influence of plant species on disease suppression by *Pseudomonas fluorescens* strain CHA0 with enhanced antibiotic production', *Plant Pathol.*, **44**, 40–50.

McGuire, R. G. (1994). 'Application of *Candida guilliermondii* in commercial citrus coatings for biocontrol of *Penicillium digitatum* on grapefruits', *Biol. Control* **4**, 1–7.

McLaren, D. L., Huang, H. C., Kozub, G. C., and Rimmer, S. R. (1994). 'Biological control of sclerotinia wilt of sunflower with *Talaromyces flavus* and *Coniothyrium minitans*', *Plant Dis.*, **78**, 231–235.

McLoughlin, T. J., Quinn, J. P., Betterman, A., and Bookland, R. (1992). '*Pseudomonas cepacia* suppression of sunflower wilt fungus and role of antigungal compounds in controlling the disease', *Appl. Environ. Microbiol.*, **58**, 1760–1763.

Mintz, A. S. and Walter, J. F. (1993). 'A private industry approach: development of GlioGard^{TM} for disease control in horticulture', in *Pest Management; Biologically Based Technologies* (eds R. D. Lumsden and J. L. Vaughn), pp. 398–403, American Chemical Society, Washington D.C.

Minuto, A., Migheli, Q., and Garabaldi, A. (1995a). 'Evaluation of antagonistic strains of *Fusarium* spp. in the biological and integrated control of Fusarium wilt of cyclamen', *Crop Prot.*, **14**, 221–226.

Mungnier, J. and Jung, G. (1985). 'Survival of bacteria and fungi in relation to water activity and the solvent properties of water in biopolymer gels', *Appl. Environ. Microbiol.*, **50**, 108–114.

Munnecke, D. E., Wilbur, W., and Darely, E. F. (1976). 'Effect of heating of drying on *Armillaria mellea* or *Trichoderma viride* and the relation to survival of *A. mellea* in soil', *Phytopathology* **66**, 1363–1368.

Minuto, A., Migheli, Q., and Garabaldi, A. (1995b). 'Integrated control of soilborne plant pathogens by solar heating and antagonistic microorganisms', *Acta. Hort.*, **382**, 138–144.

Neal, J. L., Larson, R. L., and Atkinson, T. G. (1973). 'Changes in rhizosphere populations of selected physiological groups of bacteria related to substitution of specific pairs of chromosomes in spring wheat', *Plant Soil*, **39**, 209–212.

Nelson, E. B. (1988). 'Biological control of Pythium seed rot and preemergence damping-off of cotton with *Enterobacter cloacae* and *Erwinia herbicola*', *Plant Dis.*, **72**, 140–142.

Nelson, E. B. and Maloney, A. P. (1992). 'Molecular approaches for understanding biological control mechanisms in bacteria: studies of the interaction of *Enterobacter cloacae* with *Pythium ultimum*', *Can. J. Plant Pathol.*, **14**, 106–114.

Nelson, E. B., Chao, W. L., Norton, J. M., Nash, G. T., and Harman, G. E. (1986). 'Attachment of *Enterobacter cloacae* to hyphae of *Pythium ultimum*: possible role in the biological control of *Pythium* preemergence damping off, *Phytopathology*, **76**, 327–335.

Nelson, E. B, Kuter, G. A., and Hoitink, H. A. J. (1983). 'Effect of fungal antagonists and compost age on suppression of *Rhizoctonia* damping-off in container media amended with composted hardwood bark', *Phytopathology*, **73**, 1475–1462.

O'Connell, K. P., Goodman, R. M., and Handelsman, J. (1996). 'Engineering the rhizosphere: expressing a bias', *Trends Biotechnol.*, **14**, 83–88.

Osburn, R. M., Milner, J. L., Oplinger, E. S., Smith, R. S., and Handelsman, J. (1995). 'Effect of *Bacillus cereus* UW85 on the yield of soybean at two field sites in Wisconsin', *Plant Dis.*, **79**, 551–556.

Ownley, B. H. Weller, D. M., and Aldredge, J. R. (1992). 'Relation of soil chemical and physical factors with suppression of take-all by *Pseudomonas fluorescens* 2-79', in *Plant Growth-Promoting Rhizobacteria-Progress and Prospects* (eds C. Keel, B. Koller and G. Défago), pp. 299–301, IOBC/WPRS Bull. no. 14.

Papavizas, G. C. (1985). '*Trichoderma* and *Gliocladium*: biology, ecology, and potential for biocontrol', *Annu. Rev. Phytopathol.*, **23**, 23–54.

Papavizas, G. C., and Lewis, J. A. (1989). 'Effect of *Gliocladium* and *Trichoderma* on damping-off of snap bean caused by *Sclerotium rolfsii* in the greenhouse', *Plant Pathol.*, **38**, 277–286.

Park, Y. J., Martyn, R. D., and Miller, M. E. (1996). 'dsRNA is responsible for cultural aberrations in *Monosporascus cannonballus* and hypovirulence to muskmelon', (Abstr.) *Phytopathology*, **86**, S107.

Park, C. S., Paulitz, T. C., and Baker, R. (1988). 'Biocontrol of Fusarium wilt of cucumber resulting from interactions between *Pseudomonas putida* and nonpathogenic isolates of *Fusarium oxysporum*', *Phytopathology*, **78**, 190–194.

Paulitz, T. C. (1990). 'Biochemical and ecological aspects of competition in biological control', in *New Directions in Biological Control: Alternatives for Suppressing Agricultural Pests and Diseases* (eds R. Baker and P. E. Dunn), pp. 713–724, Alan R. Liss, New York.

Paulitz, T. C. (1992). 'Biological control of damping-off diseases with seed treatments', in *Biological Control of Plant Diseases, Progress and Challenges for the Future* (eds E. C. Tjamos, G. C. Papavizas and R. J. Cook), pp. 145–156, Plenum Press, New York.

Paulitz, T. C. and Baker, R. (1987). 'Biological control of Pythium damping-off of cucumbers with *Pythium nunn*: population dynamics and disease suppression', *Phytopathology*, **77**, 335–340.

Peng, G. and Sutton, J. C. (1991). 'Evaluation of microorganisms for biocontrol of *Botrytis cinerea* in strawberry ', *Can. J. Plant Pathol.*, **13**, 247–257.

Pierson, E. A., and Weller, D. M. (1994). 'Use of mixtures of fluorescent pseudomonads to suppress take-all and improve the growth of wheat', *Phytopathology*, **84**, 940–947.

Pierson, L. S., III, and Thomashow, L. S. (1992). 'Cloning and heterologous expression of the phenazine biosynthetic locus from *Pseudomonas aureofaciens*', *Molec. Plant-Microbe Interact.*, **5**, 330–339.

Pullman, G. S., De Vay J. E., and Garber, R. H. (1981). 'Soil solarization and thermal death: A logarithmic relationship between time and temperature for four soilborne plant pathogens', *Phytopathology*, **71**, 959–964.

Raaijmakers, J. M., Weller, D. M., and Thomashow, L. S. (1997). 'Frequency of antibiotic-producing *Pseudomonas* spp. in natural environments', *Appl. Environ. Microbiol.*, **63**, 881–887.

Ramirez-Villapudua, J., and Munnecke, D. E. (1987). 'Control of cabbage yellows (*Fusarium oxysporum* f.sp. *conglutinans*) by solar heating of field soils amended with dry cabbage residues', *Plant Dis.*, **71**, 217–221.

Ramirez-Villapudu, J., and Munnecke, D. E. (1988). 'Effect of solar heating and soil amendments of cruciferous residues on *Fusarium oxysporum* f.sp. *conglutinans* and other organisms', *Phytopathology*, **78**, 289–295.

Rishbeth, J. (1975). 'Stump inoculation: a biological control of *Fomes annosus*', in *Biology and Control of Soil-Borne Pathogens*(ed. G. W. Bruehl), pp. 158–162, Am. Phytopathol. Soc., St Paul, MN.

Ristaino, J. B., Perry, K. B., and Lumsden, R. D. (1991). 'Effect of solarization and *Gliocladium virens* on sclerotia of *Sclerotium rolfsii*, soil microbiota and the incidence of southern blight of tomato', *Phytopathology*, **81**, 1117–1124.

Roberts, D. P., Marty, A. M., Dery, P. D., and Hartung, J. S. (1996). 'Isolation and modulation of growth of colonization-impaired strain of *Enterobacter cloacae* in cucumber spermosphere', *Can. J. Microbiol.*, **42**, 196–201.

Roberts, D. P., Dery, P. D., Hebbar, P. K., Mao, W., and Lomsden, R. D. (1997). 'Biological control of damping-off of cucumber caused by *Pythium ulfimun* with a root-colonization-deficient strain of *Escherichia coli*, *J. Phytopathol.*, **145**, 383–388.

Rodriguez-Kabana, R., Kloepper, J. W., Robertson, D. G., and Wells, L. W. (1992). 'Velvetbean for the management of root-knot and southern blight in peanut', *Nematropica*, **22**, 75–80.

Rojas-Rabiela, T. (1983). *La agricultura chinampera*, Universidad Autónoma Chapingo, Mexico.

Rothrock, C. S. (1992). 'Tillage systems and plant disease', *Soil Sci.*, **154**, 308–315.

Sacherer, P., Defago, G., and Haas, D. (1944). 'Extracellular protease and phospholipase C are controlled by the global regulatory gene *gac*A in the biocontrol strain *Pseudomonas fluorescens* CHA0' *FEMS Microbiol. Lett.*, **116**, 155–160.

Saksirirat, W., Boonsakdaporn, N., Sirithorn, P., and Prachinburavan, A. (1996). 'An application of the mycoparasite *Trichoderma harizianum* Rifai. in combination with mancozeb for control of tomato stem rot in northeast Thailand', in *Advances in Biological Control of Plant Diseases* (eds T. Wenhua, R. J. Cook and A. Rovira), pp. 327–329, China Agricultural University Press, Beijing, China.

Savka, M. A. and Farrand, S. K. (1992). 'Mannityl opine accumulation and exudation by transgenic tobacco', *Plant Physiol.*, **98**, 784–789.

Scher, F. M. and Baker, R. (1982). 'Effect of *Pseudomonas putida* and a synthetic iron chelator on induction of suppressiveness to Fusarium wilt pathogens', *Phytopathology*, **72**, 1567–1573.

Schroth, M. N. and Becker, J. O. (1990). 'Concepts of ecological and physiological activities of rhizobacteria related to biological control and plant growth promotion', in *Biological Control of Soil-borne Plant Pathogens* (ed. D. Hornby), pp. 389–414, CAB Intl., Oxon, UK.

Sivan, A. and Chet, I. (1993). 'Integrated control of fusarium crown and root rot of tomato with *Trichoderma harzianum* in combination with methyl bromide or soil solarization', *Crop Prot.*, **12**, 380–386.

Stabb, E. V., Jacobsen, L. M., and Handelsman, J. (1994). 'Zwittermicin A-producing strains of *Bacillus cereus* from diverse soils', *Appl. Environ. Microbiol.*, **60**, 4404–4412.

Stack, J. P., Kenerley, C. M., and Petitt, R. E. (1987). 'Influence of carbon and nitrogen sources, relative carbon and nitrogen concentrations, and soil moisture on the growth in nonsterile soil of soilborne fungal antagonists,' *Can. J. Microbiol.* **33**, 626–631.

Stapleton, J. J., and De Vay, J. E. (1984). 'Thermal components of soil solarization as related to changes in soil and root microflora and increased plant growth response', *Phytopathology*, **74**, 255–259.

Stephens, P. M., Davoren, C. W., Ryder, M. H., Doubre, B. M., and Correll, R. L. (1994). 'Field evidence for reduced severity of *Rhizoctonia* bare patch disease of wheat, due to the presence of the earthworms *Aporrectodea rosea* and *Aporrectodea trapezoides*', *Soil Biol. Biochem.*, **26**, 1495–1500.

Stosz, S. K., Fravel, D. R., and Roberts, D. P. (1996). 'In vitro analysis of the role of glucose oxidase from *Talaromyces flavus* in biocontrol of the plant pathogen *Verticillium dahliae*', *Appl. Environ. Microbiol.*, **62**, 3183–3186.

Subbarao, K. V., and Hubbard, J. C. (1996). Interactive effects of broccoli residue and temperature on *Verticillium dahliae* microsclerotia in soil and on wilt in cauliflower,' *Phytopathology*, **86**, 1303–1310.

Sutton, J. C. (1995). 'Evaluation of micro-organisms for biocontrol: *Botrytis cinerea* and strawberry, a case study', in *Advances in Plant Pathology* (eds J. H. Andrews and I. Tommerup), pp. 173–190, Academic Press, New York.

Sutton, J. C., Li, D. W., Peng, G., Yu, Y., Zhang, P., and Valdebenito-Sanhueza, R. M. (1997). '*Gliocladium roseum*: a versatile adversary of *Botrytis cinera* in crops', *Plant Dis.*, **81**, 316–328.

Taylor, A. G., Harman, G. E., and Nielsen, P. A. (1994). 'Biological seed treatments using *Trichoderma harzianum* for horticultural crops', *Hort Technol.*, **4**, 105–108.

Thomashow, L. S., and Weller, D. M. (1988).'Role of phenazine antibiotic from *Pseudomonas fluorescens* in biological control of *Geanumannomyces graminis var. tritici*', *J. Bacteriol.*, **170**, 3499–3508.

Thomashow, L. S., Weller, D. M., Bonsfall, R. F., and Pierson, L. S. III. (1990). 'Production of the antibiotic phenazine-1-carboxylic acid by fluorescent *Pseudomonas* species in the rhizosphere of wheat', *Appl. Environ. Microbiol.*, **56**, 908–912.

Thomson, S. V., Hansen, D. R., Flint, K. M., and Vandenberg, J. D. (1992). 'Dissemination of bacteria antagonistic to Erwinia amylovora by bees', *Plant Dis.*, **76**, 1052–1056.

Tjamos, E. C. and Fravel, D. R. (1995) 'Detrimental effects of sublethal heating and *Talaromyces flavus* on microsclerotia of *Verticillium dahliae*', *Phytopathology*, **85**, 388–392.

Tjamos, E. C. and Niklis, N. (1990). 'Synergism between soil solarization and *Trichoderma harzianum* preparations in controlling Fusarium wilt of beans', *Proc. 8th Conf. Mediterranean Phytopathological Union*, 145–146.

Tjamos, E. C. and Paplomatas, E. J. (1987). 'Effect of solarization on the survival of fungal antagonists of *Verticillium dahliae*', *EPPO Bull.*, **17**, 645–663.

Tjamos, E. C. and Paplomatas, E. J. (1988). 'Long-term effect of soil solarization in controlling *Verticillium* wilt of globe artichokes in Greece', *Plant Pathol.*, **37**, 507–515.

Tjamos, E. C. Papavizas, G. C., and Cook, R. J., eds (1992). *Biological Control of Plant Diseases: Progress and Challenges for the Future*, Plenum Press, New York.

Tuzun, S. and Kloepper, J. W. (1995). 'Practical application and implementation of induced resistance', in *Induced Resistance to Diseases in Plants* (eds R. Hammerschmidt and J. Kuc), pp. 152–168, Kluwer Academic Publishers, Dordrecht.

Tuzun, S., Juarez, J., Nesmith, W. C., and Kuc, J. (1992). 'Induction of systemic resistance in tobacco against metalaxyl-tolerant strains of *Peronospora tabacina* and the natural occurrence of the phenomenon in Mexico', *Phytopathology*, **82**, 425–429.

Vandermeer, J., Carney, J., Gersper, P., Perfecto, I., and Rosset, P. (1993). 'Cuba and the dilemma of modern agriculture', *Agric. Hum. Values*, **X**, 3–8.

Van Peer, R., Niemann, G. J., and Schippers, B. (1991). 'Induced resistance and phytoalexin accumulation in biological control of Fusarium wilt of carnation by *Pseudomonas* sp. strain WCS417r', *Phytopathology*, **81**, 728–734.

Villajuan-Abgona, R., Kageyama, K., and Hyakumachi, M. (1996). 'Biocontrol of Rhizoctonia damping-off of cucumber by nonpathogenic binucleate *Rhizoctonia*', *Eur. J. Plant Pathol.*, **102**, 227–235.

Vincent, M. N., Harrison, L. A., Brackin, J. M., Kovacevich, P. A., Mukerji, P., Weller, D. M., and Pierson, E. A. (1991). 'Genetic analysis of the antifungal activity of a soilborne *Pseudomonas aureofaciens* strain', *Appl. Environ. Microbiol.*, **57**, 2928–2934.

Voisard, C., Keel, C., Haas, D., and Defago, G. (1989). 'Cyanide production by *Pseudomonas fluorescens* helps suppress black root rot of tobacco under gnotobiotic conditions', *EMBO J.*, **8**, 351–358.

Washington, W. S., Shannuganathan, N., and Forbed, C. (1992). 'Fungicide control of strawberry fruit rots, and the field occurrence of resistance of *Botrytis cinerea* to iprodione, benomyl and dichlorofluamid', *Crop Prot.*, **11**, 355–360.

Watkins, J. E. and Boosalis, M. G. (1994). 'Plant disease incidence as influenced by conservation tillage systems', in *Managing Agricultural Residues* (ed. P. Unger) pp. 261–283, Lewis Publishers, Boca Raton, FL.

Wei, G., Kloepper, J. W., and Tuzun, S. (1991). 'Induction of systemic resistance of cucumber to *Colletotrichum orbiculare* by select strains of plant growth-promoting rhizobacteria', *Phytopathology*, **81**, 1508–1512.

Wei, G., Kloepper, J. W., and Tuzun, S. (1996). 'Induced systemic resistance to cucumber diseases and increased plant growth by plant growth-promoting rhizobacteria under field conditions', *Phytopathology*, **86**, 221–224.

Weller, D. M. (1983). 'Colonization of wheat roots by a fluorescent pseudomonad suppressive to take-all', *Phytopathology*, **73**, 1548–1553.

Weller, D. M. (1988). 'Biological control of soilborne plant pathogens in the rhizosphere with bacteria', *Annu. Rev. Phytopathol.*, **26**, 379–407.

Weller, D. M. and Cook, R. J. (1983). 'Suppression of take-all of wheat by seed treatments with fluorescent pseudomonads', *Phytopathology*, **73**, 463–469.

Whipps, J. M. and Gerlagh, J. M. (1992). 'Biology of *Coniothyrium minitans* and its potential for use in disease biocontrol', *Mycol. Res.*, **96**, 897–907.

Wilhite, S. E., Lumsden, R. D., and Straney, D. C. (1994). 'Mutational analysis of gliotoxin production by the biocontrol fungus *Gliocladium virens* in relation to suppression of *Pythium* damping-off', *Phytopathology*, **84**, 816–821.

Wilson, C. L. and Wisniewski, M. E., eds (1994). *Biological Control of Postharvest Diseases: Theory and Practice*, CRC Press, Boca Raton, FL.

Wilson, M. and Lindow, S. E. (1994a). 'Ecological similarity and coexistence of epiphytic ice-nucleating (Ice^+) *Pseudomonas syringae* and a non-ice-nucleating (Ice^-) biological control agent', *Appl. Environ. Microbiol.*, **60**, 3128–3137.

Wilson, M. and Lindow, S. E. (1994b). 'Coexistence among epiphytic bacterial populations mediated through nutritional resource partitioning', *Appl. Environ. Microbiol.*, **60**, 4468–4477.

Yuan, W. M. and Crawford, D. L. (1995).'Characterization of *Streptomyces lydicus* WYEC108 as a potential biocontrol agent against fungal and seed rots', *Appl. Environ. Microbiol.*, **61**, 3119–3128.

Yuen, G. Y., Craig, M. L., Kerr, E. D., and Steadman, J. R. (1994). 'Influences of antagonist population levels on the inhibition of *Sclerotinia sclerotiorum* on dry edible bean by *Erwinia herbicola*', *Phytopathology*, **84**, 495–501.

Zhou, T. and Paulitz, T. C. (1994). Induced resistance in the biocontrol of *Pythium aphanidermatum* by *Pseudomonas* spp. on cucumber', *J. Phytopathol.*, **142**, 51–63.

7 Activators for Systemic Acquired Resistance

I. YAMAGUCHI
RIKEN, Wako, Saitama, Japan

INTRODUCTION

Disease is a rare outcome in the spectrum of plant-microbe interaction. Plants have evolved a complex set of defence mechanisms that prevent infection and disease in most cases. The battery of defence reactions includes preformed physical and chemical barriers as well as induced defence mechanisms such as strengthening of cell walls and the production of defence-related molecules (Dixon and Lamb, 1990; Osbourn, 1996). Disease can result when a pathogen is able to overcome the plant defences, e.g. by either actively suppressing or outcompeting them.

The ability of a plant to respond to an infection is determined by genetic traits in both the host and the pathogen (for reviews see Dixon and Lamb, 1990; Low and Merida, 1996; Dangl *et al.*, 1996; Osbourn, 1996). Some resistance mechanisms are specific for plant cultivars and certain strains of pathogens. In these cases plant resistance genes recognize pathogen-derived molecules resulting from expression of so-called avirulence genes, which often triggers a signal cascade leading to rapid host cell death (hypersensitive response, HR, for reviews see Dangl *et al.*, 1996; Bent, 1996; Pryor and Ellis, 1993). Such 'gene-for-gene' relationships usually lead to highly efficient, but very specific, plant

Fungicidal Activity. Edited by D. H. Hutson and J. Miyamoto

resistance. In contrast, another set of plant resistance mechanisms provides broad-spectrum disease control. The mechanisms involved include preformed physical barriers (cell walls including lignin, waxes, accumulation of antimicrobial metabolites, etc.) as well as inducible mechanisms. Induction of resistance mechanisms occurs locally at the site of attempted penetration (e.g. hypersensitive response, phytoalexins, cell wall strengthening, etc.) as well as in distant (systemic) parts of the plant. This article will review the work on systemically induced resistance response (SAR) focussing on biological models of SAR, recent findings in signal transduction and on the potential for using SAR in agricultural practice.

SYSTEMIC ACQUIRED RESISTANCE (SAR)

THE BIOLOGICAL PHENOMENON

An important milestone in the development of the current understanding of the SAR phenomenon was the publication of a classic group of experiments by Frank A. Ross (1961). He demonstrated that resistance in tobacco (var. Xanthi nc) to tobacco mosaic virus (TMV) could be enhanced by a prior infection of a single leaf a few days before the subsequent challenge. Cruickshank and Mandryk (1960) showed that stem infection of tobacco with blue mold (*Peronospora tabacina*) can lead to an enhanced resistance to foliar pathogens. Later Cohen and Kuc (1981; Madamanchi and Kuc, 1991) described some carefully controlled experiments showing that SAR in tobacco induced by blue mold requires approximately 3 weeks to develop and that heat-killed conidia are unable to induce SAR. Detailed studies have shown that SAR activation in tobacco results in a significant reduction of disease symptoms by the fungi *Phytophthora parasitica*, *Cercospora nicotinae*, and as mentioned *Peronospora tabacina*. Furthermore, the plants are protected against the viruses TMV and TNV (tobacco necrosis virus) as well as against the bacteria *Erwinia carotovora* and *Pseudomonas syringae* pv. *tabaci*. However, the protection is not effective against all pathogens. For example, there is no significant protection against either *Botrytis cinerea* or *Alternaria alternata* (Vernooij *et al.*, 1995a; Ryals *et al.*, 1996).

Most interestingly, the effect of SAR induction on general plant health depends on the inoculation procedure. When conidia are injected into the cambium the SAR response was linked to severe dwarfing and premature senescence. On the other hand, infection external to the cambium leads to an increase in plant weight and leaf number (Madamanchi and Kuc, 1991; Tuzun and Kuc, 1985; Tuzun *et al.*, 1992).

In addition to tobacco, cucumber has been developed as a biological model for SAR (for a review, see Madamanchi and Kuc, 1991). SAR in cucumber can be induced by various microorganisms (e.g. TNV, *Pseudomonas lachrymans*, *P. syringae*, *Colletotrichum lagenarium*). After an incubation period of a few days, plants are protected against at least 13 diseases for up to 4–6 weeks.

SAR has also been described in other crops including potato (Stroember and Brishammer, 1991), tomato (Heller and Gessler, 1986; Kovats *et al.*, 1991), soybean (Wrather and Elrod, 1990), red clover (King *et al.*, 1964), pearl millet (Kumar *et al.*, 1993), alfalfa (O'Neill *et al.*, 1989), rice (Smith and Metraux, 1991) and others. From this wealth of information one can postulate that SAR is likely to be ubiquitous in higher plants.

An important model system to study the molecular basis of SAR is *Arabidopsis thaliana*. Turnip crinkle virus, certain *Pseudomonas syringae* pv. *tomato* (*Pst*) strains (Uknes *et al.*, 1993; Cameron *et al.*, 1994; Ryals *et al.*, 1996), or *Fusarium oxysporum* (Mauch-Mani and Slusarenko, 1994) were used for SAR-induction in *Arabidopsis* and induced plants exhibited resistance towards different pathogens. The fact that SAR could be induced in *Arabidopsis* by a certain *Pst* strain without a localized hypersensitive response (Cameron *et al.*, 1994) indicates that HR may contribute, but is not essential for SAR development. Recently, the SAR signaling pathway has been studied in great detail based on certain *Arabidopsis* mutants. The status of this work will be reviewed later.

MOLECULAR MARKERS FOR SAR

Plants often respond at the site of attempted microbial infection with a localized cell death (hypersensitive response) followed by a wide range of additional defence responses, including phytoalexin and callose formation, lignification, and cell wall cross-linking (for reviews see Dixon and Lamb, 1990; Low and Merida, 1996; Dangl *et al.*, 1996; Osbourn, 1996). These mechanisms are strictly localized and are not induced during the maintenance state of SAR (Ryals *et al.*, 1996).

Van Loon and Van Kammen (1970) as well as Gianinazzi *et al.* (1970) showed that infection of tobacco with tobacco mosaic virus (TMV) leads to the accumulation of a set of so-called 'pathogenesis-related' (PR) proteins (for a review see Bowles, 1990). Acidic, extracellular forms of these PR-proteins accumulate during the onset of resistance indicating that they may play a role in SAR. Ward *et al.* (1991) showed that nine gene families are coordinately induced both in infected, as well as in distal, untreated leaves after local infection of tobacco with TMV. The encoded proteins include acidic isoforms of PR-1 (PR-1a, PR-1b, PR-1c), β-1,3-glucanases (PR-2a, PR-2b, PR-2c), class II-chitinase (PR-3a, PR-3b), hevein-like protein (PR-4a, PR-4b), thaumatin-like protein (PR-5a, PR-5b), acidic and basic isoforms of class III chitinase, an extracellular β-1,3-glucanase (PR-Q'), the basic isoform of PR-1 and a basic protein family called SAR 8.2.

The identity and relative expression of SAR genes vary among the different species. In cucumber, chitinase is the most prominent PR-protein and the acidic PR1, the predominant PR-protein in tobacco and *Arabidopsis* is only weakly expressed. Such species-specific differences may be the result of evolutionary constraints that have selected for the most effective SAR response against the particular suite of pathogens to which an individual species is subject. Some differences may also be due to breeding processes, where goals other than disease resistance may have dominated which resulted in cultivars failing to express the full potential of the 'original' SAR response.

Since the SAR marker genes are strongly expressed during SAR, it was soon speculated that they are causally involved in the resistance response. In support of this idea, *in vitro* antimicrobial activity has been described for several PR proteins from tobacco (Schlumbaum *et al.*, 1986; Ponstein *et al.*, 1994; Bowles, 1990). Furthermore, synergistic activity has been found for chitinases and β-1, 3-glucanases both *in vitro* (Mauch *et al.*, 1988) as well as in plants over-expressing these enzymes (Zhu *et al.*, 1994). Overexpression of PR1a and SAR8.2 in tobacco resulted in significant increases in resistance towards the oomycete pathogens *Peronospora parasitica* var. *nicotianae* and *Phytophthora parasitica*, respectively (for a review see Lawton *et al.*, 1993). The results from the experiments with transgenic plants overexpressing PR-protein genes suggest that the proteins encoded are causally associated with disease resistance.

SAR in monocots is far less well studied compared with that in dicot species. Although some reports describe biologically induced SAR in monocots, none of these reports have been confirmed by independent work. Chemically induced SAR using activators originally identified in dicot plants has been reported recently in several species including wheat, barley and rice (Görlach *et al.*, 1996; Kogel *et al.*, 1994; Seguchi *et al.*, 1992). Homologues of dicot PR-genes have been identified in monocot plants including PR1. Görlach *et al.* (1996) also identified markers for chemically induced SAR in wheat. These genes encode, for example, a novel lipoxygenase and a cysteine proteinase. Three other inducible genes have unknown functions. However, since no biological model of SAR exists for wheat it cannot be confirmed that these genes are bona fide SAR genes.

THE SAR SIGNAL TRANSDUCTION CASCADE

SAR functions as a potentiator or modulator of other disease resistance mechanisms. When plants express SAR, normally compatible interactions can be converted into incompatible ones. Conversely, when SAR is inactivated (i.e. in mutants incapable of expressing SAR) a normally incompatible interaction can become compatible (Ryals *et al.*, 1996; Mauch-Mani and Slusarenko, 1996; Delaney *et al.*, 1994). Two major experimental approaches have been undertaken to study the SAR signal transduction cascade:

- elucidation of the role of salicylic acid using biochemical tools and transgenic approaches;
- dissection of the pathway and molecular cloning of important steps based on *Arabidopsis* mutants.

The Role of Salicylic Acid

White showed in 1979 that exogenously applied salicylic acid (SA) and related benzoic acid derivatives result in the accumulation of PR-proteins and protection of tobacco against TMV. In independent studies Malamy *et al.* (1990), Metraux *et al.* (1990) and Rasmussen *et al.* (1991) have shown that SA accumulates throughout the plant in tobacco and cucumber after induction of SAR by local infections. Since earlier experiments by Kuc and coworkers showed that the signal for SAR is systemically transported in the phloem sap (Madamanchi and Kuc, 1991), and because exogenously applied SA can induce SAR and synthesis of PR-proteins (White, 1979; Ward *et al.*, 1991; Malamy *et al.*, 1990) it was thought that SA may be the systemic signal for the induction of SAR throughout the plant. Raskin and coworkers showed that in tobacco the endogenous levels of SA after the onset of SAR may be sufficient to induce resistance (Enyedi *et al.*, 1992). Furthermore, Ward *et al.* (1991) confirmed that exogenously applied SA results in the induction of transcription of the entire set of SAR gene families that are also activated through the biological induction of SAR.

In order to study the role of salicylic acid in more detail, tobacco was transformed with the *nahG* gene from *Pseudomonas putida*, which encodes a salicylate hydroxylase, an enzyme that catalyzes the degradation of SA to the non-inducing metabolite catechol (Gaffney *et al.*, 1993). The *nahG*-transgenic plants were shown to express the *nahG*-gene and did not accumulate SA after the onset of SAR nor exhibit the SAR response. Mauch-Mani and Slusarenko (1996) used 2-amino-indan-2-phosphonic acid (AIP) to inhibit the activity of phenylalanine ammonia-lyase, which is thought to be involved in the biosynthesis of salicylic acid, in *Arabidopsis thaliana*. Pretreatment of *Arabidopsis* ecotype Col-O with AIP converted the interaction with *Peronospora parasitica* isolate EMWA from incompatible to compatible. The AIP effect could be suppressed by exogenously applied salicylic acid.

The early biochemical results, the AIP experiments and work with *nahG*-transgenic plants provided strong evidence that salicylic acid plays a central role both in some R-gene mediated resistance as well as SAR.

Other studies focused on the potential role of salicylic acid as the systemic signal for SAR. Metraux *et al.* (1991) have shown that the concentration of salicylic acid increases in cucumber phloem sap upon local infection with TNV. Further studies using *in vivo* labelling techniques in cucumber and tobacco indicated that salicylic acid can be transported from infected to uninfected leaves. In cucumber *c.* 50% of the salicylic acid found in uninfected leaves after induction of SAR may originate from salicylate synthesis in the primary, infected leaf. The additional 50% is most likely due to *de novo* synthesis in the non-infected leaf.

However, other experiments questioned the role of salicylic acid as a systemic signal of SAR. Rasmussen *et al.* (1991) examined this using cucumber as the experimental model, and they demonstrated that the removal of the primary leaf after SAR induction by localized *Pseudomonas syringae* infection several hours before significant accumulation of SA had no effect on SAR induction. Using *nahG*-tobacco plants Vernooij *et al.* (1995b) clearly showed that SA is not the systemic signal responsible for the induction of SAR. When scions from *nahG*-plants were grafted onto Xanthi nc rootstocks, there was no SAR induction in the *nahG* scion despite the presence of SA in the Xanthi nc rootstock after SAR induction. Conversely, the reciprocal grafts (Xanthi nc on *nahG* tobacco) showed that a local infection on the *nahG* plants led to a typical SAR response in the Xanthi scion even though no significant SA increase was detected in *nahG* rootstock.

Together these results show that the systemic signalling cannot be explained by a transport of salicylic acid alone. There may be additional factors acting synergistically with salicylic acid or the salicylic acid accumulation in the phloem is not causally involved in the SAR induction.

The biosynthesis of salicylic acid is not fully understood but the following pathway can be hypothesized. Phenylalanine is converted to trans-cinnamic acid (CA) by phenylalanine ammonia-lyase. CA is then metabolized into benzoic acid (BA), which is then hydroxylated at position 2 to salicylic acid (Yalpani *et al.*, 1993). A benzoic acid 2-hydroxylase has been characterized (Léon *et al.*, 1995) which is induced by either pathogen infection or exogenous BA application. In contrast, the detailed mechanism of conversion of CA into BA is not known yet.

It has been proposed that H_2O_2 acts as a second messenger in salicylic acid induced SAR. A 'salicylic acid-binding protein' has been purified from tobacco and was shown to be a catalase. Inhibition of this enzyme by salicylic acid causes an increase in H_2O_2. Since exogenous application of H_2O_2 was shown to induce PR1 and based on some theoretical homology to mammalian defence mechanisms, it was postulated that H_2O_2 acts as a second messenger of SAR induction (Chen and Klessig, 1991; Chen *et al.*, 1993, 1995). However, recent reports do not support this hypothesis (Bi *et al.*, 1995; Léon *et al.*, 1995; Neuenschwander *et al.*, 1995; Summermatter *et al.*, 1995):

- SAR gene expression and establishment of SAR do not correlate with an increase in H_2O_2 levels;
- exogenous application of H_2O_2 induces PR1 only at extremely high concentrations (1 M) which are also highly phytotoxic;
- in *nahG* plants even 1 M H_2O_2 did not induce SAR indicating a requirement of salicylic acid in H_2O_2-mediated SAR induction. In fact, high concentrations of H_2O_2 were shown to induce salicylic acid accumulation both in tobacco and *Arabidopsis*.

Together these experiments have shown that H_2O_2 is not the second messenger for salicylic acid-induced SAR. But has H_2O_2 any role in defense gene regulation? One can argue that very high concentrations of H_2O_2 simply cause some artefacts because of phytotoxicity. Furthermore, only millimolar concentrations of salicylic acid were shown to inhibit a range of heme-containing enzymes including catalase, ascorbate peroxidase and aconitase (Chen *et al.*, 1993; Durner and Klessig, 1995). Therefore, the reported data on catalase inhibition may be of no real biological significance.

However, the data (together with other reports) may support an alternative hypothesis. Kauss and coworkers showed that pretreatment of parsley cell cultures or cucumber cotyledons with salicylic acid dramatically increased the competence of the tissues to trigger a burst in H_2O_2 in response to elicitor treatment (Kauss *et al.*, 1992). This conditioning of cells by salicylic acid was dependant on protein synthesis and correlated with enhanced resistance of cucumber cotyledons to the fungal pathogen *Colletotrichum lagenarium*. The increase in H_2O_2 was due to *de novo* synthesis, not to an inhibition of degradation.

Furthermore, the concentrations of salicylic acid accumulated very close to infection sites are 10–100-fold higher than in systemic tissue and reach levels in the range of the reported K_d values for catalase (14 μM; Chen and Klessig, 1991) and the IC_{50} value for ascorbate peroxidase (78 μM; Durner and Klessig, 1995), respectively. Therefore, one can hypothesize that there are at least two modes of action of salicylic acid in resistance response. In systemic, non-infected tissues salicylic acid interacts with a high-affinity receptor and renders the cells competent for rapid elicitation of oxidative burst and SAR gene expression. In local, infected tissue salicylic acid reaches far higher concentrations. Here, oxidoreductases may represent a low-affinity perception mechanism for salicylic acid. Inhibition of these enzymes may lead to increases in H_2O_2 which potentiates the oxidative burst. This may trigger other local responses including the programmed cell death response (hypersensitive reaction) and defense gene expression in cells adjacent to the infection site. This would create a runaway cycle leading to high levels of both salicylic acid and H_2O_2 at the site of infection.

Interestingly, the application of sodium saccharin (1,2-benzisothiazol-3(2*H*)-one 1,1-dioxide, BIT, Figure 2), a proposed active metabolite of anti-blast chemical probenazole (Oryzemate®), to rice seedlings significantly enhanced superoxide generation in rice leaf tissues, i.e. oxidative burst, upon infection of *P. oryzae* (Sekizawa *et al.*, 1985, 1987). As the generation of superoxide can proceed to the formation of H_2O_2 and vice versa, BIT may have some similarity to salicylic acid not only in chemical structure but also in biochemical function in plant tissues.

Obviously, the identification of the high-affinity receptor for salicylic acid offers great potential both for identification of new chemical plant activators as well as for transgenic approaches. Therefore, extensive work has been done during the last few years, mainly based on *Arabidopsis* mutant analysis. Very recently this work resulted in the identification of the first element of the intracellular SAR signal transduction cascade.

Dissection of SAR Signal Transduction Cascade by Arabidopsis Mutant Analysis

In order to clone central parts of the SAR signal transduction pathway, several groups are using *Arabidopsis thaliana* mutant analysis. *Arabidopsis* is not only a well-established system for gene isolation but also well characterized regarding interaction with pathogens and non-pathogens (Dangl, 1993; Dangl *et al.*, 1996; Kunkel, 1996; Mauch-Mani and Slusarenko, 1994, 1996; Uknes *et al.*, 1992).

Three different classes of mutants have been identified (Delaney, 1997; Ryals *et al.*, 1996, 1997; Dangl *et al.*, 1996; Cao *et al.*, 1997):

(1) Mutants with constitutive expression of SAR in the absence of cell death (constitutive immunity, *cim*).

These mutants have been identified in a screening for mutants resistant against *Peronospora parasitica* and *Pseudomonas syringae* DC 3000 (Lawton *et al.*, 1993; Ryals *et al.*, 1996). The resistance of *cim3* (Ryals *et al.*, 1996) correlates with constitutive expression of PR-protein genes and elevated levels of salicylic acid. Inhibition of salicylic acid accumulation by expression of the *nahG* gene suppresses both SAR gene expression as well as resistance to *P. parasitica*. It was hypothesized that the gene mutated in *cim3* encodes an early step in the SAR signal transduction cascade, occurring after cell death but before SA accumulation (Figure 1).

(2) Mutants with constitutive expression of SAR in the presence of spontaneous lesion formation (lesion simulating disease, *lsd*, or accelerated cell death, *acd*, mutants).

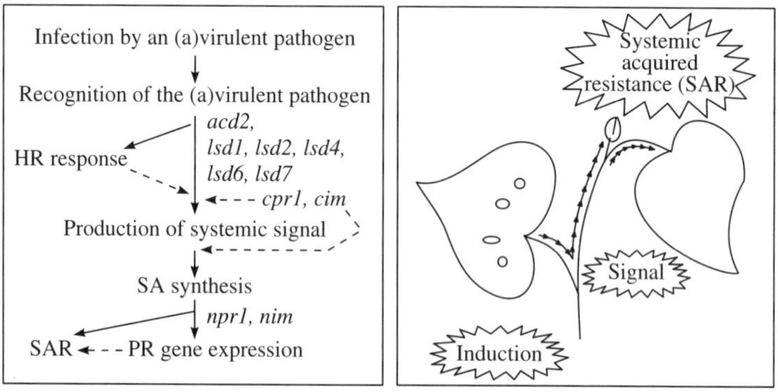

Figure 1 Proposed placement of *Arabidopsis* mutants in the SAR signalling cascade (Ryals *et al.*, 1996, 1997, Cao *et al.*, 1994, 1997)

Seven *lsd* and one *acd* mutants have been described. They are non-allelic but exhibit the same principal phenotype; they spontaneously exhibit cell death (small necrosis) followed by expression of SAR (Dangl *et al.*, 1996; Ryals *et al.*, 1996). The identification and molecular characterization of these mutants strongly indicate that plants have a genetically controlled cell death programme at least phenotypically similar to apoptosis in animals. Interestingly, the spontaneous lesion formation coupled with resistance towards pathogens has been described earlier by breeders as a useful source of resistance traits in breeding programmes (Langford, 1948; Walbot *et al.*, 1983).

In an effort to place these mutants in the SAR signal transduction pathway, crosses between the different *lsd* mutants and plants expressing *nahG* were made (Ryals *et al.*, 1996). Crosses of *lsd1*, *lsd2*, *lsd4*, *lsd6* and *lsd7* with *nahG*-plants gave progenies which were suppressed in PR-gene expression and resistance. This result is consistent with the central role of salicylic acid in the SAR signalling pathway downstream of cell death. However, interpretation of results was complicated since crosses with *lsd1*, *lsd6* and *lsd7* also exhibited suppressed lesion formation whereas the crosses with *lsd2* and *lsd4* still showed the same level of lesion as the parents. One possible explanation of the results with *lsd1*, *lsd6* and *lsd7* is that there is a salicylic acid-dependant feedback regulation of lesion formation. This hypothesis is supported by the fact that the crosses with *lsd1* and *lsd6* mutants regain lesion formation after treatment with 2,6-dichloroisonicotinic acid (INA; [II] in Figure 2), a functional analogue of salicylic acid (Weymann *et al.*, 1995).

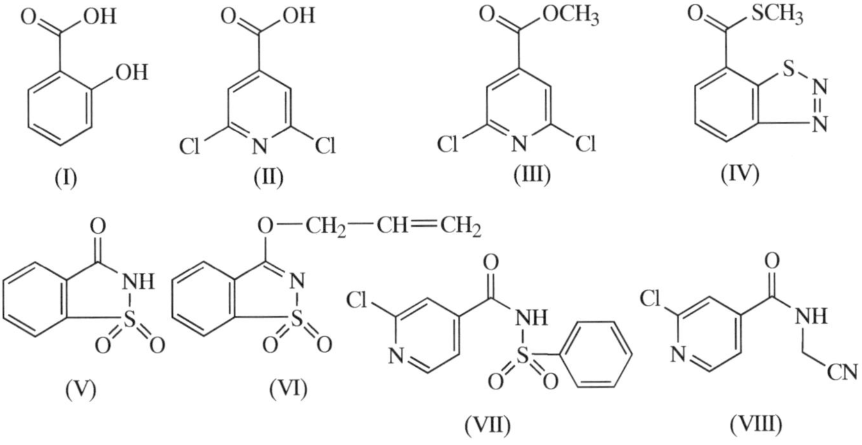

Figure 2 Chemical structures of compounds known to induce SAR. [I]: Salicylic acid (SA), [II]: 2,6-dichloroisonicotinic acid (INA), [III]: 2,6-dichloroisonicotinic acid methyl ester, [IV]: benzo(1,2,3)thiadiazole-7-carbothioic acid S-methyl ester (BTH), [V]: 1,2-benzisothiazol-3(2*H*)-one 1,1-dioxide (BIT), [VI]: probenazole (PBZ), [VII]: *N*-phenylsulfonyl-2-chloroisonicotinamide, [VIII]: *N*-cyanomethyl-2-chloroisonicotinamide (NCI)

(3) SAR compromised mutants (no immunity mutants, *nim1*; non-expressor of PR-proteins, *npr1*).

Mutants have been identified that are blocked in the SAR signalling cascade. Delaney *et al.* (1995) described six allelic recessive mutants of *Arabidopsis* which are unable to express the SAR response upon biological induction or treatment with plant activators such as salicylic acid, INA or a benzothiadiazole derivative (BTH) (Lawton *et al.*, 1996). Later one dominant mutation (*nim1-5*) was discovered by the same group (Ryals *et al.*, 1997). Similarly, Cao *et al.* (1994, 1997) described several allelic mutants which showed the same phenotype as *nim1* (called non-expressor for PR-proteins, *npr1*). *Eds* (enhanced disease susceptibility) mutants were also isolated based on their increased susceptibility towards bacterial pathogens (Glazebrook *et al.*, 1996). Certain *eds* mutants are phenotypically similar to *npr1-1*, and *eds5* as well as *eds53* have been shown to be allelic to *npr1-1* (Glazebrook *et al.*, 1996).

Nim1-plants are still able to accumulate salicylic acid upon infection. The level of free and glucose-conjugated salicylic acid is at least as high as in wild-type plants, which shows that the non-inducibility of the *nim1*-phenotype is not due to a disrupted salicylic acid synthesis pathway. Obviously, the *nim1* mutation is placed in the signal cascade downstream of salicylic acid accumulation but prior to SAR gene expression and resistance. Surprisingly, the *npr1* plants were shown to be less sensitive towards salicylic acid regarding SAR induction but far more sensitive regarding phytotoxic effects (Cao *et al.*, 1997).

The *nim1* and *npr1* gene, respectively, were recently cloned independently by Ryals *et al.* (1997) and Cao *et al.* (1997). Both groups identified the same gene using a map-based cloning strategy. Ryals *et al.* (1997) isolated and characterized five different alleles of *nim1* that show a range of phenotypes from being weakly impaired in chemically induced PR1-gene expression and fungal resistance to being very strongly blocked. Overexpression of the wild-type *NPR1/NIM1* gene complemented the *npr1/nim1*-mutation and restored the wild-type phenotype, i.e. these plants show normal SAR response both after biological or chemical activation. Cao *et al.* (1997) mentioned some preliminary experiments where overexpression of *NPR1* led to an even enhanced disease resistance against *Pseudomonas syringae* pv. *maculicola* when compared to the wild-type. Unfortunately, no detailed data have been reported so far. The results of the complementation experiments show that the *nim1* phenotype is caused by mutation in a single gene and that *nim1* is a *positive regulator* of acquired resistance responses.

The *nim1*-gene contains four exons and three introns. The encoded protein has a calculated molecular weight of 66 039 Daltons. The alignment of the amino acid sequence to those in the databases turned out to be quite challenging (Cao *et al.*, 1997; Ryals *et al.*, 1997). The protein contains four so-called ankyrin motifs. The ankyrin motif has been identified in diverse groups of proteins involved in cell structure, cell differentiation, enzymatic activities and, most interestingly, transcription regulation (Michaely and Bennet, 1992; Bork, 1993). The ankyrin motifs, which consist of 32 amino acids each, are typically involved in protein–protein interaction. A more detailed homology search revealed that *nim1* is a homologue of IκBα. These proteins are well known from mammalian systems as regulators of the NFκB transcription factor (Baeuerle and Baltimore, 1996; Baldwin, 1996). In mammalian systems, IκBα is cytosolic and able to bind NFκB. As a result NFκB cannot enter the nucleus. When the signal transduction pathway is activated, IκBα is phosphorylated at two serine residues which marks the protein for ubiquination and rapid degradation via the proteasome complex. When IκBα is degraded, NFκB is liberated and able to enter the nucleus to cause transcription. An IκBα kinase complex has been recently identified in cytoplasmatic extracts from HeLa cells (Chen *et al.*, 1996). This kinase complex may act as an integrator of multiple signal transduction pathways leading to the activation of NFκB (Lee *et al.*, 1997).

In mammals the IκBα/NFκB signal transduction can be induced by a number of different stimuli including interleukin 1, lipopolysaccharides, tumor-necrosis factor, hydrogen peroxide, virus infection, etc. (Baeuerle and Baltimore, 1996; Baldwin, 1996). Once activated, NFκB causes the transcription of a number of factors involved in inflammation and immune response including interleukin 2, 6 and 8. Most interestingly, salicylic acid is able to suppress degradation of IκBα thus leading to reduced accumulation of certain cytokinines and, as a result, reduced inflammation response (Kopp and Gosh, 1994). In transgenic mice, the knock-out of IκBα/NFκB leads to a defective immune response including increased susceptibility to viral and bacterial infections. The IκBα/NFκB system is also involved in regulation of tumor necrosis factor-induced apoptosis (Beg and Baltimore, 1996; Wang *et al.*, 1996; Van Antwerp *et al.*, 1996; Baeuerle and Baltimore, 1996; Baldwin, 1996).

The identification and characterization of *nim1* indicate that plants possess a pathway homologous to the IκBα/NFκB system known for mammals and *Drosophila* (in the fly the NFκB homologues are called *dorsal* and *dif*; see e.g. Ip *et al.*, 1993; Lemaitre *et al.*, 1996). But how could it function in SAR? In mammalian systems a stimulus such as bacterial infection will lead to an *activation* of the IκBα/NFκB system, i.e. IκBα gets phosphorylated and degraded; NFκB induces transcription of defense related genes. In plants, however, mutation of *nim1* leads to increased susceptibility—not resistance. Therefore, the transcription factor targeted by *nim1* must be a *repressor* of SAR gene expression. Infection or treatment with plant activators leads to the stabilization of the IκBα/repressor complex. This inactivation of a repressor will lead to SAR gene activation and resistance. Further studies have to identify the transcription factor targeted by *nim1* (IκBα) and the actual target of salicylic acid and other plant activators.

PLANT ACTIVATORS

In principle, the SAR system presents some interesting opportunities for the control of plant diseases in agricultural practice and for enhancing our basic knowledge of disease resistance in plants. First, disease resistance that results from the SAR response is a natural defense phenomenon. Second, the biological models of SAR in cucumber and tobacco have demonstrated that SAR can lead to long-lasting and broad-spectrum disease control. Furthermore, there is ample evidence that SAR is based on multiple mechanisms, which makes it less likely that pathogens can readily develop resistance to this control measure.

DISCOVERY AND MODE OF ACTION

SAR can be induced by microorganisms, microbial extracts or defined chemicals. The practical use of microorganisms to induce SAR, similar to the 'biological model' of SAR, seems to be feasible but is most likely restricted to selected crops grown on a small scale. In fact, no products have been introduced to the market with this mode of action yet. However, some plant growth promoting rhizobacteria (PGPR) have recently been described which induce SAR against foliar diseases in tobacco and cucumber (Wei *et al.*, 1991; Maurhofer *et al.*, 1994), and to *Fusarium* wilt of carnation (van Peer *et al.*, 1991). This strategy offers an exciting potential since disease control and increased plant health can be combined in a single seed treatment with naturally occurring microorganisms. At this time there is very little information on the biological mechanisms that are the basis for disease resistance induced by rhizobacteria. A recent report examined the induction of SAR by a *Pseudomonas fluorescens* biocontrol strain (Maurhofer *et al.*, 1994). It was demonstrated that in tobacco grown in soil inoculated with this bacterium the classical symptoms of SAR induction, including the appearance of PR-proteins and salicylic acid accumulation, were noted in the tobacco leaf tissue. This response was absent in tobacco grown in uninoculated soil, or soil inoculated with a different wild-type strain that was known not to induce SAR. Furthermore, induction of the SAR response was correlated with increased resistance to TNV.

Low molecular weight chemicals that are able to induce SAR offer a great potential for disease control in economically important crops. The following criteria need to be fulfilled before an agent can be classified as SAR-inducing or 'plant activator':

- lack of direct antimicrobial activity; no conversion of the compound *in vivo* into antimicrobial metabolites;
- the treated plants are resistant to the same spectrum of diseases as those in which SAR is induced biologically;

- induction of the same preinfectional biochemical processes as seen in systemic plant tissues after biological induction of SAR;
- plant activators are inactive in plant mutants exhibiting the *nim*-pheno–type.

It is possible that compounds that induce SAR in addition to a direct, antimicrobial activity will be discovered. A relatively simple set of experiments can be done to confirm this situation; first, pathogen strains resistant to the compound are selected *in vitro* (or *in vivo* in case of biotrophic pathogens); second, dose-response studies using sensitive and fungicide-resistant strains would indicate if the postulated resistance-induction actually plays a significant role in the disease control.

A number of well-known fungicides were described to have additional, resistance-inducing activity, e.g. fosethyl-Al (a phosphonic acid derivative, Fettouche *et al.*, 1981) or even metalaxyl (a phenylamide, Ward, 1984). In the case of fosethyl-Al, it shows weak *in vitro* activity only in phosphate enriched, but not in low-phosphate containing media (Farih *et al.*, 1981; Fenn and Coffey, 1984). In addition, metabolic inhibitors such as glyphosate decreased the effectiveness of fosethyl-Al, indicating that plant metabolism may contribute to the activity of this fungicide (Fettouche *et al.*, 1981). However, fosethyl-Al does not fulfil the criteria listed above for plant activators because fungal strains selected for insensitivity to fosethyl-Al *in vitro* are also no longer controlled *in vivo* (Fenn and Coffey, 1985; Dolan and Coffey, 1988). Furthermore, fosethyl-Al does not induce molecular markers of SAR such as chitinase in cucumber or PR1a in tobacco. (H. Kessmann, unpublished). These points show that fosethyl-Al does not induce SAR in the plant which significantly contributes to its activity.

2,6-Dichloroisonicotinic acid (INA, Figure 2 [II]) was discovered as a plant activator since it demonstrated the criteria of SAR-induction described above (for a review see Kessmann *et al.*, 1994). INA does not exhibit significant direct *in vitro* activity, but it protects cucumber against the same spectrum of diseases as does biological SAR-induction. It induces class III chitinase, a molecular marker for SAR in cucumber. INA effectively induced resistance in the field against major fungal and bacterial pathogens on various crops (Staub *et al.*, 1993; Kessmann *et al.*, 1994). Molecular studies with tobacco showed that INA induces the same set of genes as a local infection with TMV and corresponding studies with cucumber and *Arabidopsis* confirmed that INA is indeed able to mimic the biological induction of SAR (Uknes *et al.*, 1993; Ward *et al.*, 1991; Kessmann *et al.*, 1994). On barley, INA treatment led to an increased resistance against powdery mildew which looks histologically like a phenocopy of the *mlg*-mediated resistance in genetically caused powdery mildew resistance (K. Kogel, personal communication). INA induces the accumulation of thionin, a 6 kD peptide with antimicrobial properties (Wasternack *et al.*, 1994). Thionin is a member of the 'jasmonic acid induced proteins' (JIP) but other members of the JIP family are not induced, indicating a complex regulation of thionin gene expression.

Histological studies with *Arabidopsis* showed that INA treated plants respond to infection with downy mildew (*Peronospora parasitica*) with a single cell necrosis at the site of attempted penetration (Uknes *et al.*, 1993). When lower INA concentrations were used, some hyphae successfully invaded the leaves but infection also finally stops at later stages. As shown by Kauss and coworkers (Kauss *et al.*, 1992), INA is able somehow to sensitize plant tissue to respond faster to microbial attack. They showed that INA treated parsley cell cultures accumulate certain phenylpropanoids more rapidly after elicitor treatment than control cultures without INA application. Seguchi *et al.* (1992) described a similar sensitizing effect for rice using *N*-cyanomethyl-2-chloroisonicotinamide [VIII], a closely related analogue of INA. Chemically treated and *Pyricularia oryzae*-infected rice plants showed a higher increase in lipoxygenase and peroxidase activity compared to non-treated but infected controls. The molecular basis of such a sensitizing effect is unknown but may play an important role in the SAR response.

Probenazole (3-allyloxy-1,2-benzisothiazole-1,1-dioxide; PBZ, Figure 2, [VI]) and its presumed active metabolite [V] do not exhibit significant fungicidal activity *in vitro*, but give rise to resistance in rice plants against the blast disease in a race-nonspecific manner (Watanabe *et al.*, 1977). In rice plants submergibly treated with [VI], defence-related enzymes such as peroxidase (POX), lipoxygenase (LOX), phenylalanine ammonia-lyase (PAL), phospholipase A_2 (PLA_2), ACC synthase and flavo-cytochrome complex are rapidly induced or activated upon infection by *P. oryzae* (Iwata *et al.*, 1980; Shimura *et al.*, 1983; Sekizawa *et al.*, 1987). This results in the production of potential antifungal substances including α-linolenic acid and its oxygenated fatty acids as well as reactive oxygen species as O_2^- and a plant hormone, ethylene (Haga *et al.*, 1988). Reaction cascade for the induced defense mechanism was well correlated with the findings that the growth of the invading pathogens was inhibited in the infected cells; this was apparently related to the hypersensitive reaction in rice plants. PBZ was also known to exert control activity on other diseases such as bacterial spot on cucumber caused by *Pseudomonas lachrymans* and black rot on cabbage caused by *Xanthomonas campestris*.

Similar activity was observed in rice treated with *N*-cyanomethyl-2-chloroisonicotinamide [VIII] as well as *N*-phenylsulfonyl-2-chloroisonicotinamide [VII] (Yoshida *et al.*, 1990). These compounds exert potent protective efficacy for the control of blast disease and bacterial leaf blight in rice plants although they have no significant antifungal or antibacterial activities *in vitro*, as was reported for PBZ.

Though actual target sites of these chemicals for systemic acquired resistance are not yet determined, time-sequential responses in the inducible defence mechanism have been studied mostly by using [V], [VI] and [VIII] (Figure 2). In the tissues and protoplasts prepared from rice leaves treated with these

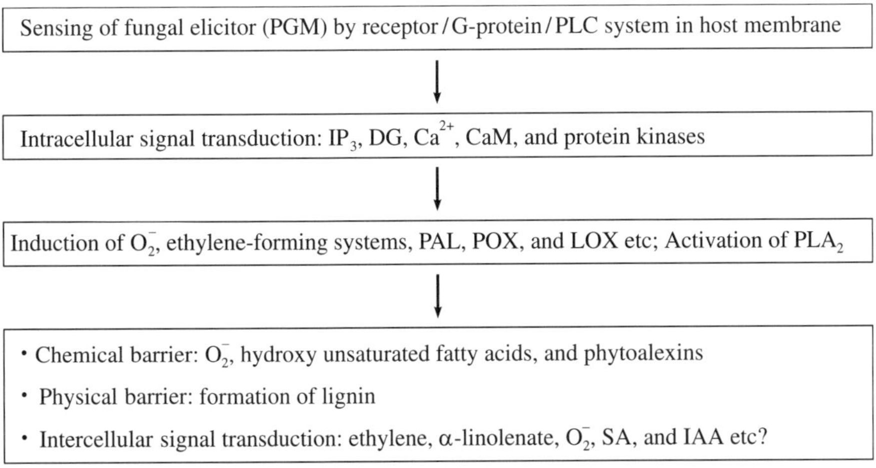

Sensing of fungal elicitor (PGM) by receptor / G-protein / PLC system in host membrane
Intracellular signal transduction: IP_3, DG, Ca^{2+}, CaM, and protein kinases
Induction of O_2^-, ethylene-forming systems, PAL, POX, and LOX etc; Activation of PLA_2
• Chemical barrier: O_2^-, hydroxy unsaturated fatty acids, and phytoalexins
• Physical barrier: formation of lignin
• Intercellular signal transduction: ethylene, α-linolenate, O_2^-, SA, and IAA etc?

Figure 3 A proposed signal induced defense mechanism in rice plant

compounds, significant O_2^- generation and α-linolenic acid release were observed at an early stage of the responses upon infection by *P. oryzae*. Prior to these reactions, however, the turnover of phosphatidylinositol (PI) in rice plant cells elicited by a fungal proteoglucomannan (PGM) was observed to be accelerated by the treatment with the compounds (Seguchi *et al.*, 1992). The importance of PI turnover in transmembrane signaling in response to external stimuli has been well established in animal cells; signal-coupled phospholipase C hydrolyzes phosphoinositide in the membrane to afford inositol 1,4,5-triphosphate (IP_3) and diacylglycerol (DG); IP_3 mobilizes Ca^{2+} from the internal Ca^{2+} pool inducing Ca^{2+}-dependent reactions, and DG activates protein kinase C. Thus the results suggest that similar signal transduction pathways may play an important role in disease resistance in higher plants and the early step(s) in the pathway appear to be affected by the compounds (Figure 3).

As for the molecular mechanism of PBZ-induced resistance, two cDNA clones of PBZ-inducible genes were recently isolated in rice by a differential screening (pPB-1 by Minami and Ando, 1994; *PBZ1* by Midoh and Iwata, 1996). *PBZ1* gene is expressed sooner by inoculation with an incompatible race of *P. oryzae* than with a compatible one. In addition, PBZ proved to induce the expression of other disease-related genes such as POX, LOX, and chitinase (Midoh and Iwata, 1997). The results suggest that *PBZ1* has an important function in disease resistance of the rice plant. However, no expression of these genes was induced by treatment with 1,2-benzisothiazol-3(2*H*)-one 1,1-dioxide (BIT, Figure 2 [V]), which was supposed to be an active form of PBZ as it is exclusively formed from PBZ in rice and BIT itself can induce resistance. Thus the action of BIT on the gene expression might be different from that of PBZ.

Recently, BIT was reported to activate GTPase activity in rice plasma membrane (Sekizawa *et al.*, 1995). Since BIT is known to stimulate cellular events through direct activation of G-proteins in mammalian taste receptor cells (Naim *et al.*, 1994), it may act on molecules related to the signal transduction pathway essential to plant recognition of pathogens as well. Further study on the molecular action mechanism of such agents is needed.

In addition to the agents described above, a wide range of chemicals and natural products have been described to have resistance-inducing properties. This list includes jasmonic acid, ethylene, β-amino acids, unsaturated fatty acids, silicon, oxalate, and phosphate (see Kessmann *et al.*, 1994). Some of these agents may induce the SAR response by causing a localized necrosis, while in other cases direct antimicrobial activity is likely. However, the fact that some agents do not specifically activate the biological SAR response in biochemical changes and spectrum of protection may suggest the possibility that other, as yet unknown, resistance mechanisms are activated.

PERFORMANCE IN PRACTICE

2,6-Dichloroisonicotinic acid (INA, Figure 2) and its derivatives were the synthetic compounds shown to activate the SAR response both in greenhouse and in field trials (Staub *et al.*, 1993). However, since INA was insufficiently tolerated by some major crop species, another class of plant activators was developed and recently introduced to the market. CGA 245704, benzo (1,2, 3)thiadiazole-7-carbothioic acid S-methyl ester (BTH, Bion®), was selected from a range of derivatives based on its excellent performance in wheat, rice, tobacco and some vegetables (Kunz *et al.*, 1997, [IV] Figure 2). BTH provides broad-spectrum protection particularly in dicot species and the spectrum of protection is identical to the spectrum observed after biological induction (Lawton *et al.*, 1996; Friedrich *et al.*, 1996). While BTH does not exhibit direct toxic activity towards pathogenic microorganisms, it induces the same set of PR-proteins in tobacco and *Arabidopsis* as observed after biological induction. In addition, BTH is inactive in *Arabidopsis nim*-mutants, which shows that a functional SAR signalling pathway is required for SAR induction by this compound (Lawton *et al.*, 1996).

In tobacco BTH is active at such low rates as 2.5 g/hill, leading to excellent protection against blue mold (*Peronospora tabacina*) and other pathogens. The activity can be further enhanced in mixtures of BTH with metalaxyl-M as well as maneb and mancozep (Figure 4). BTH protects tobacco in situations of reduced sensitivity of blue mold against phenylamide fungicides.

A unique feature of BTH is based on the fact that it can protect plants such as tomato against bacterial diseases (Figure 4b) as well as from late blight (*Phytophthora infestans*). In particular, a mixture with copper provides very good protection against bacterial spot (*Pseudomonas syringae* pv. *tomato*), bacterial speck (*Xanthomonas campestris* pv. *vesicatoria*) and late blight.

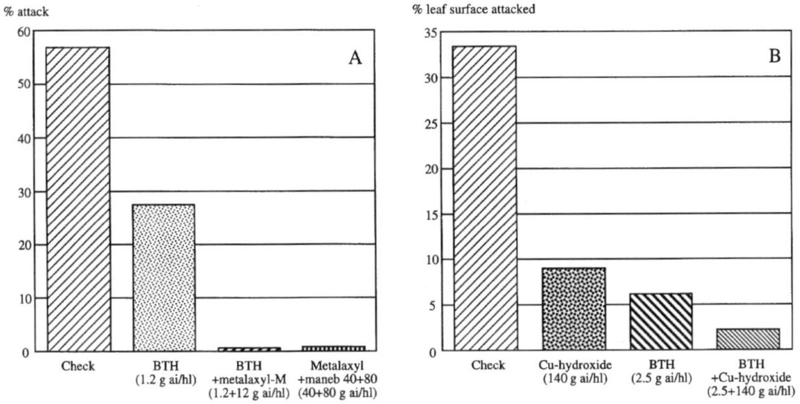

Figure 4 The plant activator BTH provides protection of tobacco (A) and tomato (B) against bluemold and bacterial spot, respectively. The performance can be further improved in mixtures with fungicides (From Novartis Crop Protection AG)

Excellent performance has also been described for lettuce and chilli against *Bremia lactucae* and *Colletotrichum* sp., respectively.

An interesting feature of plant activation is the difference between dicots and monocots. In dicot species the plant activator applied at regular intervals (7–14 days) provides broad-spectrum resistance towards fungal and bacterial pathogens. In monocot species such as wheat or rice the compound is applied only once per season to provide a long-lasting control of major pathogens. The reason for this principal difference is not understood yet.

In wheat the plant activator BTH is applied during tillering and provides protection against powdery mildew and, to a lesser extent, other diseases (brown rust and speckled leaf blotch), which lasts until ear emergence. Due to the fact that the plant activator protects the crops but does not control existing infections, it is important that it is applied before the onset of the diseases. Using the plant activator outside the application window leads to lower level of protection. However, the application window can be extended by mixtures with fungicides, e.g. cyprodinil or fenpropidin (Figure 5). In such mixtures the fungicide provides an early direct control of the disease, and the plant activator will subsequently protect the crop against further infections. Field trials with BTH (30 g ai/ha) have shown that the plant activator alone causes a yield increase of 9% (average of 7 trials) compared to untreated controls and mixtures with fungicides such as fenpropidin (375 g ai/ha) or cyprodinil (500 g ai/ha) led to 13 and 17% higher yields, respectively. For follow-up treatments, with e.g. triazole fungicides, yield increases are particularly significant and reach levels of up to 40% compared to the untreated plots.

Similar to its use in wheat, BTH was shown to provide unique long-lasting protection in rice against leaf blast caused by *Pyricularia oryzae* (teleomorph, *Magnaporthe grisea*). After being applied as a granule into the seedling box, the

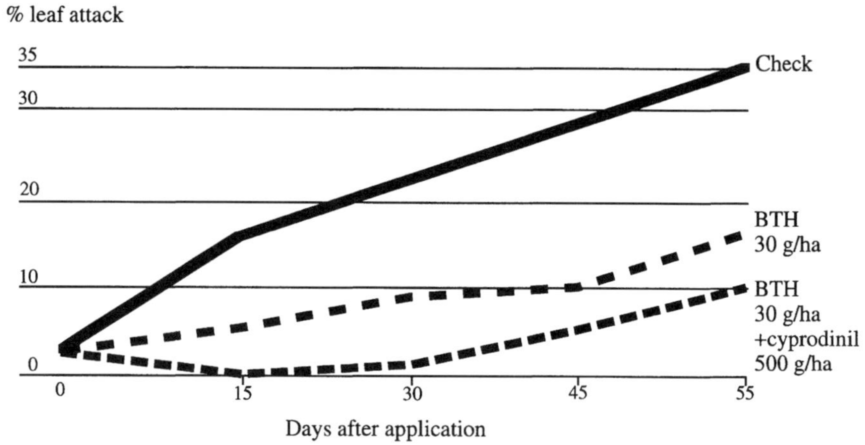

Figure 5 Protection of wheat against powdery mildew (*Erysiphe graminis*) by the plant activator BTH. BTH and the mixture with cyprodinil were applied to wheat once at early tillering stage. Data are the average of five trials in the 1994 season (From Novartis Crop Protection AG)

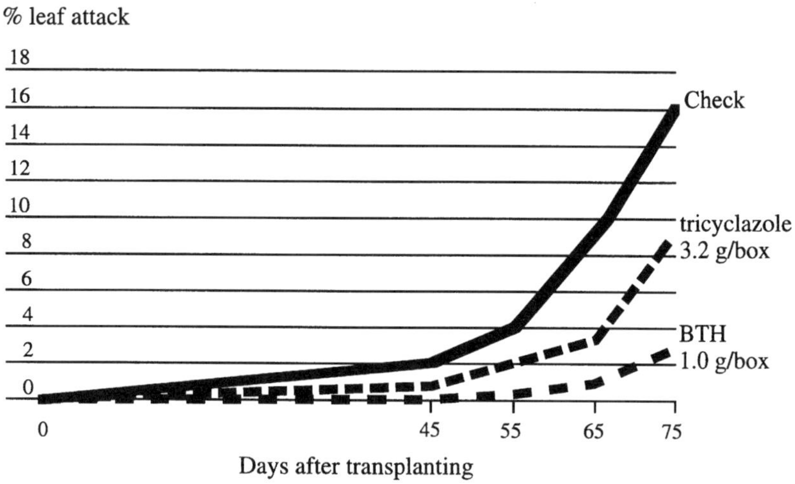

Figure 6 Long-lasting activity of BTH for rice blast control. The plant activator BTH was applied to the seedling box just prior to the transplanting. Protection of rice against *Pyricularia oryzae* by single seedling box application. Data are the average of five trials from 1993/94 seasons in Japan (From Novartis Crop Protection AG)

seedlings are transplanted into the rice fields. The young rice plants are activated at this time and protected for up to 80 days (Figure 6).

In the bioassay system for anti-blast chemicals using rice seedlings (pot test), benzisothiazoline derivatives were screened, based on the finding that soluble saccharin (sodium salt of 1,2-benzisothiazol-3(2*H*)-one 1,1-dioxide) exerts

Table 1 Diseases controlled by oryzemate[a]

Crops	Target disease	Application rate/10a	Application timing	Procedure
Rice	Blast	3–4 kg	7–10 days before onset of leaf blast and/or 3–4 weeks before heading for panicle blast	Spray application of granule formulation to the submerged paddy field
	Helminthosporium panicle blight			
	Bacterial leaf blight and grain rot	3–4 kg	7–10 days after transplantation and 3–4 weeks before heading	
Rice (seedling box)	Above diseases	(20–30 g per $30 \times 60 \times 3\,cm^3$ soil)	1–3 days before transplantation	Homogeneous application of granules into the seedling box
Cucumber	Bacterial spot	6–7.5 kg (5g/plant)	At or prior to the planting	Mixing with soil at the planting hole
Lettuce	Bacterial rot and spot	6–9 kg	At or prior to the planting	Mixing with soil
Cabbage	Black rot	6–9 kg	At or prior to the planting	Mixing with soil in the planting row
Chinese cabbage	Bacterial soft rot	6–9 kg	At or prior to the seeding or planting	Mixing with soil
Sweet pepper	Bacterial spot	(5–10 g/plant)	At or prior to the planting	Mixing with soil in the planting hole

[a] Probenazole granule (a.i. 8%). For aerial application, less amount of granule formulation (20% a.i.) is applied. (Source: Meiji Seika Kaisha, Ltd.)

some anti-b-last activity. The screening resulted in the successful development of probenazole (Oryzemate®) as a blast controlling agent (Watanabe *et al.*, 1977, 1979). Probenazole (PBZ) and its metabolites do not exhibit significant direct antifungal activity, but PBZ shows an excellent control efficacy on rice blast not only in the pot tests but also in field trials. Interestingly, in the course of study on its phytotoxicity and metabolic fate in the environment, PBZ was incidentally recognized to exert a remarkable control effect against blast disease when it was treated submergibly, suggesting that it is efficiently taken up and systemically transported in the rice plants. It was then formulated as a granule for submerged application in the paddy field and for the seedling box application. It was first registered as a blast controlling agent in Japan in 1974, and later it was found to have good control efficacy against bacterial leaf blight caused by *Xanthomonas campestris* pv. *oryzae* as well as bacterial grain rot caused by *Pseudomonas glumae* (registered in 1980 and 1982, respectively) in rice. Further, PBZ was registered against bacterial spot on cucumber (1985), bacterial rot and spot on lettuce (1989), black rot on cabbage (1990), bacterial soft rot on Chinese cabbage (1993) and bacterial spot on sweet pepper (1994), as shown in Table 1. In addition, it has been used as mixtures with insecticides such as diazinon, cartap, propoxur, monocrotophos, fenobucarb, etofenprox, carbosulfan and benfracarb, and the plant growth regulator inabenfide. It should be noted that no resistant strains of the causal pathogen *P. oryzae*, have emerged against PBZ in spite of its practical wide use for more than two decades.

CONCLUSIONS AND OUTLOOK

Plant diseases cause serious damage to crop production, particularly in humid climates. Conventional chemicals with direct fungicidal activities exert marked protective and/or curative efficacy for the control of plant diseases but they sometimes bring about adverse effects on non-target organisms. While all the modern fungicides are developed through extensive safety evaluation, there is a growing public concern about their side effects on non-target organisms and their environmental impact. In addition, some of them have met a serious problem with the emergence of resistant pathogens caused by their high selection pressure. In contrast, nonfungicidal disease controllers are considered to surpass the conventional fungicides in this aspect as they are supposed to be inherently non-cidal and to have less possibility of causing resistance. Thus there is now a keen interest in these disease controlling agents. In fact, two kinds of nonfungicidal rice blast controllers are currently on the market; melanin biosynthesis inhibitors (MBI) such as fthalide, tricyclazole and pyroquilon (Yamaguchi and Kubo, 1992) and so-called priming effectors or plant activators, e.g. probenazole and BTH, to induce host resistance against the pathogen's attack. They show high control efficacy on the diseases with a quite low toxicity to non-target organisms including mammals.

In particular, utilization of SAR as an inducible, broad-spectrum, and long-lasting disease control mechanism that is most likely present in all the higher plants, would offer a valuable new option for practical disease control as well as plant health management in general. However, our understanding of SAR is still primitive in both applied and basic aspects. For practical application, chemical agents which induce SAR with good crop tolerance in economically important crops have to be developed. These agents should not only be seen as disease control agents but also as compounds that affect plant health.

For basic research, the molecular bases of SAR, including the signalling pathway, are most interesting questions. Further, it is highly desirable that biorational design for novel, more active compounds with less adverse effects on humans and the environment should be performed. Efficient screening bioassay systems are also needed on the basis of fundamental studies on the action mechanism of the chemicals at the molecular level.

ACKNOWLEDGEMENTS

The author would like to thank Novartis Crop Protection AG and Meiji Seika Kaisha Ltd for providing important input to this review.

REFERENCES

Baeuerle, P. and Baltimore, D. (1996). 'NF-κB: ten years after', *Cell*, **87**, 13–20.

Baldwin, A. (1996). 'The NF-κB and IκB proteins: new discoveries and insights', *Ann. Rev. Immunol.*, **14**, 649–681.

Beg, A. and Baltimore, D. (1996). 'An essential role of NF-κB in preventing TNF-alpha induced cell death', *Science*, **274**, 782–784.

Bent, A. F. (1996). 'Plant disease resistance genes: function meets structure', *Plant Cell*, **8**, 1757–1771.

Bi, Y. M., Kenton, P., Darby, R., and Draper, J. (1995). 'Hydrogen peroxide does not function downstream of salicylic acid in the induction of PR protein expression', *Plant J.*, **8**, 235–245.

Bork, P. (1993). 'Hundreds of ankyrin-like repeats in functionally diverse proteins: mobile modules that cross phyla horizontally?', *Proteins: Structure, Function and Genetics*, **17**, 363–374.

Bowles, D. J. (1990). 'Defense-related proteins in higher plants', *Ann. Rev. Biochem.*, **59**, 873–907.

Cameron, R. K., Dixon, R., and Lamb, C. (1994). 'Biologically induced systemic acquired resistance in *Arabidopsis thaliana*', *Plant J.*, **5**, 715–725.

Cao, H., Bowling, S. A., Gordon, A. S., and Dong, X. (1994). 'Characterization of an *Arabidopsis* mutant that is nonresponsive to inducers of systemic acquired resistance'. *Plant Cell*, **6**, 1583–1592.

Cao, H., Glazebrook, J., Clarks, J. D., Volko, S., and Dong, X. (1997). 'The *Arabidopsis* NPR1 gene that controls systemic acquired resistance encodes a novel protein containing ankyrin repeats'. *Cell*, **88**, 57–63.

Chen, Z. and Klessig, D. (1991). 'Identification of a soluble, salicylic acid-binding protein that may function in signal transduction in the plant disease-resistance response', *Proc. Natl. Acad. Sci. USA*, **88**, 8179–8183.

Chen, Z., Silva, H., and Klessig, D. (1993). 'Active oxygen species in the induction of systemic acquired resistance by salicylic acid', *Science*, **262**, 1883–1886.

Chen, Z., Malamy, J., Henning, J., Conrath, U., Sanchezcasas, P., Silva, H., Ricigliano, J., and Klessig, D. (1995). 'Induction, modification, and transduction of the salicylic acid signal in plant defense responses', *Proc. Natl. Acad. Sci. USA*, **92**, 4134–4137.

Chen, Z. J., Parent, L., and Maniatis, T. (1996). 'Site-specific phosphorylation of 1κBa by a novel ubiquination-dependant protein kinase activity', *Cell*, **84**, 853–862.

Cohen, Y. and Kuc, J. (1981). 'Evaluation of systemic resistance to blue mold induced in tobacco leaves by prior stem inoculation with *Peronospora tabacina*', *Phytopathology*, **71**, 783–787.

Cruickshank, I. A. M., and Mandryk, A. (1960). 'The effect of stem infestations of tobacco with *Peronospora tabacina* Adam on foliage reaction to blue mold', *J. Austr. Inst. Agr. Sci.*, **26**, 369–372.

Dangl, J. L. (1993). 'Application of *Arabidopsis thaliana* to outstanding issues in plant-pathogen interactions', *Int. Rev. Cytol.*, **144**, 53–83.

Dangl, J. L., Dietrich, R. A., and Richberg, M. H. (1996). 'Death don't have no mercy: cell death programs in plant-microbe interactions', *Plant Cell*, **8**, 1793–1807.

Delaney, T. P., Uknes, S., Vernooij, B., Friedrich, L., Weymann, K., Negrotto, D., Gaffney, T., Gut-Rella, M., Kessmann, H., Ward, E., and Ryals, J. (1994). 'A central role of salicylic acid in plant disease resistance', *Science*, **266**, 1247–1250.

Delaney, T. P., Friedrich, L., and Ryals, J. (1995). '*Arabidopsis* signal transduction mutant defective in chemically and biologically induced disease resistance', *Proc. Natl. Acad. Sci. USA*, **92**, 6602–6606.

Delaney, T. P. (1997). 'Genetic dissection of acquired resistance to disease', *Plant Physiol.*, **113**, 5–12.

Dixon, R. A. and Lamb, C. (1990). 'Molecular communication in interactions between plants and microbial pathogens', *Ann. Rev. Plant Physiol. Mol. Biol.*, **41**, 339–367.

Dolan, T. C., and Coffey, M. D. (1988). 'Correlation of in vitro and in vivo behaviour of mutant strains of *Phytopthora palmivora* expressing different resistances to phosphorous acid and fosethyl-Al', *Phytopathology*, **78**, 974–978.

Durner, J. and Klessig, D. F. (1995). 'Inhibition of ascorbate peroxidase by salicylic acid and 2,6-dichloroisonicotinic acid, two inducers of plant defense responses', *Proc. Natl. Acad. Sci. USA*, **92**, 11312–11316.

Enyedi, A. J., Yalpani, N., Silverman, P., and Raskin, I. (1992). 'Signal molecules in systemic plant resistance to pathogens and pests', *Cell*, **70**, 879–886.

Farih, A., Tsao, H., and Menge, J. A. (1981). 'Fungitoxic activity of efosite-aluminium on growth, sporulation and germination of *Phytophthora parasitica* and *Phytophthora citrophora*', *Phytopathology*, **71**, 934–936.

Fenn, M. E. and Coffey, M. D. (1984). 'Studies on the in vitro and in vivo antifungal activity of fosethyl-Al and phosphorous acid', *Phytopathology*, **76**, 606–611.

Fenn, M. E. and Coffey, M. D. (1985). 'Further evidence for the direct mode of action of fosethyl-Al and phosphorous acid', *Phytopathology*, **75**, 1064–1078.

Fettouche, F., Ravise, A., and Bompeix, G. (1981). 'Suppression de la resistance induite phosethyl-AL chez la tomate a *Phytophthora capsici* avec deux inhibiteurs-glyphosate et acide a-aminooxyacetique', *Agronomie*, **9**, 826.

Friedrich, L., Lawton, K., Ruess, W., Masner, P., Specker, N., Gut-Rella, M., Meier, B., Dincher, S., Staub, T., Ukness, S., Metraux, J. P., Kessmann, H., and Ryals, J. (1996). 'A benzothiadiazole derivative induces systemic acquired resistance in tobacco', *Plant J.*, **10**, 61–70.

Gaffney, T., Friedrich, L., Vernooij, B., Negretto, D., Nye, G., Uknes, S., Ward, E., Kessmann, H., and Ryals, J. (1993). 'Requirement of salicylic acid for the induction of systemic acquired resistance', *Science*, **261**, 754–756.

Gianinazzi, S., Martin, C., and Vallee, J. C. (1970). 'Hypersensibilite aux virus, temperature et proteines solubles chez le *Nicotiana* Xanthi n.c.', *CR Acad. Sci.*, **D270**, 2382–2386.

Glazebrook, J., Rogers, E. E., and Ausubel, F. (1996). 'Isolation of *Arabidopsis* mutants with enhanced disease susceptibility by direct screening', *Genetics*, **143**, 873–882.

Görlach, J., Volrath, S., Knauf-Beiter, G., Hengy, G., Beckhove, U., Kogel, K. H., Oostendorp, M., Staub, T., Ward, E., Kessmann, H., and Ryals, J. (1996). 'Benzothiadiazole, a novel class of inducers of systemic acquired resistance, activates gene expression and disease resistance in wheat', *Plant Cell*, **8**, 629–643.

Haga, M., Haruyama, T., Kano, H., Sekizawa, Y., Urushizaki, S., and Matsumoto, K. (1988). 'Dependence on ethylene of the induction of phenylalanine ammonia-lyase activity in rice leaf infected with blast fungus', *Agric. Biol. Chem.*, **52**, 943–950.

Heller, W. E. and Gessler, C. (1986). 'Induced systemic resistance in tomato plants against *Phytophthora* infestans', *J. Phytopathol.*, **116**, 323–328.

Ip, Y., Reach, M., Engstrom, Y., Kadalayil, L., Cai, H., Gonzalez-Crespo, S., Tatei, K., and Levine, M. (1993). '*Dif*, a *dorsal* related gene that mediates an immune response in *Drosophila*', *Cell*, **75**, 753–763.

Iwata, M., Suzuki, Y., Watanabe, T., Mase, S., and Sekizawa, Y. (1980). 'Effect of probenazole on the activities of enzymes related to the resistant reaction in rice plant', *Ann. Phytopath. Soc. Japan*, **46**, 297–306.

Kauss, H., Theisinger-Hinkel, E., Mindermann, R., and Conrath, U. (1992). 'Dichlor-oisonicotinic and salicylic acid, inducers of systemic acquired resistance, enhance fungal elicitor responses in parsley cells', *Plant J.*, **2**, 655–660.

Kessmann, H., Staub, T., Hofmann, C., Maetzke, T., Herzog, J., Ward, E., Uknes, S., and Ryals, J. (1994). 'Induction of systemic acquired resistance in plants by chemicals', *Ann. Rev. Phytopathol.*, **32**, 439–459.

King, L., Hampton, R. E., and Diachun, S. (1964). 'Resistance to *Erysiphe polygoni* of red clover infected with bean yellow mosaic virus', *Science*, **146**, 1054–1055.

Kogel, K. H., Beckhove, U., Dreschers, J., Munch, S., and Romme, Y. (1994). 'Acquired resistance in barley', *Plant Physiol.*, **106**, 1269–1277.

Kopp, E., and Gosh, S. (1994). 'Inhibition of NF-κB by sodium salicylate and aspirin', *Science*, **265**, 956–959.

Kovats, K., Binder, A., and Hohl, H. R. (1991). 'Cytology of induced systemic resistance of tomato to *Phytophthora infestans*', *Planta*, **183**, 491–496.

Kumar, V. U., Meera, M. S., Hindumathy, C. K., and Shetty, H. S. (1993). 'Induced systemic resistance protects pearl millet plants against downy mildew disease due to *Sclerospora graminicola*', *Crop Protection*, **12**, 458–462.

Kunkel, B. N. (1996). 'A useful weed put to work: genetic analysis of disease resistance in *Arabidopsis thaliana*', *Trend Gen.*, **12**, 63–69.

Kunz, W., Schurter, R., and Maetzke, T. (1997). 'The chemistry of benzothiadiazole plant activators', *Pestic. Sci.*, **50**, 275–282.

Langford, A. N. (1948). 'Autogenous necrosis in tomatoes immune from *Cladosporium fulvum* Cooke', *Can. J. Res.*, **26**, 35–64.

Lawton, K., Uknes, S., Friedrich, L., Gaffney, T., Alexander, D., Goodman, R., Metraux, J. P., Kessmann, H., Ahl-Goy, P., Gut-Rella, M., Ward, E., and Ryals, J. (1993). 'The molecular biology of systemic acquired resistance', in *Mechanisms of Plant Defense Responses* (eds Fritig, B. and Legrand, M.), pp. 422–432, Kluwer Academic Publishers, Dordrecht.

Lawton, K., Friedrich, L., Hunt, M., Weymann, K., Kessmann, H., Staub, T., and Ryals, J. (1996). 'Benzothiadiazole induces disease resistance in *Arabidopsis* by activation of the systemic acquired resistance signal transduction pathway', *Plant J.*, **10**, 71–82.

Lee, F. S., Hagler, J., Chen, Z. J., and Maniatis, T. (1997). 'Activation of the 1κBa kinase complex by MEKK1, a kinase of the JNK pathway', *Cell*, **88**, 213–222.

Lemaitre, B., Nicolas, E., Michaut, L., Reichart, J. M., and Hoffmann, J. (1996). 'The dorsoventral regulatory gene cassete spatzle/toll/cactus controls the potent antifungal response in Drosophila adults', *Cell*, **86**, 973–983.

Léon, J., Lawton, M. A., and Raskin, I. (1995). 'Hydrogen peroxide stimulates salicylic acid biosynthesis in tobacco', *Plant Physiol.*, **108**, 1673–1678.

Low, P. S. and Merida, J. R. (1996). 'The oxidative burst in plant defense: function and signal transduction', *Physiol. Plant.*, **96**, 533–542.

Madamanchi, N. R., and Kuc, J. (1991). 'Induced systemic resistance in plants', in *The Fungal Spore and Disease Initiation in Plants* (eds Cole, G. T. and Hoch, H.), pp. 347–362, Plenum-Press, New York.

Malamy, J., Carr, J. P., Klessig, D., and Raskin, I. (1990). 'Salicylic acid: a likely endogenous signal in the resistance response of tobacco to viral infection', *Science*, **250**, 1002–1004.

Mauch, F., Mauch-Mani, B., and Boller, T. (1988). 'Antifungal hydrolases in pea tissue. II. Inhibition of fungal growth by combinations of chitinases and glucanases', *Plant Physiol.*, **88**, 936–942.

Mauch-Mani, B. and Slusarenko, A. J. (1994). 'Systemic acquired resistance in *Arabidopsis thaliana* induced by predisposing infection with a pathogenic isolate of *Fusarium oxysporum*', *Mol. Plant-Microbe Interact.*, **7**, 378–383.

Mauch-Mani, B., and Slusarenko, A. J. (1996). 'Production of salicylic acid precursors is a major function of phenylalanine ammonia-lyase in the resistance of *Arabidopsis to Peronospora parasitica*', *Plant Cell*, **8**, 203–212.

Maurhofer, M., Hase, C., Metraux, J. P., and Defago, G. (1994), 'Induction of systemic resistance of tobacco to tobacco necrosis virus by the root-colonizing *Pseudomonas fluorescens* strain CHA0: Influence of the gacA gene and pyoverdine production', *Phytopathology*, **84**, 139–146.

Metraux, J. P., Signer, H., Ryals, J., Ward, E. W., Wyss-Benz, M., Gaudin, J., Raschdorf, K., Schmid, E., Blum, W., and Inverardi, B. (1990). 'Increase in salicylic acid at the onset of systemic acquired resistance in cucumber', *Science*, **250**, 1004–1006.

Metraux, J. P., Ahl-Goy, P., Staub, T., Speich, J., Steinemann, A., Ryals, J., and Ward, E. (1991). 'Induced resistance in cucumber in response to 2,6-dichloroisonicotinic acid and pathogens', in *Advances in Molecular Genetics of Plant-Microbe Interaction* (eds Hennecke, H. and Verma, D. P. S.), pp. 432–439, Kluwer, Dordrecht.

Michaely, P. and Bennet, V. (1992). 'The ANK repeat: a ubiquitous motif involved in macromolecular recognition', *Trends Cell Biol.*, **2**, 127–129.

Midoh, N. and Iwata, M. (1996). 'Cloning and characterization of a probenazole-inducible gene for an intracellular pathogenesis-related protein in rice', *Plant Cell Physiol.*, **37**, 9–18.

Midoh, N. and Iwata, M. (1997). 'Expression of defense-related genes by probenazole or 1,2-benzisothiazole-3(2H)-one-1,1-dioxide', *J. Pesticide Sci.*, **22**, 45–47.

Minami, E. and Ando, I. (1994). 'Analysis of blast disease resistance induced by probenazole in rice', *J. Pesticide Sci.*, **19**, 79–83.

Naim, M., Seifert, R., Nürnberg, B., Grünbaum, L., and Schultz, G. (1994). 'Some taste substances are direct activators of G-proteins', *Biochem. J.*, **297**, 451–454.

Neuenschwander, U., Vernooij, B., Friedrich, L., Uknes, S., Kessmann, H., and Ryals, J. (1995). 'Is hydrogen peroxide a second messenger of salicylic acid in systemic acquired resistance?', *Plant J.*, **8**, 227–233.

O'Neill, N. R., Elgin, J. H., and Baker, C. J. (1989). 'Characterization of induced resistance to anthracnose in alfalfa by races, isolates and species of *Colletotrichum*', *Phytopathology*, **79**, 750–756.

Osbourn, A. E. (1996). 'Preformed antimicrobial compounds and plant defense against fungal attack', *Plant Cell*, **8**, 1821–1831.

Ponstein, A. S., Bres-Vloemans, S. A., Sela-Buurlage, M. B., Van-den-Elzen, P. J. M., Melchers, L. S., and Comelissen, B. J. C. (1994). 'A novel pathogen-and wound-inducible tobacco (*Nicotiana tabacum*) protein with antifungal activity', *Plant Physiol.*, **104**, 109–118.

Pryor, T. and Ellis, J. (1993). 'The genetic complexity of fungal resistance genes in plants', in *Advances in Plant Pathology* (eds Andrews, J. H., and Tommerup, I. C.), **Vol. 10**, pp. 281–307, Academic Press, London.

Rasmussen, J., Hammerschmidt, R., and Zook, M. N. (1991). 'Systemic induction of salicylic acid accumulation in cucumber after inoculation with *Pseudomonas syringae* pv. *syringae*', *Plant Physiol.*, **97**, 1342–1347.

Ross, F. A. (1961). 'Systemic acquired resistance induced by localized virus infection in plants', *Virology*, **14**, 340–358.

Ryals, J. A., Neuenschwander, U. H., Willits, M. G., Molina, A., Steiner, H.-Y., and Hunt, M. D. (1996). 'Systemic acquired resistance', *Plant Cell*, **8**, 1809–1819.

Ryals, J. A., Weymann, K., Lawton, K., Friedrich, L., Ellis, D., Steiner, H.-Y., Johnson, J., Delaney, T. P., Jesse, T., Vox, P., and Uknes, S. (1997). 'The *Arabidopsis* NIM 1 protein shows homology to the mammalian transcription factor inhibitor IκB'. *Plant Cell*, **9**, 425–439.

Schlumbaum, A., Mauch, F., Voegeli, U., and Boller, T. (1986), 'Plant chitinases are potent inhibitors of fungal growth', *Nature*, **324**, 365–367.

Seguchi, K., Kurotaki, M., Sekido, S., and Yamaguchi, I. (1992). 'Action mechanisms of *N*-cyanomethyl-2-chloroisonicotinamide in controlling rice blast disease', *J. Pesticide Sci.*, **17**, 107–113.

Sekizawa, Y., Haga, M., Iwata, M., Hamamoto, A., Chihara, C., and Takino, Y. (1985). 'Probenazole and burst of respiration in rice leaf tissue infected with blast fungus', *J. Pesticide Sci.*, **10**, 225–231.

Sekizawa, Y., Haga, M., Hirabayashi, E., Takeuchi, N., and Takino, Y. (1987). 'Dynamic behavior of superoxide generation in rice leaf tissue infected with blast fungus and its regulation by some substances', *Agric. Biol. Chem.*, **51**, 763–770.

Sekizawa, Y., Aoyama, H., Kimura, M., and Yamaguchi, I. (1995). 'GTPase activity in rice plasma membrance preparation enhanced by a priming effector for plant defense reactions', *J. Pesticide Sci.*, **20**, 165–168.

Shimura, M., Mase, S., Iwata, M., Suzuki, A., Watanabe, T., Sekizawa, T., Sasaki, T., Furihata, K., Seto, H., and Otake, N. (1983). 'Anti-conidial germination factors induced in the presence of probenazole in infected host leaves. 3. Structural elucidation of substances A and C', *Agric. Biol. Chem.*, **47**, 1983–1989.

Smith, J. and Metraux, J. P. (1991). '*Pseudomonas syringae* induces systemic resistance to *Pyricularia oryzae* in rice', *Physiol. Mol. Plant Pathol.*, **39**, 451–461.

Staub, T., Ahl-Goy, P., and Kessmann, H. (1993). 'Chemically induced disease resistance in plants', in *Proc. of 10th Symp. on Systemic Fungicides and Antifungal Comp.* (eds Lyr, H., and Polter, C.), **Vol. 4**, pp. 239–249, Ulmer, Stuttgart.

Stroember, A., and Brishammer, S. (1991). 'Induction of systemic resistance in potato (*Solanum tuberosum* L.) plants to late blight by local treatment with *Phytophthora infestans* (Mont.) de Bary, *Phytophthora cryptogaea* Pathyb, Laff. or dipotassium phosphate', *Potato Research*, **34**, 219–225.

Summermatter, K., Sticher, L., and Metreaux, J. P. (1995). 'Systemic responses in *Arabidopsis thaliana* infected and challenged with *Pseudomonas syringae* pv *syringae*', *Plant Physiol.*, **108**, 1379–1385.

Tuzun, S., and Kuc, J. (1985). 'A modified technique for inducing systemic resistance to blue mold and increasing growth of tobacco', *Phytopathology*, **75**, 1127–1129.

Tuzun, S., Juarez, J., Nesmith, W. C., and Kuc, J. (1992). 'Induction of systemic resistance in tobacco against metalaxyl tolerant strains of *Peronospora tabacina* and the natural occurrence of this phenomenon in Mexico', *Phytopathology*, **82**, 425–429.

Uknes, S., Mauch-Mani, B., Moyer, M., Potter, S., Williams, S., Dincher, S., Chandler, D., Slusarenko, A., Ward, E., and Ryals, J. (1992). 'Acquired resistance in *Arabidopsis*', *Plant Cell*, **4**, 645–656.

Uknes, S., Winter, A., Delaney, T., Potter, S., Ward, E., and Ryals, J. (1993). 'Biological induction of systemic acquired resistance in *Arabidopsis*', *Mol. Plant-Microbe Interact.*, **6**, 692–698.

Van Antwerp, D., Martin, S., Kafri, T., Green, D., and Verma, I. (1996). 'Suppression of TNF-alpha-induced apoptosis by NF-κB', *Science*, **274**, 787–789.

Van Loon, L. C., and Van-Kammen, A. (1970). 'Polyacrylamide disc electrophoresis of the soluble proteins from *Nicotiana tabacum* var. Samsun and Samsun NN. II. Changes in protein constitution after infection with tobacco mosaic virus', *Virology*, **40**, 199–211.

Van Peer, R., Niemann, G. J., and Schippers, B. (1991). 'Induced resistance and phytoalexin accumulation in biological control of *Fusarium* wilt of carnation by *Pseudomonas* sp. strain WCS417', *Phytopathology*, **81**, 728–734.

Vernooij, B., Friedrich, L., Ahl Goy, P., Staub, T., Kessmann, H., and Ryals, J. (1995a). '2,6-Dichloroisonicotinic acid—induced resistance to pathogens does not require the accumulation of salicylic acid', *Mol. Plant-Microbe Interact.*, **8**, 228–234.

Vernooij, B., Friedrich, L., Morse, A., Reist, R., Kolditz-Jahwar, R., Ward, E., Uknes, S. Kessmann, H., and Ryals, J. (1995b). 'Salicylic acid is not the translocated signal responsible for inducing systemic acquired resistance but is required in signal transduction', *Plant Cell*, **6**, 959–965.

Walbot, V., Hoisington, D. A., and Neuffer, M. G. (1983). 'Disease lesion mimics in maize', in *Genetic Engineering of Plants* (eds Kosuge, T., Meridith, C., and Hollaender, A.), **Vol. 3**, pp. 431–442, Plenum Press, New York.

Wang, C. Y., Mayo, M., and Baldwin, A. (1996). 'TNF-and cancer therapy-induced apoptosis: potentiation by inhibition of NF-κB', *Science*, **274**, 784–787.

Ward, E. R., Uknes, S. J., Williams, S. C., Dincher, S. S., Wiederhold, L., Alexander, D., Ahl-Goy, P., Metraux, J. P., and Ryals, J. (1991). 'Coordinate gene activity in tobacco to agents that induce systemic acquired resistance', *Plant Cell*, **3**, 1085–1094.

Ward, E. W. B. (1984). 'Suppression of metalaxyl activity by glyphosate: evidence that host defense mechanism contribute to metalaxyl inhibition of *Phytophthora megasperma* f. sp. *glycinea* in soybeans', *Physiol. Plant Pathol.*, **25**, 381–386.

Wasternack, C., Atzorn, R., Jarosch, B., and Kogel, K. H. (1994). 'Induction of thionin, the jasmonate-induced 6 kDa protein of barley by 2,6-dichloroisonicotinic acid', *J. Phytopathol.*, **140**, 280–284.

Watanabe, T., Igarashi, H., Matsumoto, K., Seki, S., Mase S., and Sekizawa, Y. (1977). 'The characteristics of probenazole (Oryzemate) for the control of rice blast', *J. Pesticide Sci.*, **2**, 291–296.

Watanabe, T., Sekizawa, Y., Shimura, M., Suzuki, Y., Matsumoto, K., Iwata, M., and Mase, S. (1979). 'Effects of probenazole (Oryzemate) on rice plants with reference to controlling rice blast', *J. Pesticide Sci.*, **4**, 53–59.

Wei, G., Kloepper, J., and Tuzun, S. (1991). 'Induction of systemic resistance in cucumber to *Colletotrichum orbiculare* by selected strains of plant-growth rhizobacteria', *Phytopathology*, **81**, 1508–1512.

Weymann, K., Hunt, M., Uknes, S., Neuenschwander, U., Lawton, K., Steiner, H. Y., and Ryals, J. (1995). 'Suppression and restoration of lesion formation in *Arabidopsis lsd* mutants', *Plant Cell*, **7**, 2013–2022.

White, R. F. (1979). 'Acetylsalicylic acid (aspirin) induces resistance to tobacco mosaic virus in tobacco', *Virology*, **99**, 410–412.

Wrather, J. A. and Elrod, J. M. (1990). 'Apparent systemic effect of *Colletotrichum truncatum* and *C. lagenarium* on the interaction of soybean and *C. truncatum*', *Phytopathology*, **80**, 472–474.

Yalpani, N., León, J., Lawton, M. A., and Raskin, I. (1993). 'Pathway of salicylic acid biosynthesis in healthy and virus-inoculated tobacco', *Plant Physiol.*, **103**, 315–321.

Yamaguchi, I. and Kubo, Y. (1992). 'Target sites of melanin biosynthesis inhibitors', in *Target Sites of Fungicide Action* (ed. Koller, W.), pp. 101–118, CRC press, New York.

Yoshida, H., Konishi, K., Nakagawa, T., Sekido, S., and Yamaguchi, I. (1990), 'Characteristics of *N*-phenylsulfonyl-2-chloroisonicotinamide as an anti-rice blast agent', *J. Pesticide Sci.*, **5**, 199.

Zhu, W., Maher, E. A., Masoud, S., Dixon, R. A., and Lamb, C. J. (1994). 'Enhanced protection against fungal attack by constitutive coexpression of chitinase and glucanase genes in transgenic tobacco', *Bio/Technology*, **12**, 807–812.

8 Novel Approaches to Disease Control

K. YONEYAMA
Meiji University, Japan

INTRODUCTION

Plant diseases are a major cause of reduction in the yields of agricultural foods and therefore great effort has been expended over many years to protect crops from these diseases. At present, several disease control measures are available to growers; these include disease-resistant cultivars, crop husbandry such as crop rotation, chemical pesticides and biological control agents. Among them, chemical pesticides or fungicides have been successfully used to control plant diseases and have led to a remarkable contribution to the increase of agricultural production. Despite the undoubted value of chemical pesticides in disease protection, their harmful impact on the environment and our health has generated many debates about limiting their extensive use in agriculture. Thus, new approaches to disease control have been sought to develop more effective means of control without detriment to the environment. One resulting approach is based on the induction of disease resistance in the host plant by

Fungicidal Activity. Edited by D. H. Hutson and J. Miyamoto

chemicals as described in a previous chapter. An alternative exciting and promising approach is molecular breeding of plant species by conferring new resistance factors into plants for self-defence against pathogens.

Progress in gene technology has allowed the understanding of the complex molecular mechanisms of plant-pathogen recognition and the natural defence strategies of host plants.This technology can also be used for the controlled and efficient production of genetically improved crop varieties far beyond the possibilities of classical breeding. The first successful attempts have been made to improve resistance against plant viruses by engineering transgenic plants and some virus-resistant transgenic crops are already at the stage of commercially useful development. In the case of bacterial and fungal diseases, the production of transgenic plants with commercially useful levels of resistance has been more limited. Nevertheless, several promising approaches in the field are now available for the development of bacterial- and fungal-resistant transgenic plants as shown in Table 1.

This chapter focuses on the genetic engineering strategies that have been used to produce transgenic plants less susceptible to bacterial and fungal diseases. Recent reviews related to genetically engineered plant improvement by Düring (1996), Panopoulos *et al.* (1996), Kahl and Winter (1995), Shah *et al.* (1995) and Herrera-Estrella and Simpson (1995) complement this review.

RESISTANCE TO PATHOGENIC TOXINS

In most plant diseases caused by microbial attack, development of disease symptoms may result from a direct or indirect effect of toxic metabolites produced by the pathogens. A number of toxins have been isolated and identified from phytopathogenic bacteria and fungi. These toxins are classified into two categories of host-specific and non-host specific toxins on the basis of the selective phytotoxicity to compatible and noncompatible hosts. A host-specific toxin selectively damages only plant varieties susceptible to the pathogens, so that the potential virulence of the pathogen is dependent on the amounts of toxin produced, and the level of disease resistance in a plant variety is correlated with that of toxin resistance. On the other hand, non-host specific toxins have a toxicity not only to host plants but also to a wide variety of non-host plants and living organisms. In this case, the toxins are not a primary determinant of pathogenicity in the pathogens, but in several cases they may play an important role as a virulence factor causing disease development or metabolic disturbance of plant cells.

In the case of non-host specific toxins that have antimicrobial activity, the toxin-producing organism must possess a mechanism to protect itself from the toxin (Durbin and Langston-Unkefer, 1988). Two of the different strategies of self-protection mechanisms developed during the evolution of toxin-producing pathogens are well known. One is based on the production of enzymes that modify a toxin to the inactive form, whereas the second is based on the production of toxin-insensitive target enzymes. In both cases, a pathogen-derived resistance gene can be used to produce transgenic plants that are less susceptible to the toxin released during the bacterial infection.

DETOXIFICATION OF PATHOGENIC TOXINS

Wildfire disease of tobacco is caused by the bacterium *Pseudomonas syringae* pv. *tabaci*, which produces the non-host specific toxin, tabtoxin. When a tobacco leaf is treated with tabtoxin, it causes chlorotic spots similar to the halos on leaves infected with the wildfire bacteria. Thus, it has been suggested that the production of disease symptoms by the attack of *P. syringae* pv. *tabaci* is directly correlated to the inhibitory effect of tabtoxin on plant cells (Braun, 1955).

Tabtoxin is a dipeptide toxin which is composed of tabtoxinine-β-lactam [2-amino-4-(3-hydroxy-2-oxoazacyclobutan-3-yl)butanoic acid] and threonine or serine(Stewart, 1971). This toxin might be synthesized in a biologically inactive form inside the bacterial cells and when released outside the cells it becomes the active toxin. Tabtoxin is readily converted to the highly active moiety of tabtoxinine-β-lactam by cleavage of threonine or serine with some amino-peptidases present in either the bacteria or the plant. The active moiety tabtoxinine-β-lactam inhibits the target enzyme glutamine synthetase, which catalyzes the synthesis of glutamine from glutamic acid in amino acid metabolism (Figure 1). This inhibition results in the abnormal accumulation in tobacco cells of ammonia causing the characteristic chlorosis (Sinden and Durbin, 1968).

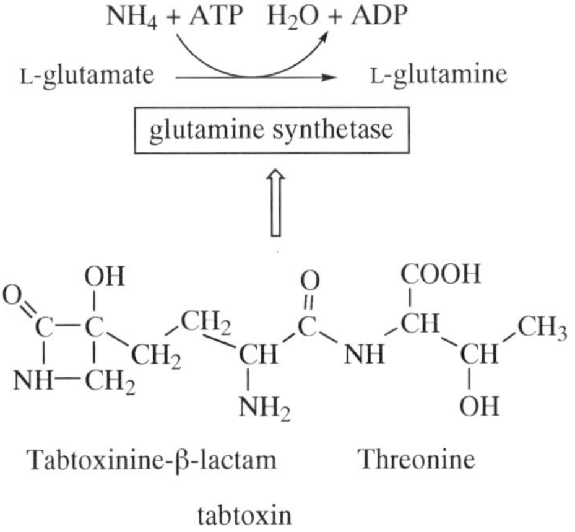

Figure 1 Structure and site of action of tabtoxin

Table 1 Transgenic plants for conferring resistance against bacterial and fungal pathogens

Gene	Source	Transgenic plants	Pathogens examined	References
Bacterial toxin tolerance				
ttr	*Pseudomonas. syringae* pv. *tabaci*	tobacco	*P. syringae* pv. *tabaci*	Anzai *et al.* (1989)
argK	*P. syringae* pv. *phaseolicola*	tobacco	*P. syringae* pv. *phaseolicola*	De La Fuente-Martinez *et al.* (1992) Hatziloukas and Panopoulos (1992)
Defence-related proteins				
Chitinase	bean	tobacco	*Rhizoctonia solani*	Broglie *et al.* (1991)
	tobacco	tobacco	*R. solani*	Vierheilig *et al.* (1993)
	rice	Indica rice	*R. solani*	Lin *et al.* (1995)
	tomato	rape	*Cylindrosporium concentricum* *Phoma lingam* *Sclerotinia sclerotiorum*	Grison *et al.* (1996)
β-1,3-Glucanase	soybean	tobacco	*Phytophthora parasitica* *Alternaria alternata*	Yoshikawa *et al.* (1993)
	tobacco	tobacco	*P. tabacina* *P. parasitica*	Lusso and Kuc (1996)
Chitinase + glucanase	rice	tobacco	*Cercospora nicotianae*	Zhu *et al.* (1994)
	barley	tobacco		Jach *et al.* (1995)
	tobacco	tomato	*Fusarium oxysporum* f. sp. *lycopersici*	Jongedijk *et al.* (1995)
PR-1	tobacco	tobacco	*Peronospora tabacina* *P. parasitica*	Alexander *et al.* (1993)
Osmotin	tobacco	potato	*Phytophthora infestans*	Liu *et al.* (1994)
Phytoalexin	grapevine	tobacco	*Botrytis cinerea*	Hain *et al.* (1993)
	grapevine	rice	*Magnaporthe grisea*	Stark-Lorenzen *et al.* (1997)

Resistance-inducing genes				
R gene *Pto*	tomato	tobacco	*P. syringae* pv. *tabaci* expressing *avrPto*	Thilmony *et al.* (1995) Rommens *et al.* (1995)
NPR1	*Arabidopsis thaliana*	NPR1-defected mutant	*P. syringae* pv. *maculicola*	Cao *et al.* (1997)
Glucose oxidase	*Aspergillus niger*	potato	*Erwinia carotovora* *P. infestans*	Wu *et al.* (1995)
Antimicrobial peptides				
Cecropin	giant silk moth	tobacco	*Burkholderia solanacearum*	Jaynes *et al.* (1993)
		tobacco	*P. syringae* pv. *tabaci*	Huang *et al.* (1997)
Attacin	cecropia moth	apple	*E. amylovora*	Norelli *et al.*(1994)
RIP	barley seed	tobacco	*R. solani*	Logemann *et al.* (1992)
Defensin	radish	tobacco	*A. aleternata* tobacco pathotype	Terras *et al.* (1995)
Thionin	barley	tobacco	*P. syringae* pv. *tabaci* *P. syringae* pv. *syringae*	Carmona *et al.* (1993)
	Arabidopsis thaliana	Arabidopsis	*F. oxysporum* f. sp. *matthiolae*	Epple *et al.* (1997)
Tachyplesin	Southeast horseshoe crab	potato	*E. carotovora*	Allefs *et al.* (1996)
Cholera toxin	subunit A1	tobacco	*P. syringae* pv. *tabaci*	Beffa *et al.* (1995)
Other origin enzymes				
Lysozyme	bacteriophage T4	potato	*E. corotovora*	Düring *et al.* (1993)
	human	tobacco	*Erysiphe cichoracearum* *P. syringae* pv. *tabaci*	Nakjima *et al.* (1997)
Chitinase	*Serratia marcescens*	tobacco	*A. alternata* tobacco pathotype	Suslow *et al.* (1988)
	Rhizopus oligosporus	tobacco	*S. sclerotiorum* *B. cinerea*	Terakawa *et al.* (1997)
Pectate lyase	*E. carotovora*	potato	*E. carotovora*	Wegener *et al.* (1996)

Although tabtoxin shows a toxic activity to a wide range of living organisms from higher plants to algae, bacteria and animals, it has no toxic effect on the producer *P. syringae* pv. *tabaci* itself. This suggests that the wildfire bacterium has a particular mechanism of resistance to tabtoxin such as inactivation of the toxin, insensitivity of the target enzyme to the toxin, or impermeability of the toxin through the cell membrane, and thereby the bacterium can be protected from the inhibitory action of tabtoxin.

The gene responsible for tabtoxin resistance from *P. syringae* pv. *tabaci* itself was cloned and named the tabtoxin resistance gene (*ttr*), in which the open reading frame of 531 bp encodes a protein of 177 amino acid residues which is molecular weight about 19 200 (Anzai *et al.*, 1990). Also, the *ttr* gene was shown to encode the tabtoxin-acetylating enzyme that inactivates tabtoxin or tabtoxinine-β-lactam by acetylation (Anzai *et al.* 1989). When transgenic tobacco plants expressing the *ttr* gene located between the 35S promoter of cauliflower mosaic virus (CaMV) and the polyadenylation signal of nopaline synthase were treated with tabtoxin or inoculated with *P. syringae* pv. *tabaci*, none of them produced the chlorotic halos typical of wildfire disease, indicating that expression of the *ttr* gene confers resistance not only to tabtoxin but also to infection by *P. syringae* pv. *tabaci*.

A major problem in this strategy is how to search for the detoxifying-enzyme genes of pathogenic toxins. A successful application for wildfire disease is the typical case that utilized a toxin-inactivating enzyme gene present in the pathogen itself. A similar way to isolate the objective genes may be used in other plant pathogens which produce specially non-host specific toxins. When plant pathogens do not possess any toxin-detoxifying enzyme genes, other biological targets for gene cloning may be available, such as toxin-degrading enzymes present in soil microorganisms, plants, algae, and other living organisms. If the detoxifying enzyme gene corresponding to a toxin of each pathogen is found, this genetic engineering strategy for disease resistance could be widely applied to fungal and bacterial diseases related to virulence-related toxins. Gene manipulation, however, must be used with care, depending on the mechanisms of action of pathogenic toxins, because some toxins interfere with the functions of plant cell membranes. One possible genetic trick to detoxify the toxins outside of the cells may be to fuse a signal peptide gene to the structure gene of an enzyme; this would lead to translocation of the enzyme to the outside of plant cells (Yoneyama and Anzai, 1993).

TOXIN-INSENSITIVE TARGET ENZYMES

The second mechanism of self-resistance is exemplified by the bacterium *P. syringae* pv. *phaseolicola*, the causal agent of halo blight disease in the common bean. *P. syringae* pv. *phaseolicola* produces the non-host specific toxin phaseolotoxin, which consists of the tripeptide (N'-sulfodiamino-phosphinyl)-ornithyl-alanyl-homoarginine (Figure 2). A hydrolysis product of phaseolo-

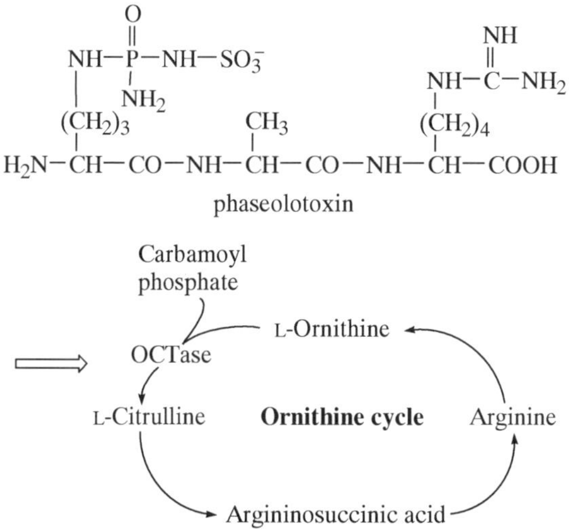

Figure 2 Structure and site of action of phaseolotoxin

toxin, octidine [N'-sulfodiamino-phosphinyl) ornithine], generated through the action of aminopeptidase, has a much higher activity. Upon infection of the bean plants with *P. syringae* pv. *phaseolicola*, local and systemic chlorosis is the direct result of phaseolotoxin action on the host plant. Phaseolotoxin specifically inhibits the enzyme ornithine carbamoyl-transferase (OCTase) involved in arginine biosynthesis, and accumulates several hundred-fold of ornithine in infected leaves (Turner and Mitchell, 1985). Since the biosynthesis of arginine in *P. syringae* pv. *phaseolicola* is presumed to follow through ornithine cycles in other prokaryotes, inhibition of OCTase by phaseolotoxin can lead to autotoxicity. However, the bacterium does not require exogenous arginine for growth in the presence of phaseolotoxin. This suggests that a self-resistance mechanism must be functioned in the cells. Further work suggested that the bacterium possesses two OCTase activities, one sensitive and another insensitive to phaseolotoxin *in vitro* (Staskawicz *et al.*, 1980). The insensitive form of the enzyme was found only in the toxin-producing strains, although the sensitive form was present in the toxin-nonproducing strains as well as the toxin-producing strains. Therefore, the toxin-insensitive OCTase is biologically significant to the self-resistance mechanism *in vivo*. The gene coding for a phaseolotoxin-insensitive OCTase (*argK*) from *P. syringae* pv. *phaseolicola* has been cloned and sequenced (Mosqueda *et al.*, 1990).

Two different groups used the same strategy to produce toxin-resistant transgenic plants (De La Fuente-Martinez *et al.*, 1992; Hatziloukas and Panopoulos, 1992). In both cases, the gene *argK* was fused to the transit peptide gene of the small subunit of ribulose-1,5-biosphosphate carboxylase under the regulation of CaMV 35S promoter, to target the resulting chimeric protein into the chloroplasts where plant OCTase is located (Shargool *et al.*, 1988). In control plants, OCTase activity was almost completely inhibited by phaseolotoxin, whereas in transgenic plants the inhibition varied from 14 to 62% depending on the level of expression of the transgene encoding the bacterial enzyme. Treatment of untransformed tobacco leaves with purified phaseolotoxin resulted in the accumulation of ornithine accompanied by leaf chlorosis. The toxin-treated area of the transgenic plants did not show any change in chlorophyll content compared with an untreated area, unlike control plants. No chlorotic symptoms appeared in the leaves from transgenic plants expressing the insensitive OCTase, even when treated with high concentrations of phaseolotoxin (De La Fuente-Martinez *et al.*, 1992; Hatziloukas and Panopoulos, 1992).

Bean plants expressing the insensitive OCTase inoculated with *P. syringae* pv. *phaseolicola* showed no watery lesions or systemic infection, whereas control plants afforded the typical symptoms of halo blight disease. In all cases, the transgenic plants responded with a hypersensitive reaction to the infection challenge (Herrera-Estrella and Simpson, 1995).

RESISTANCE VIA DEFENCE-RELATED PROTEINS

Plants recognize and resist many invading plant pathogens by inducing a rapid defence response, termed the hypersensitive response (HR). The HR results in localized cell death at the site of infection, which is thought to be responsible for limiting the spread of the pathogen. This local response often triggers nonspecific resistance throughout the plant, known as systemic acquired resistance (SAR). Once triggered, SAR provides resistance to a wide range of pathogens. In tobacco plants, it was shown that the onset of SAR correlates with the coordinated induction of the genes encoding pathogenesis-related (PR) proteins, called SAR proteins. The set of SAR markers consists of at least nine families comprising acidic forms of PR-1, β-1,3-glucanase (PR-2), class II chitinase (PR-3), hevein-like protein (PR-4), thaumatin-like protein (PR-5), and others (Ward *et al.*, 1991; van Loon *et al.*, 1994). In Arabidopsis, the SAR marker genes are PR-1, PR-2, and PR-5 (Ukness *et al.*, 1992). Expression of this set of gene families also serves as a criterion that can distinguish SAR from other resistance responses. The gene families that constitute the SAR genes apparently play an active role in the resistance process because their expression in transgenic plants can impart significant disease resistance (Broglie *et al.*, 1991; Alexander *et al.*, 1993; Liu *et al.*, 1994; Jach *et al.*, 1995).

Cloning of HR genes coding for phytoalexin synthetase and SAR genes for PR proteins from plants provides some important insights for the improvement of disease resistance in plants.

CONSTITUTIVE EXPRESSION OF PR PROTEINS

Several PR proteins possess antifungal activity *in vitro* and also possess biochemical activity as chitinases, glucanases, or other proteins. The best characterized PR proteins in plants are chitinases and β-1-3-glucanases. Four classes (I–III and V) of chitinase and three major classes (I–III) of β-1-3-glucanase are present in many plant species. Class I hydrolases are localized in plant vacuoles and have a strong inhibitory effect on fungal growth *in vitro*. In contrast, class II hydrolases, which are very similar to class I proteins on the basis of their primary structure but are localized in the extracellular space or apoplasm, have no comparable effect on fungal growth *in vivo*. Chitinases are generally found at low or basal levels in healthy plants, and their expression is drastically increased during pathogen attack. Since chitinases hydrolyze the β-1,4 linkages of the *N*-acetyl-D-glucosamine polymer, termed chitin, which is a major cell-wall component of most filamentous fungi, with the exception of the Oomycetes, their expression in plants might confer the resistance of plants to fungal pathogens that contain chitin as a cell-wall constituent. Broglie *et al.* (1991) produced transgenic tobacco plants harboring a chimeric gene encoding a bean class-1 chitinase transcriptionally driven by the 35S CaMV promoter. When these transgenic plants were grown in the presence of the soil-borne chitinous fungus *Rhizoctonia solani*, survival of the seedlings or the loss of root fresh weight in the transgenic tobacco plants was significantly reduced as compared to control plants, depending on the amount of bean chitinase expressed. Similarly, the reduction of susceptibility to *R. solani* in tobacco plants expressing the vacuolar tobacco chitinase I has been independently confirmed by Vierheilig *et al.* (1993), who also showed that deletion of the *N*-terminal chitin binding domain does not affect the ability of chitinase I to enhance resistance to *R. solani*, but deletion of the C-terminal signal for the vacuolar targeting domain does not confer any protective effect against infection of *R. solani*. Several other chitinase genes from different plant species have been expressed in transgenic plants. Lin *et al.* (1995) showed that constitutive expression of a rice chitinase gene in transgenic Indica rice plants confers increased resistance to the sheath blight pathogen, *R. solani*. The degree of resistance in the transgenic plants to this pathogen correlated with the levels of chitinase expression. Grison *et al.* (1996) produced transgenic oilseed rapes constitutively expressing a tomato class-1 chitinase and these transgenic plants also exhibited an increased tolerance to three different fungal pathogens, *Cylindrosporium concentricum, Phoma lingam, Sclerotinia sclerotiorum*, in field trials.

In addition, a β-1,3-glucanase gene has been successfully used for enhanced resistance to fungi. Yoshikawa *et al.* (1993) constructed a chimeric gene encoding a soybean glucanase cDNA gene under the control of CaMV 35S promoter, and transferred this into tobacco plants. Leaves of the transgenic plants were shown to have high levels of resistance to infection by

Phytophthora parasitica var. *nicotianae* or *Alternaria alternata* tobacco pathotype. In a similar experiment, Lusso and Kuc (1996) also showed that the constitutive expression of a β-1,3-glucanase cDNA coding for the PR-2b isoform in tobacco plants confers increased resistance to the glucan-containing fungi *Phytophthora tabacina* and *P. parasitica* var. *nicotianae*.

A further challenge has been to co-express both the chitinase and the glucanase genes in plants on the basis that these may confer higher levels of resistance to fungal pathogens than would either gene alone. The effectiveness of this strategy was demonstrated by Zhu *et al.* (1994), who showed the constitutive co-expression in transgenic tobacco of genes encoding the rice basic chitinase and the acidic glucanase gives substantially greater protection against the fungal pathogen *Cercospora nicotianae*, the causal agent of frogeye, than either transgene alone. Also, the co-expression of barley genes encoding chitinase and glucanase in transgenic tobacco was shown to be more effective in controlling fungal attack than the expression of either gene alone (Jach *et al.*, 1995). A similar result was also obtained in tomato, where co-expression of a tobacco class-I chitinase gene and a class-I β-1,3-glucanase gene in transgenic tomato led to an increased resistance to infection with *Fusarium oxysporum* f. sp. *lycopersici* (Jongedijk *et al.*, 1995).

Although the biochemical function of the PR-1 class of proteins is unknown, it is the most abundant PR protein in infected tobacco tissues. In the PR1 family, PR1a has 90% homology in sequence with PR1b and PR1c. PR1 proteins are not only induced during tobacco mosaic virus (TMV) infection in tobacco, but also by treatment with salicylic acid (SA). Since SA treatment confers resistance to TMV, several investigators have expressed the PR1a and PR1b cDNAs in transgenic tobacco plants. No resistance was found against TMV or alfalfa mosaic virus (Cutt *et al.*, 1989; Linthorst *et al.*, 1989). However, PR-1 proteins show considerable effects in transgenic plants on resistance against a variety of fungal and bacterial pathogens. Transgenic tobacco plants constitutively expressing PR-1a possessed significant tolerance to infection by two pathogenic Oomycete fungi, *Peronospora tabacina* and *P. parasitica* var. *nicotianae*, although the degree of protection in these transgenic plants was considerably less than that obtained by chemical treatment (Alexander *et al.*, 1993). It is not yet known whether PR1a exerts a direct fungicidal activity *in vivo*, or whether it slows the growth of pathogens.

One of the PR-5 proteins in tobacco, an osmotin protein, is induced by several factors, including NaCl, desiccation, ethylene, wounding, abscisic acid, viruses, fungi, and UV light. Osmotin and other osmotin-like proteins from several plant species also have antifungal activity against a variety of fungi *in vitro* (Woloshuk *et al.*, 1991; Vigers *et al.*, 1991). Transgenic potato and tobacco plants constitutively overexpressing a tobacco osmotin gene were produced by Liu *et al.* (1994), who showed that the transgenic potato led to delayed development of disease symptoms after inoculation with spore suspensions of *Phytophthora infestans*, while the transgenic tobacco did not display any change in the development of disease symptoms when challenged with either spore suspensions or fungal mycelia of *P. parasitica* var. *nicotianae*. It is interesting to note that some PR proteins may play a defensive role during fungal infection in a heterologous system but not in a homologous system.

EXPRESSION OF HETEROLOGOUS PHYTOALEXINS

Phytoalexins are antimicrobial compounds of low molecular weight that are rapidly synthesized and accumulated in plants in response to pathogens. Phytoalexin biosynthesis is induced by pathogen infection and also by a number of biotic and abiotic molecules called elicitors. Any change in the composition of the phytoalexins could add to the defence potential of the host plant. Several plants, including grapevine, synthesize the stilbene-type phytoalexin resveratrol when attacked by pathogens. Stilbenes have a fungicidal potential, and are formed in several unrelated plant species such as peanut, grapevine and pine. Stilbene biosynthesis only specifically requires the presence of stilbene synthase, and furthermore the precursor molecules for the formation of hydroxystilbenes are malonyl-CoA and *p*-coumaroyl-CoA, both present in plants. A gene derived from grapevine coding for stilbene synthase was transferred into tobacco plants. In these transgenic plants, the gene was rapidly expressed after exposure with UV light or elicitor, and resveratorl was induced in sterile roots, stems and leaves (Hain *et al.*, 1990). The transgenic tobacco plants carrying the stilbene synthase gene were more resistant to infection by *Botrytis cinerea* (Hain *et al.*, 1993). The stilbene synthase gene from grapevine was also transferred into the commercially important Japonica rice. The transgenic rice plants showed enhanced resistance to the rice blast fungus *Magnaporthe grisea* (Stark-Lorenzen *et al.*, 1997).

RESISTANCE THROUGH EXOGENOUS DEFENCE GENES

Plant resistance to diseases is determined by an interaction between a dominant or semidominant resistance (*R*) gene product in the plant and a corresponding dominant pathogen avirulence (*avr*) gene product. It has been proposed that pathogen *avr* products function as ligands and host *R* products function as receptors in an interaction leading to plant resistance to disease (Lamb, 1996). Recently, many *avr* genes have been isolated, mainly from bacteria, and *R* genes that respond to specific bacterial, fungal, or viral pathogens have been cloned from a variety of plants (see reviews by Parker and Coleman, 1997; Baker *et al.*, 1997; Hammond-Kosack and Jones, 1996). It has been shown that *R* genes can maintain their function in heterologous plant species.

HETEROLOGOUS EXPRESSION OF PLANT RESISTANCE GENES

The *Pto* gene encodes a protein kinase that confers resistance in tomato to *P. syringae* pv. *tomato* strains expressing the avirulence gene *avrPto*. Recently, it was demonstrated that the *Pto* protein kinase and its matching *Avr* product interact directly in the plant cell to initiate the plant defence response (Tang *et al.*, 1996; Scofield *et al.*, 1996). Expression of the *Pto* gene in the transgenic tobacco *Nicotiana benthamiana* or *N. tabacum* Wisconsin 38 resulted in a hypersensitive response after infection with *P. syringae* pv *tabaci* carrying *avrPto* gene (Thilmony *et al.*, 1995; Rommens *et al.*, 1995). These results indicate that disease resistance functions are conserved across species, and imply that the utility of host-specific resistance genes can be extended by transferring disease resistance genes between sexually incompatible plant species.

In current work, the Arabidopsis *NPR1* gene, which controls the onset of SAR and affects local acquired resistance, was found to encode a protein containing ankyrin repeats. When this gene was introduced into the Arabidopsis mutants with defects in the *NPR1* gene, the transgenic mutants expressing the *NPR1* gene were not only responsive to SAR induction with respect to PR-protein expression and resistance to infections, but also became more resistant to infection by *P. syringae* pv. *maculicola* in the absence of SAR induction (Cao *et al.*, 1997). It is interesting to know if the *NPR1* gene functions in heterologous plant species.

ACTIVATION OF DEFENCE SYSTEMS BY THE GLUCOSE OXIDASE GENE

Plant defence responses to pathogen infection involve the production of active oxygen species (AOS), such as the superoxide anion radical ($O_2^{\cdot-}$), hydroxyl radical ($OH^{\cdot}$), and hydrogen peroxide (H_2O_2). This process is referred to as an oxidative burst. The accumulation of AOS is one of earliest events that occurs at host-pathogen recognition. The H_2O_2 burst generated during plant–pathogen interaction is implicated not only as a local trigger for inducing HR but also as a diffusible signal for activation of cellular defence genes (Levine *et al.*, 1994). The glucose oxidase gene from *Aspergillus niger* catalyzes the oxidation of β-glucose by molecular oxygen yielding gluconic acid and H_2O_2. Glucose oxidase was identified in a number of bacterial and fungal species but not in animals or plants. Wu *et al.* (1995) produced transgenic potato plants expressing the fungal gene encoding glucose oxidase and its signal peptide, which generates H_2O_2 when glucose is oxidized. H_2O_2 levels were elevated in both leaf and tuber tissues of these plants. Expression of the fungal oxidase gene led to elevated production of H_2O_2 in transgenic potato plants. The increased level of H_2O_2 in transgenic potato plants exhibited strong resistance to a bacterial soft rot disease caused by *Erwinia carotovora* subsp. *carotovora* and enhanced resistance to potato late blight caused by *P. infestans*. Thus, the expression of an active oxygen species-generating enzyme in transgenic plants may confer broad-spectrum disease resistance via a common mechanism.

RESISTANCE VIA PEPTIDE TOXINS

Antimicrobial peptide toxins have been isolated from many different organisms. They include the cecropins, attacins, diptericins and sarcotoxins from insects (Hultmark *et al.*, 1983; Lee *et al.*, 1983; Boman and Hultmark, 1987; Hoffmann and Hoffmann, 1990; Okada and Natori, 1985), the magainins from amphibians (Zasloff, 1987), peptide toxins from microorganisms (Beffa *et al.*, 1995; Kinal *et al.*, 1995), the thionins of cereals (Garcia-Olmedo *et al.*, 1989) and other defence-related proteins from higher plants (Bowles, 1990). Several of these peptides have significant antimicrobial potential when they are expressed in transgenic plants.

INSECT-DERIVED PEPTIDE TOXINS

Many insects respond to infection by non-pathogenic bacteria with the production of potent peptide toxins which accumulate in large amounts in the insect haemolymph. The insect peptide toxins, cecropins, are a family of small, basic polypeptides with potent antibacterial activity caused by inhibiting a process involving channel formation and subsequent membrane disruption (Christensen *et al.*, 1988). Six different molecules, A–F, consisting of 35–37 amino acids have been isolated so far (Hultmark *et al.*, 1982). Activity of cecropin SB37, a homologous derivative of the naturally occurring lytic peptide cecropin B from the giant silk moth (*Hyalophora cecropia*), has been tested against a number of phytopathogenic bacteria (Nordeen *et al.*, 1992). Due to the high lytic activity of cecropins against phytopathogenic bacteria, these peptides were also utilized for enhanced defence mechanisms in genetically engineered plants. Jaynes *et al.* (1993) produced transgenic tobacco plants that harbored chimeric genes with the coding sequence of a synthetic homologue of cecropin B placed under the control of the potato proteinase inhibitor II promoter. When these plants were inoculated in their stems with *Burkholderia solanacearum*, they showed fewer wilted leaves and significantly lower mortality than control plants. When the same plants were root inoculated, however, they were equally as susceptible to infection with *B. solanacearum* as were the untransformed plants. This difference in susceptibility may be due to the tissue-specific pattern of expression of the promoter sequence used in this construct. Huang *et al.* (1997) also produced transgenic tobacco carrying a modified cecropin gene fused to a secretory sequence from barley α-amylase and regulated by the promoter and terminator from the potato proteinase inhibitor II. Transgenic plants exhibited enhanced disease resistance to the tobacco wildfire pathogen,

P. syringae pv. *tabaci*. Also, bacterial multiplication in the transgenic tobacco leaves was shown to be suppressed more than ten fold compared to control plants. However, transgenic tobacco plants expressing a chimeric gene between a signal peptide from the cecropin B gene and the mature peptide from the cecropin A gene failed to provide useful resistance to infection of *P. syringae* pv. *tabaci* (Hightower *et al.*, 1994). It is interesting that the pathogen-induced promoter and the secretory sequence were competent elements for transforming a cecropin gene into an effective disease-control gene for plants (Huang *et al.*, 1997). Care must also be taken to regulate the level of expression and the specific pattern of expression of the genes encoding cecropins, since these peptides are toxic to plant protoplasts at significantly higher concentrations than those which affect most plant bacteria.

Attacins with 180–190 amino acids are much larger than cecropins. Six different molecules, A–F, have been isolated from the cecropia moth, and these proteins are believed to be derived from two genes, processing through different steps at the N- or C-terminus. In fact, the gene corresponding to the acidic attacin E (184 amino acids) codes for a protein of 188 amino acids which include a C-terminal tetrapeptide that is missing in the mature protein. Attacins mainly act against growing Gram-negative bacteria and are not directly lytic. Rather, they disrupt the structure of the outer membrane by interfering with the synthesis of new outer membrane components (Engstrom *et al.*, 1984). One of these cDNAs coding for attacin E was coupled to plant promoters and used to transform apple (Norelli *et al.*, 1994). Transgenic plants of a susceptible apple rootstock Malling 26 possessed increased resistance compared with untransformed Malling 26 to infection by the fire blight pathogen *Erwinia amylovora* in greenhouse trials, but were still more susceptible than the most resistance rootstock Malling 7.

PLANT-DERIVED PEPTIDE TOXINS

Plant seeds have various kinds of antimicrobial toxins to antagonize pathogens. One family of the most well-known peptides is ribosome-inactivating proteins (RIPs), such as the barley seed RIP, its wheat homologue tritin and the related ricin. RIPs are *N*-glycosidases which are present in different plant species, and inhibit protein synthesis in target cells by specifically removing one highly conserved adenine residue borne on a stem-loop structure in 28S rRNA (Endo *et al.*, 1988). Two types of RIPs exist: single chain type 1 proteins, and type 2 proteins consisting of two chains, one of which bears a galactose-specific lectin domain that can bind to cell surfaces. The latter RIPs include ricin. RIPs do not inactivate self ribosomes, but show varying degrees of activity towards ribosomes of distantly related species, including fungal ribosomes. Purified barley RIP can inhibit the growth of fungi *in vitro*, and this inhibition is synergistically enhanced in the presence of cell-wall degrading enzymes such as class 1 cellulases and β-1,3-glucanases (Leah *et al.*, 1991). These characteristics of RIPs were used to produce transgenic tobacco plants that express a barley type 1 RIP cDNA under control of the wound-inducible promoter of the potato *wun1* gene (Logemann *et al.*, 1992). Analysis of these transgenic plants by western and northern blots showed that barley RIP mRNA and proteins were accumulated in leaves of these transgenic plants in response to wounding. The R1 progeny of transgenic plants producing RIP exhibited heightened protection against inoculation with the soil-borne fungus *R. solani* as judged by height differences between control and transgenic plants grown in infected soil, although direct measurements of the effect of the transgene on lesion size and fungal growth were not reported. The successful use of RIPs in disease control will depend on further experiments to examine whether expression of RIP could protect transgenic plants against other pathogenic fungi and whether the production of heterologous RIP is cytotoxic to the host plant because RIPs from different species vary considerably in their inhibition specificity. In addition, RIPs interact synergistically with cell-wall degrading enzymes, suggesting that uptake of RIPs into the fungus may be a limiting factor such that digestion of the mycelial cell wall increases effective antimicrobial activity (Leah *et al.*, 1991).

In addition to RIPs, a broad family of small antifungal peptides, termed plant defensins, were isolated from plant seeds, and found to possess antifungal activity *in vitro* against a broad spectrum of fungal pathogens. A radish gene encoding antifungal protein 2 (Rs-AFP2), which exhibits potent antifungal activity *in vitro*, was expressed in transgenic tobacco plants. A high level of resistance to the foliar pathogen *A. alternata* tobacco pathotype was observed in plants producing high levels of Rs-AFP2 (Terras *et al.*, 1995).

The thionins comprise another class of peptide toxins. These are thought to be involved in the resistance of plants to fungal and bacterial pathogens. Thionins are a family of small, cysteine-rich proteins that have toxicity against plant pathogens *in vitro*. These proteins are synthesized as precursors with a typical signal peptide. Thionins are present in the endosperm and leaves of cereals, and are induced by both biotic and abiotic stresses. Carmona *et al.* (1993) made two kinds of chimeric genes in which the genomic α-thionin gene from barley and a cDNA for the α-thionin from wheat were each placed under the control of the 35S CaMV promoter. Transgenic tobacco plants expressing high levels of the barley α-thionin showed enhanced resistance to two bacterial pathogens, *P. syringae* pv. *tabaci* and *P. syringae* pv. *syringae*, after artificial inoculation. The number of necrotic lesions and the degree of disease symptoms were reduced in leaves of transgenic R1 and R2 progeny compared to leaves of control plants. The degree of resistance also coincided with the levels of thionin expression, determined by a thionin-specific antibody. It was also found that growth of the *P. syringae* pv. *syringae* was severely inhibited but not completely arrested even in the transgenic lines with high levels of thionin expression.

In Arabidopsis, two thionin genes that are regulated differently were identified. The thionin Thi2.2 gene is expressed constitutively in the seedlings, whereas the Thi2.1 gene is inducible by methyl jasmonate, silver nitrate, and pathogenic fungi. Salicylate and ethephon have no effect, indicating that the Thi2.1 gene is inducible via a signal transduction pathway different from that for PR proteins (Epple *et al.*, 1995). Constitutive overexpression of the thionin Thi2.1 gene ehnanced the resistance of the susceptible ecotype against attack by *F. oxysporum* f. sp. *matthiolae.* Transgenic lines had a reduced loss of chlorophyll after inoculation and supported significantly less fungal growth on the cotyledons. Moreover, fungi on cotyledons of transgenic lines had more hyphae with growth anomalies, including hyperbranching, than on cotyledons of the parental line. No transcripts for pathogenesis-related PR-1, PR-5, or the pathogen-inducible plant defensin Pdfl.2 could be detected in uninoculated transgenic seedlings, indicating that all of the observed effects of the overexpressing lines are most likely the result of the toxicity of the Thi2.1 thionin. These findings strongly support the view that thionins are defence proteins (Epple *et al.*, 1997).

OTHER PEPTIDE TOXINS

Allefs *et al.* (1996) used an antimicrobial peptide, tachyplesin I gene, from Southeast horseshoe crabs. Constitutive expression of a different precursor tachyplesin I polypeptide in potato cultivars led to detectable levels of the foreign peptide only when either the sequence coding for the full-size precursor or the mature tachyplesin polypeptide were fused to the barley-hordothionin signal peptide, but not the signal peptide from tobacco PR-S protein. The transgenic potato plants with detectable levels of tachyplesin were shown to have enhanced resistance to soft rot caused by *E. carotovora* sp. *atroseptica.*

Another interesting toxin is a cholera toxin (CTX), which is a multimeric protein consisting of A1, A2 and five B subunits. The A1 subunit catalyzes the ADP-ribosylation of Ga, which irreversibly blocks the GTPase activity of GTP-binding proteins (G-proteins) leading to the activation of the downstream signalling pathways (Simon *et al.*, 1991). CTX can mimic the effects of red light on *Cab-1* gene expression which is mediated by phytochrome (Neuhaus *et al.*, 1993; Bowler *et al.*, 1994). Transgenic tobacco plants expressing the A1 subunit of CTX driven by the light-inducible wheat *Cab-1* promoter showed greatly reduced susceptibility to the bacterial pathogen *P. syringae* pv. *tabaci*, and accumulated high levels of SA and constitutively expressed PR protein genes encoding PR-1 and the class II isoforms of PR-2 and PR-3. In contrast, the class I isoforms of PR-2 and PR-3, known to be induced in tobacco by stress, by ethylene treatment and as part of the hypersensitive response to infection, were not induced. This means that tissues of transgenic plants expressing CTX mimic the state of SAR in the absence of infection. G proteins modified by CTX could generate a non-systemic signal which acts upstream of SA induction (Beffa *et al.*, 1995).

Ustilago maydis, a fungal pathogen of maize, is the causative agent of corn smut. Some natural isolates of *U. maydis* are persistently infected by a double-strand RNA virus (UmV). UmV is non-infectious and propagates only along with the fungus. Some UmV strains encode secreted polypeptide toxins capable of killing other susceptible strains of *U. maydis*. One of these toxins, the KP6 killer toxin, was produced in transgenic tobacco plants expressing the viral toxin cDNA under the control of CaMV promoter. The two components of the KP6 toxin, designated, α and β with activity and specificity identical to those found in the toxin secreted by *U. maydis* cells, were isolated from the intercellular fluid of the transgenic tobacco plants. The β polypeptide from tobacco was identical in size and *N*-terminal sequence to the *U. maydis* KP6 β polypeptide, indicating that the KP6 preprotoxin is expressed and processed to its active mature form in tobacco plants (Kinal *et al.*, 1995; Park *et al.*, 1996). It will be interesting to know whether the systemic production of this viral killer toxin in maize plants confers resistance to *U. maydis* strains lacking tolerance to killer toxins.

RESISTANCE VIA ANTIMICROBIAL ENZYMES

Several genes of antibacterial and antifungal enzymes have been cloned from microorganisms and animals. Most of them have cell-wall degrading activity against bacteria and fungi, such as the lysozymes from various organisms and phage (Jolles and Jolles, 1984), and the chitinases from fungi (Terakawa *et al.*, 1997). These enzymes also have significant antimicrobial activity in transgenic plants.

LYSOZYMES

Lysozymes have a specific hydrolytic activity against the bacterial cell-wall peptidoglycan. They are β-1, 4-*N*-acetyl muramidases, cleaving the glycosidic bonds between the C–1 of *N*-acetyl muramic acid and the C–4 of *N*-acetyl-glucosamine in peptidoglycan. Lysozyme activity has been reported in a wide variety of organisms, including phages, mammals and plants. Although the role of plant lysozymes in the mechanisms of resistance against phytobacteria is unclear, their participation in plant defence responses is highly conceivable.

Expression of foreign lysozyme genes in transgenic plants has been proposed as a mechanism to enhance resistance against bacterial pathogens. The lysozyme genes used in the transgenic plants were derived from hen egg albumin (Trudel *et al.*, 1992), bacteriophage T4 (Düring *et al.*, 1993), and human (Nakajima *et al.*, 1997). Expression of the hen-egg lysozyme from a full-length cDNA, including its own signal peptide, in transgenic tobacco plants showed significant levels of expression, but only a small proportion of the total lysozyme produced was secreted into the intercellular spaces. The data do not indicate whether the transgenic plants exhibited resistance to bacterial infection (Trudel *et al.*, 1992).

Bacteriophage T4 lysozyme is the most active lysozyme reported to date, and also is active not only against Gram-negative but also Gram-positive bacteria. This enzyme was introduced into potato plants, in which the lysozyme-coding sequence was fused to a barley α-amylase signal peptide under the control of 35S CaMV promoter. The transgenic potato plants expressing the chimeric gene resulted in effective resistance against infection by *E. carotovora* under laboratory and greenhouse conditions, even though the levels of expression and secretion were low. Unfortunately, in none of these experiments was the growth of *E. carotovora* determined in infected tissues.

The human lysozyme gene, which is assembled by the stepwise ligation of chemically synthesized oligonucleotides, was introduced into tobacco. The lysozyme gene was highly expressed under the control of CaMV 35S promoter, and the gene product accumulated in the transgenic tobacco plants. Interestingly, the transgenic tobacco plants showed enhanced resistance against the fungus *Erysiphe cichoracearum*, both conidia formation and mycelial growth were reduced, and the size of the colony was diminished. Growth of the plant bacterium *P. syringae* pv. *tabaci* was also strongly retarded in the transgenic tobacco, and the chlorotic halo of the disease symptom was reduced to 17% of that observed in control tobacco. Thus, introduction of a human lysozyme gene seems to be an effective approach to protect the plants from both fungal and bacterial diseases.

Although the level of expression of foreign lysozyme obtained in these reports was low, the degree of resistance is quite promising. The design of better gene constructs could lead to higher levels of expression and consequently to more effective resistance.

MICROBE-DERIVED ENZYMES

Certain kinds of bacteria and fungi produce antifungal enzymes such as chitinase and cellulase. Microbial chitinases as well as plant chitinases also are useful tools for conferring disease resistance into plants. Several genes encoding chitinases have been isolated from various kinds of microorganisms. The bacterial chitinase gene from *Serratia marcescens* (*chi-A*) was successfully introduced into tobacco plants (Jones, 1988; Suslow *et al.*, 1988). Transgenic tobaccco plants highly expressing the *chi-A* gene showed more resistance to infection by *A. alternata* tobacco pathotype than did control plants. Fungal chitinases also are effective in plant defence systems. Two kinds of chitinases, chitinase I and II, from the filamentous fungus *Rhizopus oligosporus* are quite similar in their sequences to those of yeast rather than those from bacteria. The gene encoding chitinase I, *chi1*, which is involved in cell autolysis of *R. oligosporus*, was introduced into tobacco plants. In the transgenic homozygous progeny, chitinase activity in the young leaves was three- to fourfold higher than that in control plants. Transgenic tobacco leaves infected with the fungal pathogens, *S. sclerotiorum* and *B. cinerea*, revealed remarkable suppression in the development of disease symptoms as compared with control leaves (Terakawa *et al.*, 1997). Although the precise mechanism of disease resistance of chitinase-transformed plants is not clear, microbe-derived chitinase genes will be useful in engineering plants with enhanced protection to fungal pathogens.

BACTERIAL PECTOLYTIC ENZYMES

Degradation of the plant cell wall by the bacterial pectolytic enzymes leads to the formation of sugar oligomers of different sizes. The degradation products, unsaturated oligogalacturonates (OGs), elicit numerous physiological responses in plants and induce a proteinase inhibitor in plants, suggesting that the released OGs function as elicitors for a plant defence response. Therefore, expression of pectolytic enzyme genes in transgenic plants may lead to a resistance of soft-rot disease caused by *E. carotovora*. Wegener *et al.* (1996) constructed a chimeric gene by fusing pectate lyase PL3 gene of *E. carotovora* subsp. *atroseptica* to the promoter of the potato patatin B33 gene or to the 35S promoter of CaMV. Transgenic potato plants harboring the 35S-PL3 chimeric gene exhibited constitutive expression of PL3 in the leaves and tubers, and provided more resistance to tuber tissue maceration by *E. carotovora* subsp. *atroseptica* or its enzymes. Wounding also induced the transcription of the plant defence related gene encoding phenylalanine ammonia-lyase (PAL) to a high level in tubers of a PL3-expressing transgenic potato. It seems to be involved in the elicitation process of the PAL mRNA transcription. However, this approach may be limited to a narrow range of host-pathogen systems.

CONCLUSIONS

The molecular breeding of plants by the introduction of disease resistance genes is an exciting and, in the long term, promising means of disease control. Recent progress in genetic engineering and in the understanding of molecular mechanisms of pathogenicity and natural or induced plant resistance have provided a significant stimulus for the introduction of novel forms of transgenic resistance in plants. Genes from different organisms including bacteria, fungi and plants have been used to enhance disease resistance in plants. Some genes were shown to be effective against different pathogens, whereas others are specific only to a single pathogen or a pathovar of a given pathogen. The majority of these approaches are based on constitutively expressing single proteins, such as toxin-detoxifying enzymes, toxin-insensitive target enzymes, PR proteins, anti-microbial proteins, etc. Transgenic plants resistant to pathogenic toxins, in which resistance genes are derived from the pathogens themselves, are less susceptible to the toxin released during the bacterial infection. These transgenic plants offer good prospects for resistance to single pathogens, but not to multiple pathogens. Other strategies for the introduction of antimicrobial proteins into plants, such as chitinase, glucanase, PR proteins, and other antimicrobial proteins, have also been shown to be effective, but it seems likely that combinations of such factors will often be needed to engineer levels of protection that will be useful in the field. Such combinatorial deployment of antimicrobial genes, each giving partial protection, may be desirable as exemplified with increased resistance in the combination of β-1,3-glucanase and chitinase. More refined strategies are needed to mimic and modify naturally evolved defence reactions of plants, thereby conferring a more durable resistance to a broad range of pathogens. The progression of such strategies will be dependent on information on the molecular analysis of interactions between pathogen *avr* genes and plant *R* genes, and on the genetic dissection of *R* gene-mediated induction of the HR and SAR host defences.

Presently, the transgenic plants produced so far are still in the laboratory stage. Further development of genetically modified plants could provide environmentally safe and economically feasible disease control and will provide the farmer with new prospective and additional options for disease control in crops.

REFERENCES

Alexander, D., Goodman, R. M., Gut-Rella, M., Glascock, C., Weymann, K., Friedrich, L., Maddox, D., Ahl-Goy, P., Luntz, T., Ward, E., and Ryals, J. (1993). 'Increased tolerance to two oomycete pathogens in transgenic tobacco expressing pathogenesis-related protein la', *Proc. Natl. Acad. Sci. USA*, **90**, 7327–7331.

Allefs, J. J., de Jong, E. R., Florack, D. E. A., Hoogendoorn, J., and Stiekema, W. J. (1996). '*Erwinia* soft rot resistance of potato cultivars expressing antimicrobial peptide tachyplesin, I', *Mol. Breed.*, **2**, 97–105.

Anzai, H., Yoneyama, K., and Yamaguchi, I. (1989). 'Transgenic tobacco resistant to a bacterial disease by the detoxification of a pathogenic toxin', *Mol. Gen. Genet.*, **219**, 492–494.

Anzai, H., Yoneyama, K., and Yamaguchi, I. (1990). 'The nucleotide sequence of tabtoxin resistance gene (*ttr*) of *Pseudomonas syringae* pv. *tabaci*', *Nucl. Acids Res.*, **18**, 1890.

Baker, B., Zambryski, P., Staskawicz, B., and Dinesh-Kumar, S. P. (1997). 'Signaling in plant-microbe interactions', *Science*, **276**, 726–733.

Beffa, R., Szell, M., Meuwly, P., Pay, A., Vögeli-Lange, R., Metraux, J. P., Neuhaus, G., Meins, F. Jr., and Nagy, F. (1995). 'Cholera toxin elevates pathogen resistance and induces pathogenesis-related gene expression in tobacco', *EMBO J.*, **14**, 5753–5761.

Boman, H. G., and Hultmark, D. (1987). 'Cell-free immunity in insects', *Ann. Rev. Microbiol.*, **41**, 103–126.

Bowler, C., Neuhaus, G., Yamagata, H., and Chua, N. H. (1994). 'Cyclic GMP and calcium mediate phytochrome phototransduction', *Cell*, **77**, 73–81.

Bowles, D. J. (1990). 'Defense-related proteins in higher plants', *Ann. Rev. Biochem.*, **59**, 873–907.

Braun, A. C. (1955). 'A study on the mode of action of the wildfire toxin', *Phytopathology*, **45**, 659–664.

Broglie, K., Chet, I., Holliday, M., Cressman, R., Biddle, P., Knowlton, C., Mauvais, C. J., and Broglie, R. (1991). 'Transgenic plants with enhanced resistance to the fungal pathogen *Rhizoctonia solani*', *Science*, **254**, 1194–1197.

Cao, H., Glazebrook, J., Clarke, J. D., Volko, S., and Dong, X. (1997). 'The Arabidopsis *NPRI* gene that controls systemic acquired resistance encodes a novel protein containing ankyrin repeats', *Cell*, **88**, 57–63.

Carmona, M. J., Molina, A., Fernandez, J. A., Lopez-Fando, J. J., and Garcia-Olmedo, F. (1993). 'Expression of the α-thionin gene from barley in tobacco confers enhanced resistance to bacterial pathogens', *Plant J.*, **3**, 457–462.

Christensen, B., Fink, J., Merrifield, R. B., and Mauzerall, D. (1988). 'Channel-forming properties of cecropins and related model compounds incorporated into planar lipid membranes', *Proc. Natl. Acad. Sci. USA*, **85**, 5072–5076.

Cutt, J. R., Harpster, M. H., Dixon, D. C., Carr, J. P., Dunsmuir, P., and Klessig, D. F. (1989). 'Disease response to tobacco mosaic virus in transgenic plants that constitutively express the pathogenesis-related PR1b gene', *Virology* , **173**, 89–97.

De La Fuente-Martinez, J. M., Mosquerda-Cano, G., Alvarez-Morales, A., and Herrera-Estrella, L. (1992). 'Expression of a bacterial phaseolotoxin-resistant ornithyl transcarbamylase in transgenic tobacco confers resistance to *Pseudomonas syringae* pv. *phaseolicola*', *Bio/Technology*, **10**, 905–909.

Durbin, R. D. and Langston-Unkefer, P. J. (1988). 'The mechanism for self-protection against bacterial phytotoxins', *Ann. Rev. Phytopathol.*, **26**, 313–329.

Düring, K. (1996). 'Genetic engineering for resistance to bacteria in transgenic plants by introduction of foreign genes', *Mol. Breeding*, **2**, 297–305.

Düring, K., Porsch, P., Fladung, M., and Lörz, H. (1993). 'Transgenic potato plants resistant to the phytopathogenic bacterium *Erwinia carotovora*', *Plant J.*, **3**, 587–598.

Endo, Y., Tsurugi, K., and Ebert, R. F. (1988). 'The mechanism of action of barley toxin: A type I ribosome-inactivating protein with RNA *N*-glucosidase activity', *Biochem. Biophys. Acta*, **954**, 224–226.

Engstrom, P., Carlsson, A., Engstrom, A., Tao, Z. J., and Bennich, H. (1984). 'The antibacterial effect of attacins from the silk moth *Hyalophora cecropia* is directed against the outer membrane of *Escherichia coli*', *EMBO J.*, **3** 3347–3351.

Epple, P., Apel, K., and Bohlmann, H. ((1995). 'An *Arabidopsis thaliana* thionin gene is inducible via a signal transduction pathway different from that for pathogenesis-related proteins', *Plant Physiol.*, **109**, 813–820.

Epple, P., Apel, K., and Bohlmann, H. (1997). 'Overexpression of an endogenous thionin enhances resistance of Arabidopsis against *Fusarium oxysporum*', *Plant Cell*, **9**, 509–520.

Garcia-Olmedo, F., Rodriguez-Palenzuela, P., Hernandez-Lucas, C., Ponz, F., Marana, C., Carmona, M. J., Lopez-Fando, J., Fernandez, J. A., and Carbonero, P. (1989). 'The thionins, a protein family that includes prothionins, viscotoxins and crambins', *Oxf. Surv. Plant Mol. Cell Biol.*, **6**, 31–60.

Grison, R., Grezes-Besset, B., Schneider, M., Lucante, N., Olsen, L., Leguay, J. J., and Toppan, A. (1996). 'Field tolerance to fungal pathogens of *Brassica napus* constitutively expressing a chimeric chitinase gene', *Nat. Biotechnol.*, **14**, 643–646.

Hain, R., Bieseler, B., Kindl, H., Schröder, G., and Stöcker, R. (1990). 'Expression of a stilbene synthase gene in *Nicotiana tabacum* results in synthesis of the phytoalexin resveratrol', *Plant Mol. Biol.*, **15**, 325–335.

Hain, R., Reif, H., Krause, E., Langebartels, R., Kindl, H., Vornam, B., Wiese, W., Schmelzer, E., Schreier, P. H., Stöcker, R. H., and Stenzel, K. (1993). 'Disease resistance results from foreign phytoalexin expression in a novel plant', *Nature*, **361**, 153–156.

Hammond-Kosack, K. E., and Jones, J. D. G. (1996). 'Resistance gene-dependent plant defense responses', *Plant Cell*, **8**, 1773–1791.

Hatziloukas, E., and Panopoulos, N. J. (1992). 'Origin, structure, and regulation of *argK*, encoding the phaseolotoxin-resistant ornithine carbamoyltransferase in *Pseudomonas syringae* pv. *phaseolicola*, and functional expression of *argK* in transgenic tobacco,' *J. Bacteriol.*, **174**, 5895–5909.

Herrera-Estrella, L., and Simpson, J. (1995). 'Genetically engineered resistance to bacterial and fungal pathogens', *World J. Microbiol. Biotech.*, **11**, 383–392.

Hightower, R., Baden, C., Penzes, E., and Dunsmuir, P. (1994). 'The expression of cecropin peptide in transgenic tobacco does not confer resistance to *Pseudomonas syringae* pv. *tabaci*', *Plant Cell Rep.*, **13**, 295–299.

Hoffmann, J. A. and Hoffmann, D. (1990). 'The inducible antibacterial peptides of dipteran insects', *Res. Immunol.*, **141**, 910–918.

Huang, Y., Nordeen, R. O., Di, M., Owens, L. D., and McBeath, J. H. (1997). 'Expression of an engineered cecropin gene cassette in transgenic tobacco plants confers disease resistance to *Pseudomonas syringae* pv. *tabaci*', *Phytopathology*, **87**, 494–499.

Hultmark, D., Engstrom, A., Bennich, H., Kapur, R., and Boman, H. G. (1982). 'Insect immunity: Isolation and structure of cecropin D and four minor antibacterial components from Cecropia pupae', *Eur. J. Biochem.*, **127**, 207–217.

Hultmark, D., Engstrom, A., Andersson, K., Steiner, H., Bennich, H., and Boman, H. G. (1983). 'Insect immunity, Attacins, a family of antibacterial proteins from *Hyalophora cecropia*', *EMBO J.*, **2**, 571–576.

Jach, G., Görnhardt, B., Mundy, J., Logemann, J., Pinsdorf, E., Leah, R., Schell, J., and Maas, C. (1995). 'Enhanced quantitative resistance against fungal disease by combinatorial expression of different barley antifungal proteins in transgenic tobacco', *Plant J.*, **8**, 97–109.

Jaynes, J. M., Nagpala, P., Destefano-Beltran, L., Huang, J. H., Kim, J., Denny, T., and Cetiner, S. (1993). 'Expression of a cecropin B lytic peptide analog in transgenic tobacco confers enhanced resistance to bacterial wilt caused by *Pseudomonas solanacearum*', *Plant Sci.*, **89**, 43–53.

Jolles, P. and Jolles, J. (1984). 'What's new in lysozyme research? Always a model system, today as yesterday', *Mol. Cell Biochem.*, **63**, 165–189.

Jones, J. D. (1988). 'Expression of bacterial chitinase protein in tobacco leaves using two photosynthetic gene promoters', *Mol. Gen. Genet.*, **212**, 536–542.

Jongedijk, E., Tigelaar, H., van Roekel, J. S. C., Bres-Vloemans, S. A., Dekker, I., van den Elzen, P. J. M., Cornelissen, B. J. C., and Melchers, L. S. (1995). 'Synergistic activity of chitinases and β-1,3-glucanases enhances fungal resistance in transgenic tomato plants, *Euphytica*, **85**, 173–180.

Kahl, G. and Winter, P. (1995). 'Plant genetic engineering for crop improvement', *World J. Microbiol. Biotechnol.*, **11**, 449–460.

Kinal, H., Park, C., Berry, J. O., Koltin, Y., and Bruenn, J. A. (1995). 'Processing and secretion of a virally encoded antifungal toxin in transgenic tobacco plants, evidence for a Kex2p pathway in plants', *Plant Cell*, **7**, 677–688.

Lamb, C. (1996). 'A ligand-receptor mechanism in plant-pathogen recognition', *Science*, **274**, 2038–2040.

Leah, R., Tommerup, H., Svendsen, I., and Mundy, J. (1991). 'Biochemical and molecular characterization of three barley seed proteins with anti-fungal properties,' *J. Biol. Chem.*, **266**, 1564–1573.

Lee, J. Y., Edlund, T., Ny, T., Faye, I., and Boman, H. G. (1983). 'Insect immunity. Isolation of cDNA clones corresponding to attacins and immune protein P4 from *Hyalophora cecropia*', *EMBO J.*, **2**, 577–581.

Levine, A., Tenhaken, R., Dixon, R., and Lamb, C. (1994). 'H_2O_2 from the oxidative burst orchestrates the plant hypersensitive disease resistance', *Cell*, **79**, 583–593.

Lin, W., Anuratha, C. S., Datta, K., Potrykus, I., Muthukrishnan, S., and Datta, S. K. (1995). 'Genetic engineering of rice for resistance to sheath blight', *Bio/Technology*, **13**, 686–691.

Linthorst, H. J., Meuwissen, R. L., Kauffman, S., and Bol, J. F. (1989). 'Constitutive expression of pathogenesis-related proteins PR-1, GRP, and PR-S in tobacco has no effect on virus infection', *Plant Cell*, **1**, 285–291.

Liu, D., Raghothama, K. G., Hasegawa, P. M., and Bressan, R. A. (1994). 'Osmotin overexpression in potato delays development of disease symptoms, *Proc. Natl. Acad. Sci. USA*, **91**, 1888–1892.

Logemann, J., Jach, G., Tommerup, H., Mundy, J., and Schell, J. (1992). 'Expression of a barley ribosome-inactivating protein leads to increased fungal protection in transgenic tobacco plants', *Bio/Technology*, **10**, 305–308.

Lusso, M. and Kuc, J. (1996). The effect of sense and antisense expression of the PR-N gene for β-1,3-glucanase on disease resistance of tobacco to fungi and viruses', *Physiol. Mol. Plant pathol.*, **49**, 267–283.

Mosqueda, G., Van den Broeck, G., Saucedo, O., Bailey, A. M., Alvalez-Morales, A., and Herrera-Estrella, L. (1990). 'Isolation and characterization of the gene from *Pseudomonas syringae* pv. *phaseolicola* encoding the phaseolotoxin-insensitive ornithine carbamoyltransferase', *Mol. Gen. Genet.*, **222**, 461–466.

Nakajima, H., Muranaka, T., Ishige, F., Akutsu, K., and Oeda, K. (1997). 'Fungal and bacterial disease resistance in transgenic plants expressing human lysozyme', *Plant Cell Rep.*, **16**, 674–679.

Neuhaus, G., Bowler, C., Kern, R., and Chua, N. H. (1993). 'Calcium/calmodulin-dependent and -independent phytochrome signal transduction pathways', *Cell*, **73**, 937–952.

Nordeen, R. O., Sinden, S. L., Jaynes, J. M., and Owens, L. D. (1992). 'Activity of cecropin sb37 against protoplasts from several plant species and their bacterial pathogens', *Plant Sci.*, **82**, 101–107.

Norelli, J. L., Aldwinckle, H. S., Destefano-Beltran, L., and Jaynes, J. M. (1994). 'Transgenic 'Malling 26' apple expressing the attacin E gene has increased resistance to *Erwinia amylovora*', *Euphytica*, **77**, 123–128.

Okada, M. and Natori, S. (1985). 'Primary structure of sarcotoxin I, an antibacterial protein induce in the hemolymph of *Sacophaga peregrina* (fresh fly) larvae', *J. Biol. Chem.*, **260**, 7174–7177.

Panopoulos, N. J., Hatziloukas, E., and Afendra, A. S. (1996). 'Transgenic crop resistance to bacteria', *Field Crops Res.*, **45**, 85–97.

Park, C., Berry, J. O., and Bruenn, J. O. (1996). 'High-level secretion of a virally encoded anti-fungal toxin in transgenic tobacco plants', *Plant Mol. Biol.*, **39**, 359–366.

Parker, J. E. and Coleman, M. J. (1997). 'Molecular intimacy between proteins specifying plant-pathogen recognition,' *Trends Biochem. Sci.*, **22**, 291–296.

Rommens, C. M., Salmeron, J. M., Oldroyd, G. E. D., and Staskawicz, B. J. (1995). 'Intergeneric transfer and functional expression of the tomato disease resistance gene *Pto*', *Plant Cell*, **7**, 1537–1544.

Scofield, S. R., Tobias, C. M., Rathjen, J. P., Chang, J. H., Lavelle, D. T., Michelmore, R. W., and Staskawicz, B. J. (1996). 'Molecular basis of gene-for-gene specificity in bacterial speck disease of tomato', *Science*, **274**, 2063–2065.

Shah, D. M., Rommens, C. M. T., and Beachy, R. N. (1995). 'Resistance to diseases and insects in transgenic plants, progress and applications to agriculture', *Trends Biotechnol.*, **13**, 362–368.

Shargool, P. D., Jain, J. C., and McKay, G. (1988). 'Ornithine biosynthesis and arginine biosynthesis and degradation in plant cells,', *Phytochemistry*, **27**, 1571–1574.

Simon, M. I., Strathmann, M. P., and Gautam, N. (1991). 'Diversity of G proteins in signal transduction', *Science*, **252**, 802–808.

Sinden, S. L. and Durbin, R. D. (1968). 'Glutamine synthetase inhibition: Possible mode of action of wildfire toxin from *Pseudomonas tabaci*', *Nature*, **219**, 379–380.

Stark-Lorenzen, P., Nelke, B., Hänbler, G., Mühlbach, H. P., and Thomzik, J. E. (1997). 'Transfer of a grapevine stilbene synthase gene to rice (*Oryza sativa* L.)', *Plant Cell Rep.*, **16**, 668–673.

Staskawicz, B. J., Panopoulos, N. J., and Hoogenraad, N. J. (1980). 'Phaseolotoxin-insensitive ornithine carbamoyltransferase of *Pseudomonas syringae* pv. *phaseolicola*: Basis for immunity to phaseolotoxin', *J. Bacteriol.*, **142**, 720–723.

Stewart, W. (1971). 'Isolation and proof of structure of wildfire toxin', *Nature*, **229**, 174–178.

Suslow, T. V., Matsubara, D., Jones, J., Lee, R., and Dunsmui, P. (1988). 'Effect of expression of bacterial chitinase on tobacco susceptibility to leaf brown spot', *Phytopathology*, **78**, 1556.

Tang, X., Frederick, R. D., Zhou, J., Halterman, D. A., Jia, Y., and Martin, G. B. (1996). 'Initiation of plant disease resistance by physical interaction of AvrPto and Pto kinase', *Science*, **274**, 2060–2063.

Terakawa, T., Takaya, N., Horiuchi, H., Koike, M., and Takagi, M. (1997). 'A fungal chitinase gene from *Rhizopus oligosporus* confers antifungal activity to transgenic tobacco', *Plant Cell Rep.*, **16**, 439–443.

Terras, F. R. G., Eggermont, K., Kovaleva, V., Raikhel, N. V., Osborn, R. W., Kester, A., Rees, S. B., Torrekens, S., Van Leuven, F., Vanderleyden, J., Cammue, B. P. A., and Broekaert, W. F. (1995). 'Small cysteine-rich antifungal proteins from radish, their role in host defense, *Plant Cell*, **7**, 573–588.

Thilmony, R. L., Chen, Z., Bressan, R. A., and Martin, G. B. (1995). 'Expression of the tomato *Pto* gene in tobacco enhances resistance to *Pseudomonas syringae* pv. *tabaci* expressing *avrPto*', *Plant Cell*, **7**, 1529–1536.

Trudel, J., Potvin, C., and Asselin, A. (1992). 'Expression of active hen egg white lysozyme in transgenic tobacco', *Plant Sci.*, **87**, 55–67.

Turner, J. G. and Mitchell, R. E. (1985). 'Association between symptom development and inhibition of ornithine carbamoyltransferase in bean leaves treated with phaseolotoxin', *Plant Physiol.*, **79**, 468–473.

Uknes, S., Mauch-Mani, B., Moyer, M., Potter, S., Williams, S., Dincher, S., Chandler, D., Slusarenko, A., Ward, E., and Ryals, J. (1992). 'Acquired resistance in Arabidopsis', *Plant Cell*, **4**, 645–656.

van Loon, L. C., Pierpoint, W. S., Boller, T., and Camejero, V. (1994). 'Recommendations for naming plant pathogenesis-related proteins', *Plant Mol. Biol. Rep.*, **12**, 245–264.

Vierheilig, H., Alt, M., Neuhaus, J. M., Boller, T., and Wiemken, A. (1993). 'Colonization of transgenic *Nicotiana sylvestris* plant, expressing different forms of *Nicotiana tabacum* chitinase, by the root pathogen *Rhizoctonia solani* and by the mycorrhizal symbiont *Glomus mosseae*', *Mol. Plant-Microbe Interact.*, **6**, 261–264.

Vigers, A. J., Roberts, W. K., and Selitrennikoff, C. P. (1991). 'A new family of antifungal proteins', *Mol. Plant-Microbe Interact.*, **4**, 315–323.

Ward, E. R., Uknes, S. J., Williams, S. C., Dincher, S. S., Wiederhold, D. L., Alexander, D. C., Ahl-Goy, P., Metraux, J. P., and Ryals, J. A. (1991). 'Coordinate gene activity in response to agents that induce systemic acquired resistance', *Plant Cell*, **3**, 1085–1094.

Wegener, C., Bartling, S., Olsen, O., Weber, J., and von Wettstein, D. (1996). 'Pectate lyase activation in transgenic potatoes confers pre-activation of defense against *Erwinia carotovora*', *Physiol. Mol. plant Pathol.*, **49**, 359–376.

Woloshuk, C. P., Meulenhoff, J. S., Sela-Buurlage, M., van den Elzen, P. J., and Cornelissen, B. J. (1991). 'Pathogen-induced proteins with inhibitory activity toward *Phytophthora infestans*', *Plant Cell*, **3**, 619–628.

Wu, G., Shortt, B. J., Lawrence, E. B., Levine, E. B., Fitzsimmons, K. C., and Shahl, D. M. (1995). 'Disease resistance conferred by expression of a gene encoding H_2O_2-generating glucose oxidase in transgenic potato plants', *Plant Cell*, **7**, 1357–1368.
Yoneyama, K. and Anzai, H. (1993). 'Transgenic plants resistant to diseases by the detoxification of toxins', in *Biotechnology in Plant Disease Control*, ed. Chet, I., John Wiley and Sons, New York, pp. 115–137.
Yoshikawa, M., Tsuda, M., and Takeuchi, Y. (1993). 'Resistance to fungal diseases in transgenic tobacco plants expressing the phytoalexin elicitor-releasing factor, β-1,3-endoglucanase, from soybean', *Naturwissenschaften*, **80**, 417–420.
Zasloff, M. (1987). 'Magainins, a class of antimicrobial peptides from Xenopus skin, isolation, characterization of two active forms, and partial cDNA sequence of a precursor', *Proc. Natl. Acad. Sci. USA*, **84**, 5449–5453.
Zhu, Q., Maher, E. A., Masoud, S., Dixon, R. A., and Lamb, C. J. (1994). 'Enhanced protection against fungal attack by constitutive co-expression of chitinase and glucanase genes in transgenic tobacco', *Bio/Technology*, **12**, 807–812.

Index